AF388939

Combining Sensors and Multibody Models for Applications in Vehicles, Machines, Robots and Humans

Combining Sensors and Multibody Models for Applications in Vehicles, Machines, Robots and Humans

Editors

Javier Cuadrado
Miguel A. Naya

MDPI • Basel • Beijing • Wuhan • Barcelona • Belgrade • Manchester • Tokyo • Cluj • Tianjin

Editors
Javier Cuadrado
University of La Coruña
Spain

Miguel A. Naya
University of La Coruña
Spain

Editorial Office
MDPI
St. Alban-Anlage 66
4052 Basel, Switzerland

This is a reprint of articles from the Special Issue published online in the open access journal *Sensors* (ISSN 1424-8220) (available at: https://www.mdpi.com/journal/sensors/special_issues/multibody_sensors).

For citation purposes, cite each article independently as indicated on the article page online and as indicated below:

LastName, A.A.; LastName, B.B.; LastName, C.C. Article Title. *Journal Name* **Year**, *Volume Number*, Page Range.

ISBN 978-3-0365-2357-6 (Hbk)
ISBN 978-3-0365-2358-3 (PDF)

Contents

About the Editors

Javier Cuadrado graduated in Mechanical Engineering from the University of Navarra (Spain) in 1990. Afterwards, he worked as assistant professor at the School of Engineering of the same university and, under the Spanish Ministry of Education and Science, developed his doctorate under the supervision of Prof. Javier Garcia de Jalon at the department of Applied Mechanics of CEIT (Spain), a research center settled in the Basque Country, closely related to the University of Navarra. He received his Ph.D. degree in Mechanical Engineering from that university in 1993, having specialized in the computational analysis of multibody systems. In 1994, he joined the Mechanical Engineering department of the University of La Coruña (Spain) as Associate Professor, becoming a Full Professor in 2000. In March 2002, he founded the Laboratory of Mechanical Engineering (http://lim.ii.udc.es) in this department.

Miguel A. Naya graduated in Industrial Engineering in 1996 from the University of Vigo (Spain). He began his Ph.D. studies at University of La Coruña in 1996 where he was granted permission to work in the research project "Methods for Intelligent Real-TimeSimulation of Multibody Dynamics", funded by the U.S. Army Research Office and led by Prof. Eduardo Bayo. He became Assistant Professor in Mechanical Engineering at the same university in 1998 and, after obtaining his Ph.D. degree in June 2007, achieved the position of Associate Professor in 2009. He was Head of the Department of Industrial Engineering of the University of La Coruña between 2009 and 2013, and Secretary of this Deparment from 2013 to 2017. Currently, he is Head of the Department of Naval and Industrial Engineering of the University of La Coruña. His research is focused on the application of multibody dynamics to the simulation and control of cars.

Preface to "Combining Sensors and Multibody Models for Applications in Vehicles, Machines, Robots and Humans"

Measuring and simulating sensors and physical models represent two ways of obtaining information from a dynamical system. Each of them has advantages and disadvantages, but the combination of both leads to more consistent results. This general idea can be particularized to mobile mechanical systems, also known as multibody systems. In this type of system, with popular examples including vehicles, machines, robots and humans, there is a tradition of using simplified models along with sensors to estimate the states, inputs and/or parameters of the system. However, in the last decade, research activity has been devoted to substituting the simplified models with detailed multibody models, which can be useful in some applications. Therefore, the aim of this Special Issue has been to gather contributions that illustrate the interest in this recent trend.

Javier Cuadrado, Miguel A. Naya
Editors

*sensors*

Editorial

Editorial of Special Issue "Combining Sensors and Multibody Models for Applications in Vehicles, Machines, Robots and Humans"

Javier Cuadrado * and Miguel Á. Naya

Laboratory of Mechanical Engineering, University of La Coruña, 15403 Ferrol, Spain; miguel.naya@udc.es
* Correspondence: javier.cuadrado@udc.es

Citation: Cuadrado, J.; Naya, M.Á. Editorial of Special Issue "Combining Sensors and Multibody Models for Applications in Vehicles, Machines, Robots and Humans". *Sensors* 2021, 21, 6345. https://doi.org/10.3390/s21196345

Received: 14 September 2021
Accepted: 18 September 2021
Published: 23 September 2021

Publisher's Note: MDPI stays neutral with regard to jurisdictional claims in published maps and institutional affiliations.

The combination of physical sensors and computational models to provide additional information about system states, inputs and/or parameters, in what is known as virtual sensing, is becoming more and more popular in many sectors, such as the automotive, aeronautics, aerospatial, railway, machinery, robotics and human biomechanics sectors. While, in many cases, control-oriented models, which are generally simple, are the best choice, multibody models, which can be much more detailed, may have applications, for example, during the design stage of a new product.

This Special Issue sought to attract works dealing with the many challenges that must be overcome when developing multibody-based virtual sensors. These challenges include the selection of the fusion algorithm and its parameters, the coupling or the independence between the fusion algorithm and the multibody formulation, the magnitudes to be estimated, the stability and accuracy of the adopted solution, optimization of the computational cost, real-time issues and implementation on embedded hardware. Studies of the application of multibody-based virtual sensors to, for example, vehicles, heavy machinery, mobile or humanoid robots, assistive orthotic and prosthetic devices or the measurement and analysis of human movement were also welcome.

In the next paragraphs, a brief description of the content of each contribution forming the Special Issue is provided.

The first work [1] focuses on the measurement of human motion and employs an extended Kalman filter to carry out the optical motion capture, the problem being to adjust the filtering level of the algorithm, for which the accelerometers contained in inertial measurement units are used.

The second work [2] addresses the measurement of railway track geometric irregularities by means of a system included on board, composed of a kinematic model of the track and a set of sensors.

The objective of the third work [3] is to avoid the danger of reaching singular positions in a parallel robot for rehabilitation. For this purpose, a control system is developed based on computer vision and the multibody model of the robot.

The fourth work [4] is not centered on a particular field of application, but on the general problem of state and input estimation, its contribution being a method to extract the matrices required for a discrete Kalman filter which provides stable and accurate results, demonstrated in a slider–crank mechanism.

The fifth work [5] foresees naval applications in the future, although it tackles the general problem of state, input and parameter estimation, proposing that the model equations are based on Pontryagin's treatment of Hamiltonian systems, and that feedback is used to re-initialize the initial values of a reformulated two-point boundary value problem.

The sixth work [6] deals with human-in-the-loop applications with haptic feedback, providing examples in the automotive and musical fields. Sensors are used to characterize and validate the multibody model, to measure the system kinematics and dynamics within the human-in-the-loop process, and to validate the haptic device behavior.

The field of application of the seventh work [7] is machinery. More specifically, the work seeks to determine the characteristic curves of a directional control valve in a hydraulic system. For this purpose, it employs an augmented discrete extended Kalman filter based on a multibody model of the mechanism and the lumped fluid theory for a fluid power system, along with a curve-fitting method to describe the characteristic curves of the valve.

The eighth work [8] looks at the general problem of state and force estimation, considering the automotive field as a clear target of application. The authors apply an adaptive method to estimate the noise covariance matrices of a state and input estimator based on multibody models.

The ninth work [9] intends to estimate the sideslip angle in road vehicles, and proposes an alternative to traditional methods of state estimation by representing the problem as a probabilistic graphical model which can be optimized by several methods.

The tenth work [10] is a spatial application consisting of the design of a six-legged integrated lander and rover for repetitive lunar landings. The possibilities of successful landing despite the failure of some integrated drive units of the system are explored, assuming that at least three legs of the hexapod are operative.

In summary, it can be seen that the papers gathered in the Special Issue contributed either by proposing solutions to the general problem of state, input and/or parameter estimation [4,5,7–9], and/or by suggesting applications of the combined use of sensors and multibody models to different fields, such as automotive [6,8,9], railway [2], naval [5], spatial [10], machinery [4,7], robotics [3], biomechanics [1] and music [6] applications, thus showing the theoretical challenges and practical interest of this research topic.

Finally, we wish to thank the authors, reviewers and journal staff for their commitment and effort, which made it possible to complete this Special Issue on time.

Funding: This research received no external funding.

Conflicts of Interest: The authors declare no conflict of interest.

References

1. Cuadrado, J.; Michaud, F.; Lugrís, U.; Soto, M.P. Using Accelerometer Data to Tune the Parameters of an Extended Kalman Filter for Optical Motion Capture: Preliminary Application to Gait Analysis. *Sensors* **2021**, *21*, 427. [CrossRef] [PubMed]
2. Escalona, J.; Urda, P.; Muñoz, S. A Track Geometry Measuring System Based on Multibody Kinematics, Inertial Sensors and Computer Vision. *Sensors* **2021**, *21*, 683. [CrossRef] [PubMed]
3. Pulloquinga, J.L.; Escarabajal, R.J.; Ferrandiz, J.; Valles, M.; Mata, V.; Urizar, M. Vision-Based Hybrid Controller to Release a 4-DOF Parallel Robot from a Type II Singularity. *Sensors* **2021**, *21*, 4080. [CrossRef] [PubMed]
4. Adduci, R.; Vermaut, M.; Naets, F.; Croes, J.; Desmet, W. A Discrete-Time Extended Kalman Filter Approach Tailored for Multibody Models: State-Input Estimation. *Sensors* **2021**, *21*, 4495. [CrossRef] [PubMed]
5. Sands, T. Virtual Sensing of Motion Using Pontryagin's Treatment of Hamiltonian Systems. *Sensors* **2021**, *21*, 4603. [CrossRef] [PubMed]
6. Docquier, N.; Timmermans, S.; Fisette, P. Haptic Devices Based on Real-Time Dynamic Models of Multibody Systems. *Sensors* **2021**, *21*, 4794. [CrossRef] [PubMed]
7. Khadim, Q.; Kiani-Oshtorjani, M.; Jaiswal, S.; Matikainen, M.; Mikkola, A. Estimating the Characteristic Curve of a Directional Control Valve in a Combined Multibody and Hydraulic System Using an Augmented Discrete Extended Kalman Filter. *Sensors* **2021**, *21*, 5029. [CrossRef] [PubMed]
8. Rodríguez, A.; Sanjurjo, E.; Pastorino, R.; Naya, M.A. Multibody-Based Input and State Observers Using Adaptive Extended Kalman Filter. *Sensors* **2021**, *21*, 5241. [CrossRef] [PubMed]
9. Leanza, G.; Reina, J.L.; Blanco-Claraco, J.L. A Factor-Graph-Based Approach to Vehicle Sideslip Angle Estimation. *Sensors* **2021**, *21*, 5409. [CrossRef] [PubMed]
10. Yin, K.; Zhou, S.; Sun, Q.; Gao, F. Lunar Surface Fault-Tolerant Soft-Landing Performance and Experiment for a Six-Legged Movable Repetitive Lander. *Sensors* **2021**, *21*, 5680. [CrossRef] [PubMed]

Article

Using Accelerometer Data to Tune the Parameters of an Extended Kalman Filter for Optical Motion Capture: Preliminary Application to Gait Analysis

Javier Cuadrado, Florian Michaud, Urbano Lugrís * and Manuel Pérez Soto

Laboratory of Mechanical Engineering, University of La Coruña, 15403 Ferrol, Spain; javier.cuadrado@udc.es (J.C.); florian.michaud@udc.es (F.M.); manuel.perez.soto@udc.es (M.P.S.)
* Correspondence: urbano.lugris@udc.es

Abstract: Optical motion capture is currently the most popular method for acquiring motion data in biomechanical applications. However, it presents a number of problems that make the process difficult and inefficient, such as marker occlusions and unwanted reflections. In addition, the obtained trajectories must be numerically differentiated twice in time in order to get the accelerations. Since the trajectories are normally noisy, they need to be filtered first, and the selection of the optimal amount of filtering is not trivial. In this work, an extended Kalman filter (EKF) that manages marker occlusions and undesired reflections in a robust way is presented. A preliminary test with inertial measurement units (IMUs) is carried out to determine their local reference frames. Then, the gait analysis of a healthy subject is performed using optical markers and IMUs simultaneously. The filtering parameters used in the optical motion capture process are tuned in order to achieve good correlation between the obtained accelerations and those measured by the IMUs. The results show that the EKF provides a robust and efficient method for optical system-based motion analysis, and that the availability of accelerations measured by inertial sensors can be very helpful for the adjustment of the filters.

Keywords: Kalman filter; motion capture; gait analysis; inertial sensor

Citation: Cuadrado, J.; Michaud, F.; Lugrís, U.; Pérez Soto, M. Using Accelerometer Data to Tune the Parameters of an Extended Kalman Filter for Optical Motion Capture: Preliminary Application to Gait Analysis. *Sensors* **2021**, *21*, 427. https://doi.org/10.3390/s21020427

Received: 30 October 2020
Accepted: 4 January 2021
Published: 9 January 2021

Publisher's Note: MDPI stays neutral with regard to jurisdictional claims in published maps and institutional affiliations.

1. Introduction

Human motion capture during gait provides a way to understand the principles of the natural mode of locomotion of the human being. Technological advances have changed its practice and improved its accuracy along history [1,2]. Recent developments in microelectromechanical systems (MEMS) have caused a renewed interest in the use of inertial measurement units (IMUs) to perform three-dimensional (3D) human movement reconstruction [3–10]. However, getting orientation from IMUs presents accuracy and consistency issues [11–16], especially in the presence of environmental ferromagnetic disturbances or when measuring fast complex movements over long periods of time [17]. This is why, although the performance of inertial sensors has improved in the last decade, optical motion capture remains as the preferred method to perform precise biomechanical studies. In fact, as pointed out in [18,19], IMU-based methods for motion capture and reconstruction are usually validated against optical methods, which remain as the golden standard reference. The problem with optical motion capture systems is that it is very difficult to ensure that all markers are visible to the cameras all the time and, moreover, other reflective objects present in the capture zone can be incorrectly identified as markers. In general, obtaining the skeletal motion involves some manual post-processing of the captured data, so the technique is not straightforward [20,21]. This problem can be overcome by using an extended Kalman filter (EKF) [22], as will be described later in this paper.

The typically high-frequency noise harmonics present in the recorded marker trajectories are hardly perceptible at displacement level. However, after numerical differentiation, the amplitude of each harmonic increases with its harmonic number; for velocities, it increases linearly, and, for accelerations, the increase is proportional to the square of the harmonic number [2]. For this reason, filtering is required when trying to obtain velocities and accelerations by numerically differentiating position data. The problem here lies on the choice of the cutoff frequency of the filter, since it is difficult to achieve a value that filters out most of the noise, without also removing relevant motion information [23–27]. From what has been said, it is deduced that different filtering quality is demanded for the animation of virtual characters, where only configuration-level information is needed, and for biomechanical analysis including dynamics, where velocity- and acceleration-level information is also required.

Numerous filtering approaches can be found in the literature [2,23,28,29]. While the Butterworth filter has generally been preferred, because impulsive-type inputs are rarely seen in human movement data [28,30,31], recent studies have applied the Kalman filter [32] (most commonly used in the literature for inertial sensors), thus improving the accuracy of estimated joint kinematics and computed orientation data [22,33,34]. However, beyond the choice of a filtering algorithm, the main problem remains in the tuning of its parameters. In almost all the filtering studies for gait analysis, the smoothing level is based on the author's decision of how much noise is acceptable. A common criterion is to establish an error threshold at position level, and then setting the cutoff frequency accordingly [22,23]. Regardless of the method used, there is no way of assessing the accuracy of the obtained accelerations by relying on position data alone. In order to provide an objective filter-tuning procedure, this work proposes to compare the filtered accelerations from the optical capture with their experimental measurement from the inertial sensors in the case of gait. In [24], a similar procedure was applied in the case of jumping, but it was not successful due to the overshoot provided by the accelerometers in their horizontal measurements.

IMUs are capable of estimating their own orientation within an Earth-fixed frame by using sensor fusion algorithms, such as Madgwick's algorithm [35] or the EKF [36]. These algorithms provide an estimate of the orientation by combining the information from the triaxial accelerometers, gyroscopes and magnetometers present in the IMU. Because IMUs show limitations to give an accurate orientation [13] (closely related to sensor calibration, magnetometer sensitivity, and presence of accelerations other than gravity) and, moreover, their Earth-based global frame will in general differ from that of the optical motion capture system, a preliminary test was performed with nine IMUs. This test was carried out in order to assess the instrumental errors associated, select the most accurate units, and determine the global frame offset corresponding to each one [11]. Second, the gait analysis of a healthy subject was conducted. The motion was recorded by both the optical and the inertial techniques, using the seven most accurate IMUs among the nine previously tested. The human motion was then reconstructed by using both the classic Vaughan's method [37], which does not impose the kinematic constraints and is similar to those proposed in [19] and the EKF introduced in this paper, which allows automatic marker labeling, is robust to short marker occlusions and imposes kinematic constraints, even in real time, so the local accelerations measured by the IMUs could be used to tune the filters applied to the optical motion capture data.

The remaining of the paper is organized as follows. Section 2 describes the two experiments carried out, sensor test and gait analysis. It points out the errors that may be incurred by inertial sensors, and proposes a way to minimize their influence. Then, it explains the two motion reconstruction methods applied and compared in this work, with a detailed description of the EKF, and shows the procedure to obtain the accelerations of the IMU attachment points from the optical system-based analysis, so that they can be compared with the accelerations measured by the IMUs. Section 3 presents the results of both the preliminary test and the gait analysis, showing the errors of the inertial sensors in orientations and accelerations, and the effect of the filter parameters adopted for the motion reconstructions methods in the accuracy of the accelerations obtained from the optical system recordings. Finally, Section 4 discusses the results and points out the limitations of the study, while Section 5 draws the conclusions of the work.

2. Materials and Methods

2.1. Preliminary Test

IMUs provide the measured accelerations expressed in their local reference frames, while the optical motion capture system provides marker positions within its own fixed reference frame. Therefore, in order to establish a comparison between accelerations coming from both techniques, it is necessary to express them in the same reference frame, and this implies obtaining the transformation matrix between each IMU and the fixed frame used by the optical motion capture system.

As mentioned above, an IMU can use sensor fusion algorithms to estimate its orientation within an Earth-fixed frame. This Earth-fixed frame is usually defined as NED (North-East-Down) or NWU (North-West-Up), and will be probably rotated with respect to the reference frame used by the motion capture lab. Therefore, the first step is to determine the offset between both reference frames, for which two methods can be applied: (i) the first option is to carry out a preliminary IMU calibration process, as the spot check proposed in [11]; (ii) the second alternative is to attach three markers to each IMU, so the local frames can be obtained directly from the optical motion capture system [38]. Since the second method requires a large number of markers, thus making the motion capture process more involved and error-prone, the calibration approach has been chosen in this work.

2.1.1. Experimental Data Collection

Nine IMUs (STT-IWS, STT Systems, San Sebastián, Spain) sampling at 100 Hz were fixed on a flat rigid wooden plate (with no ferromagnetic disturbances), equally spaced and accurately aligned to each other. Four reflective markers were also attached to four of the sensors, as illustrated in Figure 1.

Figure 1. Calibration setup composed by nine IMUs and four markers on a rigid plate.

The optical motion capture system was formed by 18 infrared cameras (OptiTrack FLEX 3, Natural Point, Corvallis, OR, USA), also sampling at 100 Hz. Starting with the plate on the floor, where it was kept for 5 s, it was manually moved around for 30 s and, finally, put again in the original place during 5 s. Data from both the IMU set and the optical system were recorded and the plate orientation during the motion was obtained from: (i) each IMU, based on gravity, magnetic North and gyroscope integration within the commercial software iSen provided by the manufacturer; (ii) the optical system, by rigid-body motion reconstruction based on the trajectories of the reflective markers 1, 2 and 3.

2.1.2. Sensor Orientation and Geomagnetic Frame of Reference

Figure 2 shows the three reference frames involved in the problem. The first reference frame is the global reference frame of the motion capture lab, obtained after calibration of the optical system, and it is noted with subscript O (after optical). This reference frame is fixed and common for all IMUs. The second reference frame is the global, Earth-fixed reference frame of each inertial sensor, and it is noted with subscript E (after Earth-fixed), and superscript *i* denoting the IMU number. Although this frame should be the same for all the IMUs, their inherent errors in determining gravity and magnetic North directions lead to discrepancies among sensors. The third reference frame is the local reference frame of each inertial sensor, and it is noted with subscript I (after inertial). In the calibration setup, the local reference frame is the same for all the IMUs, and it coincides with the local reference frame of the wooden plate. Note that the axes of this reference frame have a bar on them, meaning that they are moving axes, rigidly attached to the wooden plate.

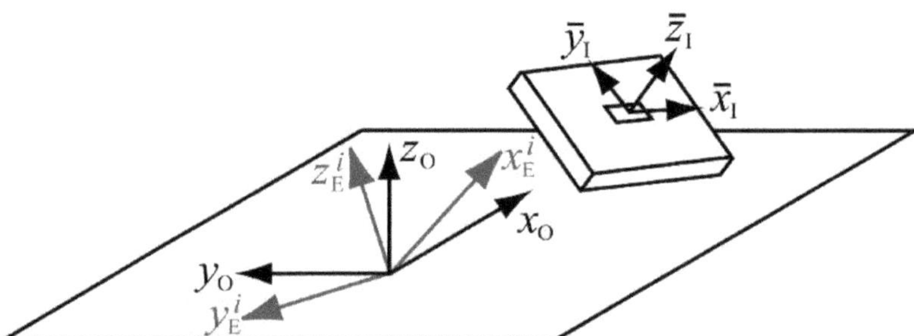

Figure 2. The three reference frames involved in the calibration: fixed global reference frame of the optical system (subscript O); Earth-fixed global reference frame of each IMU (in grey, subscript E and superscript i); moving local reference frame of all the IMUs and the wooden plate (subscript I).

If $\mathbf{R}_{\mathrm{OI}}$ is the variable rotation matrix that transforms the components of a vector expressed in the reference frame I into the components of the same vector expressed in the reference frame O, $\mathbf{R}_{\mathrm{OE}}^{i}$ is the constant rotation matrix that does the same between frames E and O for the inertial sensor i, and $\mathbf{R}_{\mathrm{EI}}^{i}$ is the variable rotation matrix that makes the same between frames I and E for that sensor, the following relation can be stated at any instant of the plate motion:

$$\mathbf{R}_{\mathrm{OI}} = \mathbf{R}_{\mathrm{OE}}^{i}\mathbf{R}_{\mathrm{EI}}^{i} \tag{1}$$

At any time point, the trajectories of the markers measured by the optical system provide $\mathbf{R}_{\mathrm{OI}}$, while the sensor fusion algorithm from the ith IMU provides $\mathbf{R}_{\mathrm{EI}}^{i}$. Therefore $\mathbf{R}_{\mathrm{OE}}^{i}$ can be derived from Equation (1) as,

$$\mathbf{R}_{\mathrm{OE}}^{i} = \mathbf{R}_{\mathrm{OI}}\mathbf{R}_{\mathrm{EI}}^{i\,\mathrm{T}} \tag{2}$$

Each $\mathbf{R}_{\mathrm{OE}}^{i}$ matrix must be constant, since it represents a rotation between two fixed frames. However, if it is calculated for all the time points of the recorded motion, the obtained values will not be completely constant, due to sensor and estimation errors. In order to find a unique matrix for each IMU, an average rotation matrix is calculated and taken as its effective $\mathbf{R}_{\mathrm{OE}}^{i}$. Since rotation matrices are orthogonal, care must be taken when averaging them, so that orthogonality is preserved. The method followed here consisted of extracting the roll, pitch and yaw angles from each rotation matrix at every time point, averaging them, and using these values to build back the corresponding effective rotation matrix. This calibration procedure to get the $\mathbf{R}_{OE}^i$ matrices yields different results for different days due to magnetic changes, so it should be ideally performed right before using the IMUs.

Once the IMUs were properly calibrated, the orientation error provided by each of them along the motion of the wooden plate was obtained. For each time point, the trajectories of the markers measured by the optical system provided matrix $\mathbf{R}_{OI}$, while data from ith IMU provided $\mathbf{R}_{EI}^i$. Since the constant matrix $\mathbf{R}_{OE}^i$ had been obtained for each IMU in the calibration process described before, it can be written from Equation (1),

$$\mathbf{R}_{OI}^i = \mathbf{R}_{OE}^i \mathbf{R}_{EI}^i \tag{3}$$

where $\mathbf{R}_{OI}^i$ is the rotation matrix between frames I and O provided by the ith inertial sensor. Ideal IMUs would provide the same matrix $\mathbf{R}_{OI}^i$ for all the sensors, and it would be coincident with matrix $\mathbf{R}_{OI}$ provided by the optical system. However, due to errors in the IMUs, such matrices differ, and the orientation error committed by each IMU at each time point can then be obtained by calculating the roll, pitch and yaw angles of $\mathbf{R}_{OI}^i$, and comparing them with the roll, pitch and yaw angles of $\mathbf{R}_{OI}$, taken as reference. This was done for the nine IMUs, the results being shown in Section 3.

Once the method to obtain the orientation error of each inertial sensor has been described, the objective of comparing the accelerations provided by the optical and the inertial systems is addressed. The optical system provides the trajectories of the markers, based on which the position history of any point of the body can be obtained through a motion reconstruction method. Then, double differentiation of the position history yields velocity and acceleration histories of the point considered. Positions, velocities and accelerations are expressed in the global reference frame of the motion capture lab (previously denoted by O). On the other hand, IMU accelerometers measure a combination of the gravitational and translational accelerations. As reported by Woodman [21], it is necessary to have very accurate rotation sensors in inertial navigation systems, because knowing the precise orientation of the body allows to properly subtract the gravitational acceleration from the measurement, in order to find the translational acceleration. Each IMU provides its acceleration expressed in its local reference frame (previously denoted by I). Therefore, to compare accelerations obtained through the optical and inertial techniques, it is necessary to express them in the same reference frame and to take into account the gravity constant, which is present in the inertial case.

To highlight all the mentioned issues, the acceleration of point 4 in Figure 1 was obtained in three different ways. First, since point 4 had a marker on it, the marker trajectory was filtered by means of a 8 Hz forward-backward 2nd order Butterworth filter, then it was differentiated twice with respect to time, and the gravity constant (9.81 m/s^2) was added to the vertical component of the resulting acceleration; the presence of the marker attached at the point made it unnecessary the use of any motion reconstruction method, thus eliminating a source of error for the optical system. Second, the acceleration provided by the IMU at point 4 was expressed in frame O by multiplying it by matrix $\mathbf{R}_{OI}^5$ (orientation provided by the IMU #5, attached to that point). Third, the acceleration provided by the IMU at point 4 was expressed in frame O by multiplying it by matrix $\mathbf{R}_{OI}$ (orientation provided by the optical system after the mentioned filtering of the marker trajectory). The resulting accelerations and their comparison are shown in Section 3.

2.2. Gait Analysis

2.2.1. Experimental Data Collection

A healthy adult male, 24 years old, 70 kg, and 175 cm, performed a complete gait cycle. Both 36 reflective markers in all his body segments for optical motion capture (same equipment as that described in Section 2.1) and 7 IMUs (the best seven among the nine tested in the preliminary test) at pelvis, thighs, shanks and feet for inertial motion capture were attached to the subject's body, as can be seen in Figure 3. One additional marker was attached to each IMU so as to determine its local position within the corresponding segment during a static pose recording.

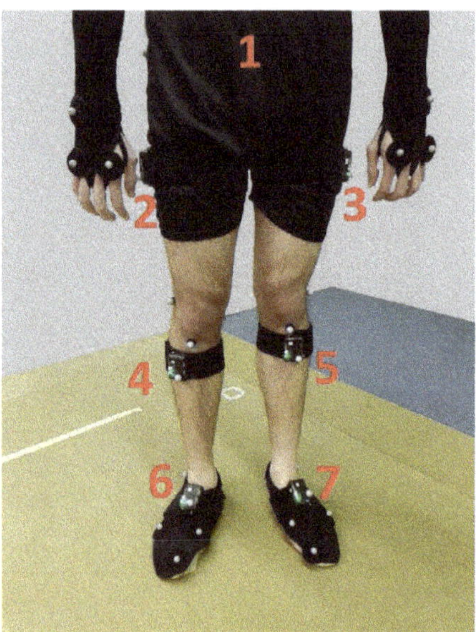

Figure 3. Markers and IMUs (red numbers) attached to the subject's body for gait analysis.

2.2.2. Skeletal Model and Kinematics

The human body is modeled as a three-dimensional multibody system formed by rigid bodies, as shown in Figure 4. The model consists of 18 anatomical segments [39]: two hind feet, two forefeet, two shanks, two thighs, pelvis, torso, neck, head, two arms, two forearms, and two hands. The segments are linked by ideal spherical joints (black dots in Figure 4b), thus defining a model with 57 degrees of freedom (DOF). The axes of the global reference frame are defined as follows: x-axis in the antero-posterior direction, y-axis in the medio-lateral direction, and z-axis in the vertical direction.

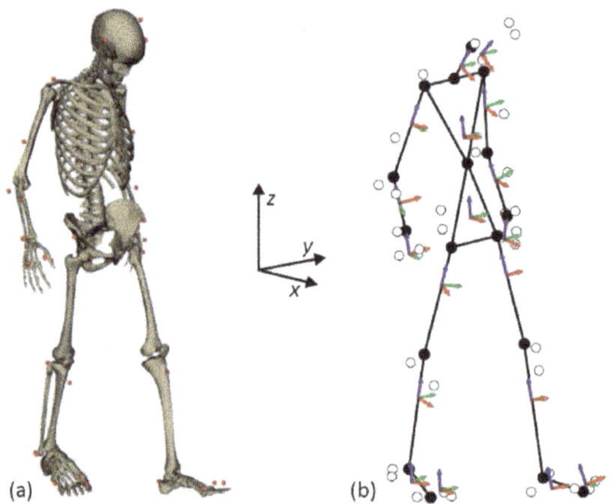

Figure 4. 3D human model: (**a**) graphical output; (**b**) multibody model showing the segments: joints (black dots), and marker locations (white dots).

2.2.3. Motion Reconstruction from Motion Capture Data

Optical motion capture records the motion of entities (markers) that are external to the body, and the objective is to use the marker data to determine the positions and orientations of the body segments. The traditional approach for accomplishing this is to use the method described by Vaughan [37]: (i) select three non-collinear entities, which can be either markers or already located joints, within each body segment; (ii) define an orthogonal reference frame for the corresponding segment, based on the three selected entities; (iii) use correlation equations, based on archived anthropometric data and body measurements, to estimate the position and orientation of the body segment. When applying this method, marker trajectories are previously filtered with a low-pass filter (forward-backward 2nd order Butterworth filter), whose cutoff frequency must be selected by the analyst.

Another commonly used approach is to solve a weighted optimization problem, in order to fit the skeletal model to the measured markers, as done in the OpenSim software [40]. The fitting is carried out in two steps. First, a reference skeletal model, with virtual markers fixed to the anatomical points, is scaled in order to match the markers from a static capture, taken in a reference pose. Then, a second optimization problem finds the positions and orientations of the scaled body segments that best track the motion capture data. This last optimization uses an independent set of positions and orientations as design variables, which can be filtered and differentiated afterwards to find velocities and accelerations.

These methods present important drawbacks. In the first one, the local frames are obtained directly from the markers, which are not rigidly attached to the bones, so the obtained skeletal motion is not consistent with the rigid body constraints, i.e., the distances between joints do not remain constant. This can be addressed by enforcing kinematic consistency in a post-processing stage [41]. Another problem, common to both methods, is that they require clean capture data: marker trajectories must not contain any gaps, and the markers need to be properly labeled at every time step, something that is not always guaranteed by the motion capture system. The procedure used for fixing this problem usually involves some manual gap filling and marker labeling [20] and, in some cases of severe marker loss, it is not even possible to salvage a take. The main consequence of these drawbacks is the impossibility of knowing if a motion capture take has been successful until all post-processing has been carried out.

2.2.4. Extended Kalman Filter for Motion Reconstruction

In order to overcome these drawbacks, a motion capture algorithm based on the extended Kalman filter has been developed. The filter uses a purely kinematic model for the plant, while the markers act as position sensors. The kinematic model mostly coincides with that described in Figure 4 but, in order to avoid the need for additional markers, the spherical joint at the base of the neck has been substituted by a universal joint, and metacarpophalangeal joints are modeled here as revolute pairs. Therefore, the kinematic model used in the Kalman filter has 52 degrees of freedom instead of 57. Since the Kalman filter requires using independent state variables, the position of the model must be defined by a set of independent coordinates, including 3 base body translations (pelvis), 2 relative angles at the toes, 2 relative angles at the base of the neck, and 45 Euler angles representing the absolute orientation of the remaining bodies.

The Kalman filter is based on a discrete white noise acceleration model (DWNA) [42], in which the plant is considered as a discrete-time state-space system,

$$\mathbf{x}_{k+1} = \mathbf{F}\mathbf{x}_k + \mathbf{\Gamma}\mathbf{a}_k \tag{4}$$

where $\mathbf{x}_{k+1}$ and $\mathbf{x}_k$ are the state vector at time instants $k+1$ and k respectively, $\mathbf{F}$ is the state propagation matrix, $\mathbf{a}_k$ is the process noise vector, and $\mathbf{\Gamma}$ is the noise gain matrix. The DWNA is a second-order kinematic model, so the state vector contains the 52 degrees of freedom, $\mathbf{q}$, along with their first time derivatives, $\dot{\mathbf{q}}$. Accelerations are introduced in the system through the process noise vector $\mathbf{a}$. This vector contains the 52 independent accelerations, being each of them a discrete-time zero-mean white sequence. Therefore, they are assumed to be constant along every time step, and their values are random variables with a zero-mean normal distribution of variance σ_a^2. This variance has dimensions of squared acceleration for the translational DOFs, and squared angular acceleration for the angular ones. In order to reduce the number of parameters, in this work the same numerical value will be used for all of them.

Taking into account that accelerations are assumed to remain constant along each time step, the state transition in Equation (4) particularized to any given DOF i is,

$$\begin{bmatrix} q_{k+1}^i \\ \dot{q}_{k+1}^i \end{bmatrix} = \begin{bmatrix} 1 & \Delta t \\ 0 & 1 \end{bmatrix} \begin{bmatrix} q_k^i \\ \dot{q}_k^i \end{bmatrix} + \begin{bmatrix} \frac{1}{2}\Delta t^2 \\ \Delta t \end{bmatrix} a_k^i \; i = 1, \dots, 52 \tag{5}$$

where Δt is the sampling period, which is fixed to 10 ms (the motion capture cameras used in this work have a maximum frame rate of 100 Hz). As can be seen in Equation (5), the state propagation and noise gain matrices defined for each DOF only depend on the time step Δt, so they are constant and equal for all of them. Therefore, matrices $\mathbf{F}$ and $\mathbf{\Gamma}$ for the whole system are the result of assembling these individual matrices, following the structure of the state vector $\mathbf{x}$.

The process noise covariance matrix $\mathbf{Q}^i$ has also the same form for all DOFs [42],

$$\mathbf{Q}^i = \begin{bmatrix} \frac{1}{4}\Delta t^4 & \frac{1}{2}\Delta t^3 \\ \frac{1}{2}\Delta t^3 & \Delta t^2 \end{bmatrix} \sigma_a^2 \tag{6}$$

The $\mathbf{Q}$ matrix for the whole system is the result of assembling these individual matrices, as done for the state transition and noise gain matrices.

The observation function $\mathbf{h}(\mathbf{x})$ provides the observation vector $\mathbf{z}$, which in this case contains the absolute x, y and z coordinates of the 36 optical markers, as a function of the state vector $\mathbf{x}$,

$$\mathbf{z}_k = \mathbf{h}(\mathbf{x}_k) + \mathbf{w}_k \tag{7}$$

The additive term $\mathbf{w}_k$ represents the noise introduced by the motion capture system, along with the skin motion artifact. Since the latter is correlated to the skeletal motion, modeling the sensor noise as a random variable following a Gaussian distribution is not strictly correct, so the Kalman filter will not be optimal. All sensors are considered independent and equally affected by noise, so the observation noise covariance matrix $\mathbf{R}$ is a diagonal matrix, whose diagonal elements are all equal to the sensor noise variance σ_s^2, which has dimensions of squared length.

In order to compute the absolute marker positions from the system states, a recursive kinematic relationship can be established, as shown in Figure 5. The absolute position $\mathbf{z}_i$ of a marker i, which is attached to body b (right hand in Figure 5), can be obtained from the following recursive relationships,

$$\begin{aligned} \mathbf{z}_i &= \mathbf{r}_b + \mathbf{A}_b \overline{\mathbf{m}}_i \\ \mathbf{r}_b &= \mathbf{r}_{b-1} + \mathbf{A}_{b-1} \overline{\mathbf{r}}_b \end{aligned} \tag{8}$$

where $\mathbf{r}_b$ is the absolute position of the proximal joint of body b, $\mathbf{A}_b$ is the rotation matrix of the same body, which depends on its three orientation angles, and $\overline{\mathbf{m}}_i$ is the local position vector of marker i within the local frame of body b. In the observation model, the markers are considered as rigidly attached to the skeleton, so $\overline{\mathbf{m}}_i$ is a constant vector. The vector $\mathbf{r}_b$ itself can be obtained in a recursive way from the position vector $\mathbf{r}_{b-1}$ and orientation matrix $\mathbf{A}_{b-1}$ of the preceding body in the kinematic chain, knowing that $\overline{\mathbf{r}}_b$ is the position of the proximal joint of body b in the local frame of $b - 1$, which, due to the rigid body assumption, is considered constant. This recursive process starts at the pelvis, whose position vector is contained directly in $\mathbf{q}$ and, consequently, in $\mathbf{x}$.

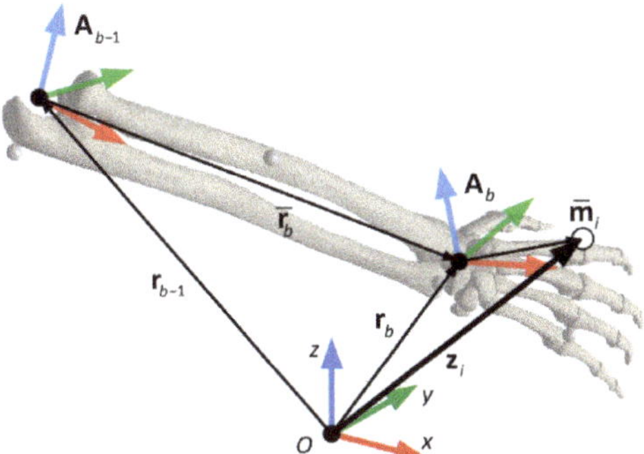

Figure 5. Kinematic description of the observation function h(x).

The local position vectors $\bar{\mathbf{r}}_b$ and $\bar{\mathbf{m}}_i$ must be scaled prior to running the Kalman filter, in order for the model to adjust to the experimental data. This is performed by solving, at a reference pose, a nonlinear least squares optimization problem, in which the design variables are a set of scale factors $\mathbf{k}$ and the skeletal degrees of freedom $\mathbf{q}$, being the objective function the quadratic error between measured and estimated marker positions,

$$\min_{\mathbf{q},\mathbf{k}} f(\mathbf{q},\mathbf{k}) = [\mathbf{h}_a(\mathbf{x},\mathbf{k}) - \mathbf{z}]^T [\mathbf{h}_a(\mathbf{x},\mathbf{k}) - \mathbf{z}] \tag{9}$$

where $\mathbf{h}_a(\mathbf{x},\mathbf{k})$ is an augmented version of the observation function that also takes the scale factors as input variables. The resulting scale factors are then used to scale the $\bar{\mathbf{r}}_b$ and $\bar{\mathbf{m}}_i$ vectors that will be used in $\mathbf{h}(\mathbf{x})$. It has been found that the Levenberg-Marquardt algorithm works very well for this problem, converging in a very robust way even from rough initial estimates.

The Kalman filter algorithm follows a recursive predictor-corrector scheme. It uses the current estimate of the state vector, $\hat{\mathbf{x}}_k$, along with the sensor measurements, $\mathbf{z}_{k+1}$, to obtain an optimal estimate $\hat{\mathbf{x}}_{k+1}$ at the next time step. In the predictor stage, the state estimate is updated by means of the state transition matrix $\mathbf{F}$, leading to the so-called *a priori* estimate $\hat{\mathbf{x}}_{k+1}^-$. The estimate covariance matrix $\mathbf{P}$ is updated accordingly, by using matrices $\mathbf{F}$ and $\mathbf{Q}$,

$$\begin{aligned} \hat{\mathbf{x}}_{k+1}^- &= \mathbf{F}\hat{\mathbf{x}}_k \\ \mathbf{P}_{k+1}^- &= \mathbf{F}\mathbf{P}_k\mathbf{F}^T + \mathbf{Q} \end{aligned} \tag{10}$$

The state estimate at the first time step, $\hat{\mathbf{x}}_0$, will contain the initial independent positions, $\mathbf{q}_0$, along with the corresponding velocities, $\dot{\mathbf{q}}_0$. The positions are obtained after solving the initial marker labeling problem, which will be described later. The initial velocities are unknown, as well as the value of $\mathbf{P}_0$, so they are both set to zero, but they converge quickly to their correct values after a short transient.

The corrector stage uses the sensor measurements $\mathbf{z}_{k+1}$ to find the optimal *a posteriori* estimate $\hat{\mathbf{x}}_{k+1}$, as well as its corresponding covariance matrix, $\mathbf{P}_{k+1}$,

$$\begin{aligned} \mathbf{K}_{k+1} &= \mathbf{P}_{k+1}^- \mathbf{H}_{k+1}^T \left(\mathbf{H}_{k+1} \mathbf{P}_{k+1}^- \mathbf{H}_{k+1}^T + \mathbf{R} \right)^{-1} \\ \hat{\mathbf{x}}_{k+1} &= \hat{\mathbf{x}}_{k+1}^- + \mathbf{K}_{k+1} \left[\mathbf{z}_{k+1} - \mathbf{h}(\mathbf{x}_{k+1}^-) \right] \\ \mathbf{P}_{k+1} &= (\mathbf{I} - \mathbf{K}_{k+1}\mathbf{H}_{k+1})\mathbf{P}_{k+1}^- \end{aligned} \tag{11}$$

In these equations, $\mathbf{H}_{k+1}$ is the Jacobian matrix of the observation function, evaluated at $\hat{\mathbf{x}}_{k+1}^-$. This matrix can be computed very efficiently due to the recursive nature of $\mathbf{h}(\mathbf{x})$. Moreover, it is quite sparse, due to the usage of absolute angles as state variables: most rotation matrices will only depend on three angles, greatly simplifying their derivatives. In addition, the gradient used in the scale optimization problem shown in Equation (9) mostly coincides with this matrix, which is very convenient from the implementation point of view.

Since this algorithm is recursive and each step can be evaluated very efficiently, it can be used for real-time motion reconstruction and visualization, as opposed to the previously mentioned methods, which provide skeletal motion after post-processing the captured data. In order to achieve on-the-fly motion reconstruction, the marker labeling and occlusion issues must be addressed.

The problem of initial marker labeling is addressed by using a simple heuristic method to identify the markers. The procedure consists of checking their relative positions at the initial time step, according to a reference pose. Then, the same Levenberg-Marquardt optimization algorithm previously used for scaling the model is used here to fit the DOFs to the measured markers. This time, the objective function uses the regular observation function $\mathbf{h}(\mathbf{x})$ with the scale factors already applied to the local position vectors, so only the positions $\mathbf{q}_0$ are considered as design variables. If the objective function value after the optimization (i.e., the fitting error) is below a certain threshold, the marker order is considered valid, and the iterative process of the Kalman filter can begin.

During the execution of the Kalman filter, marker labeling must be carried out on the fly, between the predictor in Equation (10) and the corrector in Equation (11). This is because, for several reasons, the raw measurement vector $\mathbf{z}^r_{k+1}$ obtained from the cameras cannot be directly used within the corrector. First, the markers are provided as an unsorted list by the cameras. Second, some markers may be missing due to occlusions. Third, other bright objects present during the motion capture can be incorrectly identified as markers. Therefore, the raw measurement $\mathbf{z}^r_{k+1}$ must be correctly labeled and sorted, the missing markers need to be identified, and all spurious markers have to be discarded, in order to get the "clean" measurement vector $\mathbf{z}_{k+1}$. After the Kalman filter predictor has computed the a priori state estimate $\mathbf{x}^-_{k+1}$, the observation function $\mathbf{h}(\mathbf{x})$ is evaluated at that point to obtain the corresponding set of estimated marker positions $\hat{\mathbf{z}}_{k+1}$. Ideally, these estimated markers would coincide with the measured ones $\mathbf{z}^r_{k+1}$, and this fact can be used to identify the measured markers by using a simple, nearest-neighbor approach. First, a matrix of squared cross-distances $\mathbf{D}$ is built, such that

$$D_{ij} = \left(\hat{\mathbf{z}}_i - \mathbf{z}^r_j \right)^T \left(\hat{\mathbf{z}}_i - \mathbf{z}^r_j \right) \tag{12}$$

where $\hat{\mathbf{z}}_i$ contains the estimated x, y and z coordinates of marker i, and $\mathbf{z}^r_j$ is the position vector of measured marker j. By setting a maximum search distance, estimated markers that do not have a measured one close enough are considered as missing, and the remaining ones are assigned to their closest measured counterparts. Any marker from $\mathbf{z}^r_{k+1}$ remaining unassigned, after all estimated markers have been either paired to their measured counterparts or marked as missing, are regarded as spurious, so they are discarded. In order to avoid resizing vectors and matrices at runtime, missing markers are set to zero in $\mathbf{z}$, and the same is done to their corresponding rows in $\mathbf{H}$, so they do not affect the correction.

The EKF can provide a smoothing effect depending on the tuning of its parameters, so in this case there is no need for filtering the marker trajectories. If the sensor noise variance σ_s^2 is fixed to a constant value, the smoothing can be controlled by the process noise variance, i.e., the acceleration variance σ_a^2. Low values of the variance limit the accelerations the system can reach at every time step, thus having a smoothing effect on the resulting position histories, while high values of the variance allow for larger accelerations, so that the system can follow the sensors (i.e., the markers) more closely, at the expense of introducing sensor noise into the reconstructed motion. In this work, the accelerations are obtained by further filtering the independent positions, and differentiating them twice to obtain velocities and accelerations. There exist higher order state-space models that include accelerations in the state vector, but they present two major issues when used with position sensors only: the resulting accelerations are noisy and delayed, and some unwanted oscillations may appear in the resulting motion for certain values of the filter parameters.

The analyst has in this case two parameters for tuning the obtained accelerations: the process noise variance and the cutoff frequency of the Butterworth filter. In order to find their optimum values, the use of accelerometers can be of great help.

2.2.5. Calculation of the Accelerations

In order to make the accelerations obtained from the optical system directly comparable to those obtained by the inertial sensors, it was necessary to transform the former into the local axes of the corresponding IMUs, and to add the gravity effect to them. Such an acceleration obtained from the optical system will be called hereafter the virtual acceleration, as it comes from a virtual accelerometer. It is possible to do the opposite, i.e., to rotate the IMU measurements to the global frame of reference instead, subtracting the acceleration of gravity afterwards. However, this procedure involves mixing data from both systems (to rotate the IMU accelerations to the global frame), so the first alternative seems more appropriate.

Figure 6 shows a segment or body of the multibody model of the human skeleton, where the black dots are joints connecting the segment with its neighbors, and the white dots are the markers attached to the segment. The small rectangle represents the IMU attached to the segment, which in turn has a marker attached to it, as also shown in Figure 3. The local reference frame of the body is denoted by B, a moving frame rigidly attached to the body, and its origin is defined in frame O by the position vector $\mathbf{r}_B$ while the local position of the IMU in frame B is given by the constant vector $\bar{\mathbf{r}}_i$ (i is the number of the IMU attached to that particular body). The following equation can be written,

$$\mathbf{r}_i = \mathbf{r}_B + \mathbf{R}_{OB}\bar{\mathbf{r}}_i \tag{13}$$

where $\mathbf{r}_i$ is the position vector of the IMU in frame O and $\mathbf{R}_{OB}$ is the rotation matrix between frames B and O. Then, $\bar{\mathbf{r}}_i$ can be worked out as,

$$\bar{\mathbf{r}}_i = \mathbf{R}_{OB}^{T}(\mathbf{r}_i - \mathbf{r}_B) \tag{14}$$

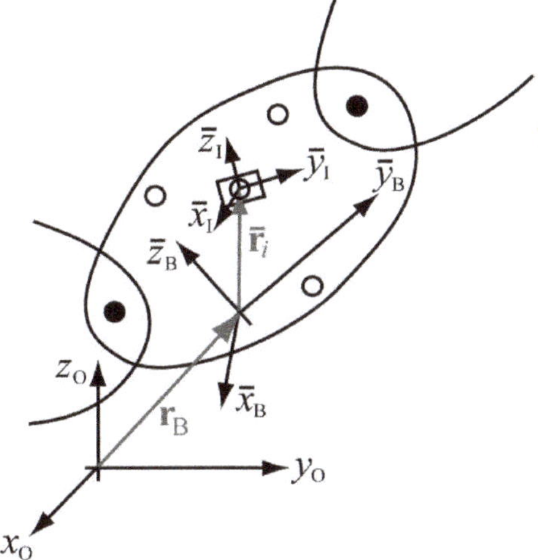

Figure 6. Kinematics at segment level to get the virtual acceleration (acceleration of the point of the body where the IMU is attached) from the optical motion capture recordings. The black dots represent joints connecting the segment with its neighbors. The white dots represent the markers attached to the segment.

Regarding the orientations, the following relation stands,

$$\mathbf{R}_{\mathrm{OB}}\overline{\mathbf{R}}_{\mathrm{BI}} = \mathbf{R}_{\mathrm{OE}}^{i}\mathbf{R}_{\mathrm{EI}}^{i} \tag{15}$$

being $\overline{\mathbf{R}}_{\mathrm{BI}}$ the unknown constant rotation matrix between frames B and I. Superscript i in the rotation matrices of the right-hand side refers to the number of the IMU attached to that particular body. From Equation (15), the constant rotation matrix $\overline{\mathbf{R}}_{\mathrm{BI}}$ can be worked out as,

$$\overline{\mathbf{R}}_{\mathrm{BI}} = \mathbf{R}_{\mathrm{OB}}^{\mathrm{T}}\mathbf{R}_{\mathrm{OE}}^{i}\mathbf{R}_{\mathrm{EI}}^{i} \tag{16}$$

The two constant terms worked out in Equation (14) and Equation (16), respectively, must be determined in order to obtain, later on, the virtual acceleration. To this end, a capture of a static pose of the subject is recorded by both the optical and inertial systems. From the positions of the markers, $\mathbf{r}_i$, $\mathbf{r}_{\mathrm{B}}$ and $\mathbf{R}_{\mathrm{OB}}$ can be derived, so that $\overline{\mathbf{r}}_i$ is calculated from Equation (14). On the other hand, the constant matrix $\mathbf{R}_{\mathrm{OE}}^{i}$ had been obtained in the calibration process carried out during the preliminary test, while $\mathbf{R}_{\mathrm{EI}}^{i}$ can be derived from the orientation provided by the IMU, so that $\overline{\mathbf{R}}_{\mathrm{BI}}$ is calculated using Equation (16). It must be noted that this is the only point, along the process of getting the virtual acceleration, in which the orientation provided by the IMU is used. However, this does not induce a significant error, since the estimated orientations are much more accurate in static conditions.

Once the constant terms $\overline{\mathbf{r}}_i$ and $\overline{\mathbf{R}}_{\mathrm{BI}}$ have been determined in the described preprocess, the history of the virtual acceleration can be derived from the info recorded by the optical system. At each time point, the global acceleration of the point where the IMU is attached, expressed in frame O, can be obtained by differentiating Equation (13) twice with respect to time,

$$\ddot{\mathbf{r}}_i = \ddot{\mathbf{r}}_{\mathrm{B}} + \ddot{\mathbf{R}}_{\mathrm{OB}}\overline{\mathbf{r}}_i \tag{17}$$

where $\ddot{\mathbf{r}}_{\mathrm{B}}$ and $\ddot{\mathbf{R}}_{\mathrm{OB}}$ are calculated as the second derivative with respect to time of the position data obtained from the optical motion capture. The virtual acceleration, still expressed in frame O, is obtained by including the gravity effect into the acceleration given by the Equation (17),

$$\mathbf{a}_i = \ddot{\mathbf{r}}_i + \mathbf{g} \tag{18}$$

being $\mathbf{g}$ the gravity vector (9.81 m/s^2 in the positive vertical direction, as it would be perceived by the IMU). To get the virtual acceleration, vector $\mathbf{a}_i$ must be expressed in the local frame of the IMU, I,

$$\overline{\mathbf{a}}_i = \mathbf{R}_{\mathrm{OI}}^{\mathrm{T}}\mathbf{a}_i \tag{19}$$

with,

$$\mathbf{R}_{\mathrm{OI}} = \mathbf{R}_{\mathrm{OB}}\overline{\mathbf{R}}_{\mathrm{BI}} \tag{20}$$

where $\mathbf{R}_{\mathrm{OB}}$ is calculated from the optical motion capture. Compacting Equations (17)–(20) into a single expression, the virtual acceleration can be written as,

$$\overline{\mathbf{a}}_i = \overline{\mathbf{R}}_{\mathrm{BI}}^{\mathrm{T}}\mathbf{R}_{\mathrm{OB}}^{\mathrm{T}}\left(\ddot{\mathbf{r}}_{\mathrm{B}} + \ddot{\mathbf{R}}_{\mathrm{OB}}\overline{\mathbf{r}}_i + \mathbf{g}\right) \tag{21}$$

Therefore, the acceleration directly measured by the IMU can now be compared to the virtual acceleration provided by the Equation (21) from the measurements of the optical system, and the filtering parameters applied in the latter can be adjusted so as to yield the optimal correlation. The error was measured as the root-mean-square error (RMSE) between the histories of the two accelerations compared, the results being shown in Section 3.

3. Results

3.1. Preliminary Test and Calibration

As explained in Section 2, the orientation error committed by the *i*th IMU, $i = 1, \dots, 9$, at each time point, was obtained by calculating the roll, pitch and yaw angles of $\mathbf{R}^i_{OI}$, and comparing them with the roll, pitch and yaw angles of $\mathbf{R}_{OI}$, taken as reference. Figure 7 shows the error incurred by each IMU in roll, pitch and yaw angles, along the time of the calibration experiment. Maximum errors of 19° in yaw (around the vertical axis) with respect to the reference (optical system) were found, while mean error differences of up to 4° were detected among IMUs. Similar results were obtained for $\mathbf{R}^i_{OE}$, the rotation matrix between frames E and O for the inertial sensor *i*: differences of up to 8° in yaw were detected among IMUs.

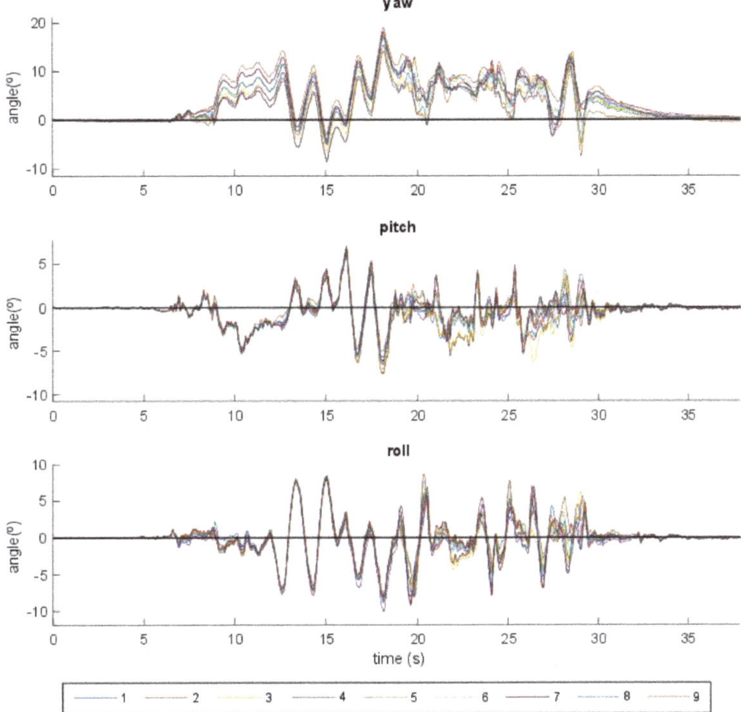

Figure 7. Orientation errors in roll, pitch and yaw incurred by the nine IMUs with respect to the optical system (reference).

As explained in the last paragraph of Section 2.1.2, the acceleration of point 4 of the wooden plate was obtained in three different ways: (i) from the optical system, using the marker attached to point 4; (ii) from the inertial system, using the orientation provided by the inertial system; (iii) from the inertial system, using the orientation provided by the optical system. A forward-backward 2nd order Butterworth filter with a cutoff frequency of 8 Hz was applied to the optically captured trajectories of the markers, while no filtering was applied to the inertial measurements. Figure 8 gathers the global components, expressed in frame O, of the three accelerations. While the x- and z-components are similar, significant discrepancies are observed between 15 and 20 s for the y-component, with a maximum error of 1.9 m/s² when using the orientation provided by the IMU. Moreover, the accelerometer shows some peaks that are not captured by the optical system, for instance when the plate touches the ground after the 30 s mark. Due to the low sampling rate of the optical system, it cannot capture high-frequency events such as impacts, regardless of the filter cutoff frequency.

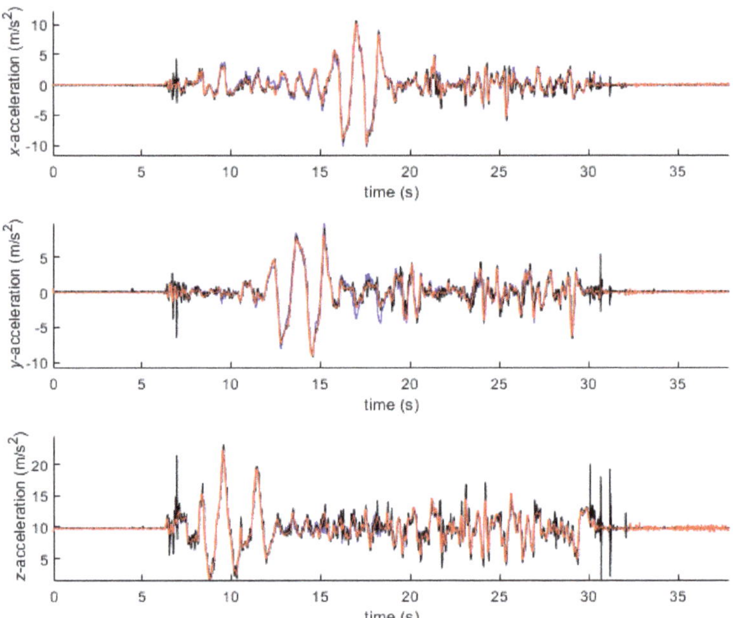

Figure 8. Comparison of the global acceleration of point 4 of the wooden plate, obtained by three methods: from the optical system (red); from the inertial system with the orientation provided by the inertial system (blue); from the inertial system with the orientation provided by the optical system (black).

3.2. Gait Analysis

This Section is devoted to gather the results obtained when comparing the accelerations provided by each of the seven IMUs during the gait analysis described in Section 2.2, with the so-called virtual accelerations obtained from the optical motion capture. As explained in the mentioned Section, the trajectories of the markers recorded by the optical system should be processed by a motion reconstruction method, which includes filtering of the recorded data. Therefore, the results from each of the two reconstruction methods proposed in Section 2.2 are shown in what follows.

3.2.1. Vaughan's Method

The virtual accelerations obtained after the application of the forward-backward 2nd order Butterworth filter with different cutoff frequencies were compared with those directly measured by the IMUs. Figure 9 shows the three components of the accelerations at the seven segments analyzed for cutoff frequencies of 6, 12 and 40 Hz, while Figure 10 provides more detail for the left foot. Table 1 gathers the RMSE of the optical-system based accelerations, with cutoff frequencies ranging between 6 and 40 Hz, with respect to those directly measured by the inertial system.

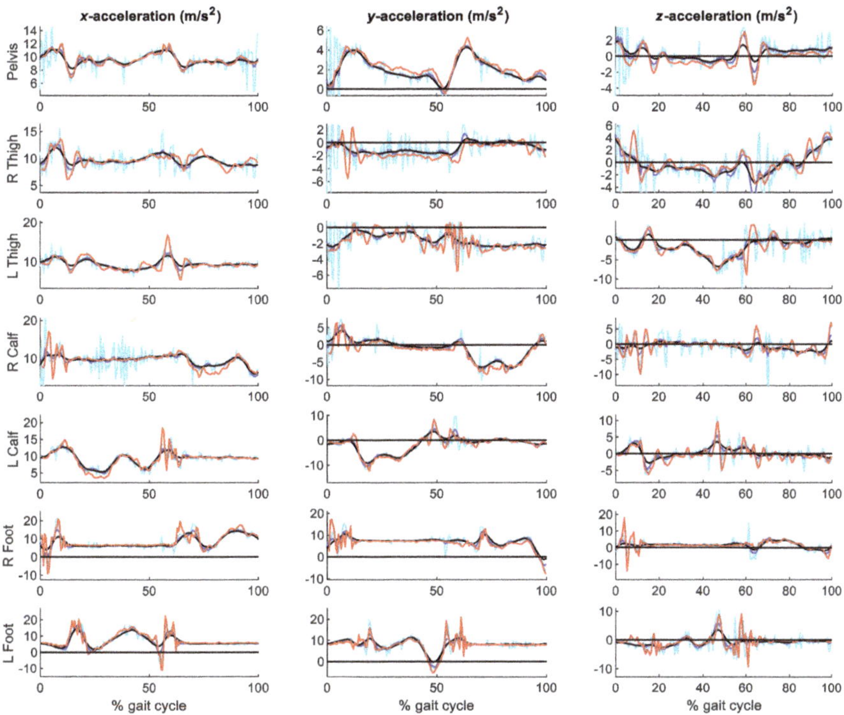

Figure 9. Accelerations obtained from the optical system with Vaughan's method, for cutoff frequencies of 6 Hz (black), 12 Hz (blue) and 40 Hz (cyan), respectively, vs. accelerations measured by the IMUs (red).

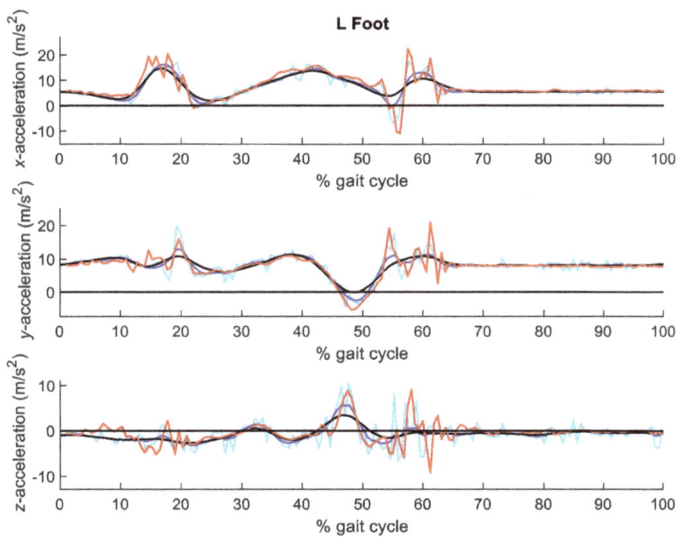

Figure 10. Detail of accelerations at the left foot obtained from the optical system with Vaughan's method, for cutoff frequencies of 6 Hz (black), 12 Hz (blue) and 40 Hz (cyan), respectively, vs. accelerations measured by the IMUs (red).

Table 1. RMSE of the accelerations obtained from the optical system through Vaughan's method with different cutoff frequencies, with respect to the accelerations measured by the IMUs, taken as reference. The row with the lowest mean RMSE is highlighted in red.

Cutoff Freq. (Hz)	RMSE (m/s^2)							
	Pelvis	R Thigh	L Thigh	R Tibia	L Tibia	R Foot	L Foot	Mean
6	0.626	1.034	1.123	1.583	1.502	2.743	2.485	1.585
8	0.578	1.005	1.071	1.538	1.448	2.640	2.405	1.336
10	0.559	0.997	1.043	1.515	1.428	2.571	2.362	1.309
12	0.559	1.003	1.031	1.508	1.426	2.529	2.339	1.299
15	0.583	1.034	1.036	1.521	1.445	2.504	2.328	1.306
20	0.680	1.138	1.086	1.607	1.510	2.526	2.354	1.363
25	0.840	1.303	1.181	1.771	1.602	2.590	2.422	1.464
30	1.045	1.517	1.314	2.002	1.706	2.679	2.526	1.599
40	1.531	2.035	1.654	2.602	1.931	2.897	2.815	1.933

It can be seen that the influence of the filtering parameter is significant. For high cutoff frequencies (above 20 Hz), the accelerations were too noisy, with peak errors over 5 m/s^2. Conversely, for low cutoff frequencies (below 8 Hz), the accelerations were too smooth, not reaching the experimental peak measurements of the inertial sensors. As opposed to the preliminary test, some acceleration peaks can be captured by the optical system at high cutoff frequencies, due to the softer contacting materials involved in this case, but at the cost of very noisy accelerations along the whole capture. The lowest errors were obtained for a cutoff frequency of 12 Hz, as highlighted in Table 1.

3.2.2. Extended Kalman Filter

The accelerations obtained after the application of different values of the process noise standard deviation σ_a (with the sensors noise standard deviation σ_s fixed to 0.001 m), along with the application of the forward-backward 2nd order Butterworth filter with different cutoff frequencies to the position data, were compared to those directly measured by the IMUs. The units for σ_a are omitted in what follows for the sake of brevity, since they depend on the associated degree of freedom: for translational DOFs, σ_a is expressed in m/s^2 whereas for rotational DOFs it is in rad/s^2.

Figure 11 shows the three components of the accelerations at the seven segments analyzed for combinations of σ_a and cutoff frequencies of 0.1/30 Hz, 1/20 Hz and 50/6 Hz, while Figure 12 provides more detail for the left foot. Table 2 gathers the RMSE of the optical-system based accelerations with σ_a ranging between 0.1 and 50 m/s^2 (or rad/s^2, depending on the corresponding coordinate), and cutoff frequencies ranging between 6 and 30 Hz, with respect to those directly measured by the inertial system.

It can be seen in Figures 11 and 12 that the accelerations obtained with the EKF were smoother than those obtained with Vaughan's method, and that the experimental peak measurements of the inertial sensors were better dissociated from the noise peaks. Moreover, Table 2 presents lower values of the RMSEs. The best results were obtained for a process noise standard deviation of 1 m/s^2 combined with a 20 Hz Butterworth filter.

In addition to obtaining better accelerations, it should be noted that the EKF automatically ensures kinematic consistency, whereas Vaughan's method shows joint distance variations above 1 cm along the gait cycle, so the resulting motion would require further post-processing depending on the intended application.

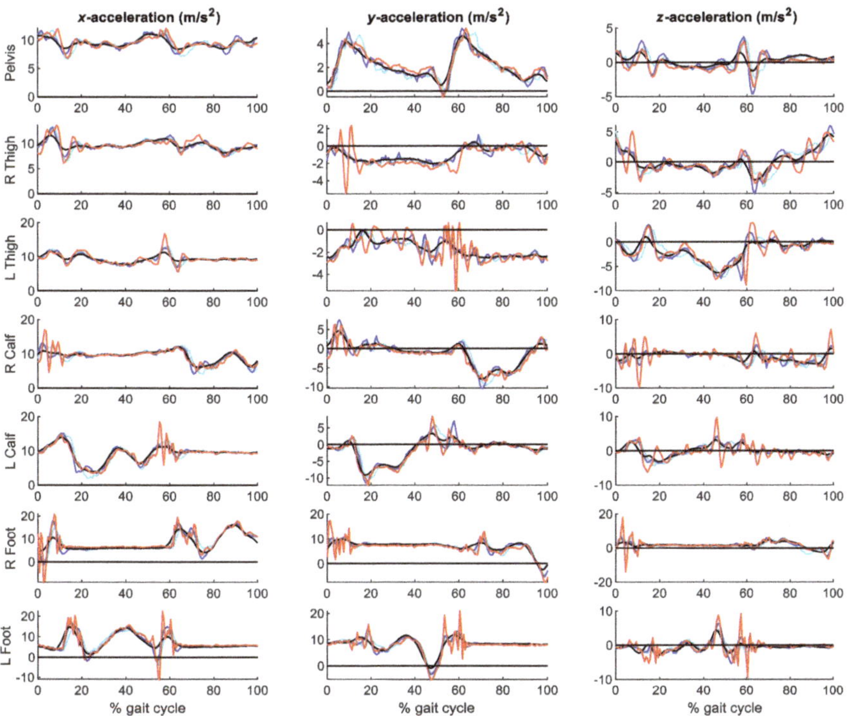

Figure 11. Accelerations obtained from the optical system with the EKF for combined process noise variances and cutoff frequencies of 0.1/30 Hz (black), 1/20 Hz (blue) and 50/6 Hz (cyan), respectively, vs. accelerations measured by the IMUs (red).

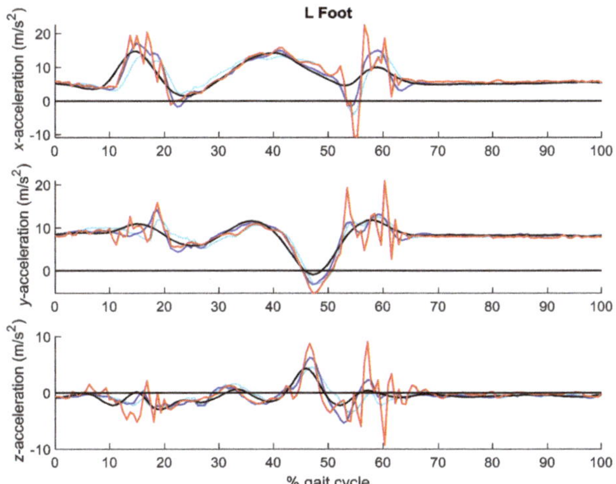

Figure 12. Detail of accelerations at the left foot obtained from the optical system with the EKF-based method for combined process noise standard deviations and cutoff frequencies of 0.1/20 Hz (black), 1/15 Hz (blue) and 50/6 Hz (cyan), respectively, vs. accelerations measured by the IMUs (red).

Table 2. RMSE of the accelerations obtained from the optical system through the EKF-based method with different combinations of process noise standard deviations and cutoff frequencies, with respect to the accelerations measured by the IMUs, taken as reference. The row with the lowest mean RMSE is highlighted in red.

Acc. Std. (m/s² or rad/s²)	Cutoff Freq. (Hz)	RMSE (m/s²)							
		Pelvis	R Thigh	L Thigh	R Tibia	L Tibia	R Foot	L Foot	Mean
0.1	6	0.717	1.047	1.195	1.663	1.624	2.649	2.430	1.618
0.1	10	0.679	1.020	1.155	1.679	1.618	2.530	2.338	1.377
0.1	15	0.676	1.018	1.136	1.678	1.614	2.433	2.274	1.354
0.1	20	0.682	1.022	1.133	1.674	1.611	2.377	2.229	1.341
0.1	25	0.690	1.028	1.135	1.672	1.612	2.341	2.198	1.335
0.1	30	0.698	1.035	1.140	1.672	1.615	2.318	2.176	1.332
0.5	6	0.606	1.011	1.108	1.513	1.442	2.553	2.274	1.313
0.5	10	0.545	0.969	1.030	1.514	1.412	2.340	2.105	1.239
0.5	15	0.557	0.965	1.017	1.524	1.415	2.183	2.004	1.208
0.5	20	0.582	0.977	1.034	1.530	1.424	2.098	1.943	1.198
0.5	25	0.608	0.996	1.058	1.538	1.439	2.049	1.907	1.199
0.5	30	0.634	1.018	1.083	1.553	1.458	2.018	1.887	1.207
1	6	0.604	1.006	1.099	1.484	1.433	2.565	2.267	1.307
1	10	0.538	0.963	1.026	1.454	1.383	2.330	2.072	1.221
1	15	0.562	0.961	1.033	1.457	1.379	2.161	1.955	1.188
1	20	0.601	0.978	1.067	1.466	1.393	2.074	1.890	1.183
1	25	0.639	1.004	1.105	1.483	1.417	2.026	1.857	1.191
1	30	0.679	1.036	1.143	1.511	1.446	1.999	1.843	1.207
10	6	0.632	0.987	1.132	1.473	1.439	2.580	2.324	1.321
10	10	0.576	0.926	1.086	1.393	1.343	2.344	2.131	1.225
10	15	0.606	0.928	1.115	1.374	1.300	2.174	2.002	1.187
10	20	0.665	0.973	1.168	1.397	1.307	2.084	1.934	1.191
10	25	0.739	1.041	1.228	1.452	1.342	2.037	1.909	1.218
10	30	0.823	1.124	1.292	1.538	1.391	2.014	1.913	1.262
50	6	0.633	0.989	1.134	1.478	1.442	2.578	2.330	1.323
50	10	0.574	0.933	1.088	1.401	1.348	2.343	2.142	1.229
50	15	0.603	0.943	1.119	1.386	1.307	2.177	2.018	1.194
50	20	0.666	1.000	1.179	1.424	1.319	2.091	1.955	1.204
50	25	0.753	1.085	1.253	1.507	1.359	2.047	1.937	1.243
50	30	0.859	1.190	1.336	1.632	1.416	2.032	1.953	1.302

4. Discussion and Limitations of the Study

This work proposes both an extended Kalman filter that facilitates optical motion capture, and an objective filter-tuning procedure that improves the resulting accelerations in gait analysis by using accelerometer data. First, a preliminary test including nine IMUs was carried out to assess the errors incurred by the inertial sensors in the measured orientations and accelerations. Second, the gait analysis of a healthy subject was performed. Both optical motion capture and inertial motion capture (using the seven most accurate IMUs out of the nine tested in the spot check) were recorded. The motion was then reconstructed by the classic Vaughan's method (filtering the marker trajectories with a Butterworth filter) and by the proposed EKF (applying a process noise variance and filtering the marker trajectories with a Butterworth filter), and the accelerations measured by the IMUs were used to tune the parameters of the filters for both methods.

As observed earlier in [11–13], the preliminary test highlighted the IMUs limitation to yield an accurate orientation. These errors depend on the calibration of the accelerometers and magnetometers, and on the algorithm used to estimate the orientations. Brodie [13] showed that it is possible to reduce the errors by substituting the commercial algorithm implemented in the inertial sensors by an improved one, which is consistent with the experience of the authors using other algorithms [35,36], but even in this case their accuracy remains limited. Therefore, although the performance of the IMUs has been improved in the last decade, optoelectronic systems are still used as the golden standard reference [18,19,43].

For this reason, it was decided to reduce the use of the orientations provided by the IMUs to a minimum for the gait analysis, taking as reference the local accelerations measured by the IMUs, and applying all the required transformations in the optical methods so as to obtain the accelerations, denoted as virtual accelerations, which are directly comparable to those provided by the inertial sensors.

This decision was enforced after observing, in the preliminary test, the effect of the orientation errors incurred by the IMUs on the global accelerations. As reported by Woodman [44], it is necessary to have very accurate rotation sensors in inertial navigation systems because the precise orientation of the body must be known in order to mathematically calculate the gravitational acceleration to find the translational acceleration. As observed in Figure 8, the gravitational acceleration was incorrectly estimated and appeared as translational acceleration perpendicular to the gravitational vector. To alleviate this problem, the orientations obtained from the optical system could be used instead but, as they are sensitive to the filter tuning, the resulting global accelerations from the IMUs would be distorted too.

The virtual accelerations obtained by Vaughan's method were very sensitive to the filtering applied to the trajectories of the markers. Best matches with experimental values were observed for cutoff frequencies ranging between 10 and 15 Hz. Bartlett [45] stated that cutoff frequencies between 4 and 8 Hz are often used in filtering movement data, while the OpenSim software [40] recommends to use a cutoff frequency of 6 Hz. However, it was observed that by using low cutoff frequencies, the accelerations were too smooth and the peaks measured by the IMUs were not reached. Schreven et al. [31] found that filtering the data with a cutoff frequency of 6 Hz decreases the accuracy of the reconstructed kinematics and, hence, can affect the accuracy of the joint moments obtained from inverse dynamics, as shown in [46].

Regarding the EKF method, apart from its robustness and simplicity of use, it showed a better accuracy in the resulting accelerations. The best filtering was obtained for a plant noise variance of 1 m/s^2 (or rad/s^2, depending on the corresponding coordinate) and a cutoff frequency of 20 Hz. Noise was eliminated, peaks measured by the IMUs were almost reached, and the resulting RMSEs were better than those incurred by Vaughan's method. Moreover, the EKF offered consistent kinematics by providing constant lengths of the body segments along the motion. Vaughan's method is similar to those proposed in [19] and, like them, does not impose the kinematic constraints to compute the joint kinematics from the marker trajectories. Therefore, it would require an additional step to correct these inconsistencies before dynamic analysis.

Although gait may be perceived as a smooth activity, acceleration peaks due to foot impact are observed in Figures 10 and 12, captured by the inertial system. In fact, they were also captured by the optical system when sampling at 100 Hz. Focusing on the acceleration peaks due to left foot landing, happening at around 60% of the gait cycle, it can be seen in Figure 10 that filtering with a cutoff frequency of 40 Hz already allows to capture them, but at the cost of keeping a lot of noise in the rest of the signal. On the other hand, using a cutoff frequency under 30 Hz provides a much cleaner signal, but notably oversmooths the impact peaks. Therefore, the procedure proposed in this paper can be useful for other researchers to evaluate existing filtering methods, design new ones and chose the best filtering parameters, but also to select the best capture frequency for their applications, because they will be able to distinguish between peaks due to noise and peaks due to actual motion.

The conducted study has been based on the results obtained from the gait analysis of one single subject. Although it could be expected that the frequency content of the motion signals is more dependent on the type of activity than on the particular subject performing it, tests including a greater number of subjects would be advisable in order to confirm the presented conclusions. This has been the reason to include the word 'preliminary' in the title.

5. Conclusions

The conclusion is twofold. First, when performing motion capture and analysis using a marker-based optical system, the extended Kalman filter significantly streamlines the motion capture and reconstruction process, since it facilitates automatic marker labeling, and manages occlusions and reflections in a robust and efficient way. Second, the availability of accelerations measured by inertial sensors can be very helpful for the tuning of the filters, no matter which motion reconstruction method is used. Consequently, the reliability of the obtained accelerations is improved.

Author Contributions: F.M. designed and performed the experiments with the supervision of U.L. and J.C.; U.L. implemented the EKF algorithm; F.M. analyzed the data. F.M., U.L. and J.C. wrote the manuscript; M.P.S. helped F.M. to perform the experiments. All authors have read and agreed to the published version of the manuscript.

Funding: This work was funded by the Spanish MCI under project PGC2018-095145-B-I00, co-financed by the EU through the EFRD program, and by the Galician Government under grant ED431C2019/29. Moreover, F. Michaud would like to acknowledge the support of the Spanish MCI by means of the doctoral research contract BES-2016-076901, co-financed by the EU through the ESF program.

Institutional Review Board Statement: Ethical review and approval were waived for this study, due to the non-invasive and non-dangerous character of the experiments.

Informed Consent Statement: Informed consent was obtained from all subjects involved in the study.

Data Availability Statement: The data presented in this study are available on request from the corresponding author. The data are not publicly available due to privacy restrictions.

Acknowledgments: Authors would like to acknowledge the different funding supports. Moreover, authors would like to acknowledge Mario Lamas for his voluntary participation in this project.

Conflicts of Interest: No conflicts of interest lie with any of the authors.

References

1. Baker, R. The history of gait analysis before the advent of modern computers. *Gait Posture* **2007**, *26*, 331–342. [CrossRef] [PubMed]
2. Winter, D.A. *Biomechanics and Motor Control of Human Movement*, 4th ed.; John Wiley & Sons, Inc.: Hoboken, NJ, USA, 2009.
3. Bachmann, E.R.; Yun, X.; McGhee, R.B. Sourceless tracking of human posture using small inertial/magnetic sensors. In Proceedings of the 2003 IEEE International Symposium on Computational Intelligence in Robotics and Automation. Computational Intelligence in Robotics and Automation for the New Millennium (Cat. No.03EX694), Kobe, Japan, 16–20 July 2003; Volume 2, pp. 822–829.
4. Fong, D.; Chan, Y.Y. The Use of Wearable Inertial Motion Sensors in Human Lower Limb Biomechanics Studies: A Systematic Review. *Sensors* **2010**, *10*, 11556–11565. [CrossRef] [PubMed]
5. Sabatini, A.M.; Martelloni, C.; Scapellato, S.; Cavallo, F. Assessment of Walking Features From Foot Inertial Sensing. *IEEE Trans. Biomed. Eng.* **2005**, *52*, 486–494. [CrossRef] [PubMed]
6. Liu, T.; Inoue, Y.; Shibata, K. Development of a wearable sensor system for quantitative gait analysis. *Measurement* **2009**, *42*, 978–988. [CrossRef]
7. Teufl, W.; Lorenz, M.; Miezal, M.; Taetz, B.; Fröhlich, M.; Bleser, G. Towards inertial sensor based mobile gait analysis: Event-detection and spatio-temporal parameters. *Sensors* **2018**, *19*, 38. [CrossRef]
8. Okkalidis, N.; Camilleri, K.P.; Gatt, A.; Bugeja, M.K.; Falzon, O. A review of foot pose and trajectory estimation methods using inertial and auxiliary sensors for kinematic gait analysis. *Biomed. Tech.* **2020**, *65*, 653–671. [CrossRef]
9. Lambrecht, S.; del-Ama, A.J. Human movement analysis with inertial sensors. *Biosyst. Biorobotics* **2014**, *4*, 305–328.
10. Blair, S.J. Biomechanical Considerations in Goal- Kicking Accuracy: Application of an Inertial Measurement System. Ph.D. Thesis, College of Sport and Exercise Science Institute for Health and Sport (IHES), Melbourne, Australia, 2019.
11. Picerno, P.; Cereatti, A.; Cappozzo, A. A spot check for assessing static orientation consistency of inertial and magnetic sensing units. *Gait Posture* **2011**, *33*, 373–378. [CrossRef]
12. Seaman, A.; McPhee, J. Comparison of optical and inertial tracking of full golf swings. *Procedia Eng.* **2012**, *34*, 461–466. [CrossRef]
13. Brodie, M.A.; Walmsley, A.; Page, W. The static accuracy and calibration of inertial measurement units for 3D orientation. *Comput. Methods Biomech. Biomed. Eng.* **2008**, *11*, 641–648. [CrossRef]
14. Lebel, K.; Boissy, P.; Nguyen, H.; Duval, C. Inertial measurement systems for segments and joints kinematics assessment: Towards an understanding of the variations in sensors accuracy. *Biomed. Eng. Online* **2017**, *16*, 56. [CrossRef] [PubMed]

15. Robert-Lachaine, X.; Mecheri, H.; Larue, C.; Plamondon, A. Validation of inertial measurement units with an optoelectronic system for whole-body motion analysis. *Med. Biol. Eng. Comput.* **2017**, *55*, 609–619. [CrossRef] [PubMed]
16. Poitras, I.; Dupuis, F.; Bielmann, M.; Campeau-Lecours, A.; Mercier, C.; Bouyer, L.J.; Roy, J. Validity and reliability ofwearable sensors for joint angle estimation: A systematic review. *Sensors* **2019**, *19*, 1555. [CrossRef] [PubMed]
17. Rezaei, A.; Cuthbert, T.J.; Gholami, M.; Menon, C. Application-based production and testing of a core–sheath fiber strain sensor for wearable electronics: Feasibility study of using the sensors in measuring tri-axial trunk motion angles. *Sensors* **2019**, *19*, 4288. [CrossRef] [PubMed]
18. Weygers, I.; Kok, M.; Konings, M.; Hallez, H.; de Vroey, H.; Claeys, K. Inertial sensor-based lower limb joint kinematics: A methodological systematic review. *Sensors* **2020**, *20*, 673. [CrossRef]
19. Pacher, L.; Chatellier, C.; Vauzelle, R.; Fradet, L. Sensor-to-segment calibration methodologies for lower-body kinematic analysis with inertial sensors: A systematic review. *Sensors* **2020**, *20*, 3322. [CrossRef]
20. Holden, D. Robust solving of optical motion capture data by denoising. *ACM Trans. Graph.* **2018**, *37*, 165. [CrossRef]
21. Ghorbani, S.; Etemad, A.; Troje, N.F. Auto-labelling of Markers in Optical Motion Capture by Permutation Learning. *Lect. Notes Comput. Sci.* **2019**, *11542*, 167–178.
22. Lugrís, U.; Vilela, R.; Sanjurjo, E.; Mouzo, F.; Michaud, F. Implementation of an Extended Kalman Filter for robust real-time motion capture using IR cameras and optical markers. In Proceedings of the IUTAM Symposium on Intelligent Multibody Systems—Dynamics, Control, Simulation, Sozopol, Bulgaria, 11–15 September 2017; pp. 3–4.
23. Skogstad, S.A.v.; Nymoen, K.; Høvin, M.E.; Holm, S.; Jensenius, A.R. Filtering Motion Capture Data for Real-Time Applications. In Proceedings of the 13th International Conference on New Interfaces for Musical Expression, Daejeon, Korea, 27–30 May 2013; pp. 142–147.
24. Bisseling, R.W.; Hof, A.L. Handling of impact forces in inverse dynamics. *J. Biomech.* **2006**, *39*, 2438–2444. [CrossRef]
25. Liu, X.; Cheung, Y.M.; Peng, S.J.; Cui, Z.; Zhong, B.; Du, J.X. Automatic motion capture data denoising via filtered subspace clustering and low rank matrix approximation. *Signal Process.* **2014**, *105*, 350–362. [CrossRef]
26. Skurowski, P.; Pawlyta, M. On the noise complexity in an optical motion capture facility. *Sensors* **2019**, *19*, 4435. [CrossRef] [PubMed]
27. Burdack, J.; Horst, F.; Giesselbach, S.; Hassan, I.; Daffner, S.; Schöllhorn, W.I. Systematic Comparison of the Influence of Different Data Preprocessing Methods on the Performance of Gait Classifications Using Machine Learning. *Front. Bioeng. Biotechnol.* **2020**, *8*, 260. [CrossRef] [PubMed]
28. Sinclair, J.; Taylor, P.J.; Hobbs, S.J. Digital filtering of three-dimensional lower extremity kinematics: An assessment. *J. Hum. Kinet.* **2013**, *39*, 25–36. [CrossRef] [PubMed]
29. Molloy, M.; Salazar-Torres, J.; Kerr, C.; McDowell, B.C.; Cosgrove, A.P. The effects of industry standard averaging and filtering techniques in kinematic gait analysis. *Gait Posture* **2008**, *28*, 559–562. [CrossRef] [PubMed]
30. Cappello, A.; la Palombara, P.F.; Leardini, A. Optimization and smoothing techniques in movement analysis. *Int. J. Biomed. Comput.* **1996**, *41*, 137–151. [CrossRef]
31. Schreven, S.; Beek, P.J.; Smeets, J.B.J. Optimising filtering parameters for a 3D motion analysis system. *J. Electromyogr. Kinesiol.* **2015**, *25*, 808–814. [CrossRef]
32. Kalman, R.E. A New Approach to Linear Filtering and Prediction Problems. *J. Basic Eng.* **1960**, *82*, 35–45. [CrossRef]
33. De Groote, F.; de Laet, T.; Jonkers, I.; de Schutter, J. Kalman smoothing improves the estimation of joint kinematics and kinetics in marker-based human gait analysis. *J. Biomech.* **2008**, *41*, 3390–3398. [CrossRef]
34. Cerveri, P.; Pedotti, A.; Ferrigno, G. Robust recovery of human motion from video using Kalman filters and virtual humans. *Hum. Mov. Sci.* **2003**, *22*, 377–404. [CrossRef]
35. Madgwick, S.O.H.; Harrison, A.J.L.; Vaidyanathan, R. Estimation of IMU and MARG orientation using a gradient descent algorithm. In Proceedings of the 2011 IEEE International Conference on Rehabilitation Robotics, Zurich, Switzerland, 29 June–1 July 2011; pp. 1–7.
36. Sabatini, A.M. Quaternion-Based Extended Kalman Filter for Determining Orientation by Inertial and Magnetic Sensing. *IEEE Trans. Biomed. Eng.* **2006**, *53*, 1346–1356. [CrossRef]
37. Vaughan, C.L.; Davis, B.L.; O'Connor, J.C. *Dynamics of Human Gait*, 2nd ed.; Kiboho Publishers: Cape Town, South Africa, 1999.
38. Cordillet, S.; Bideau, N.; Bideau, B.; Nicolas, G. Estimation of 3D knee joint angles during cycling using inertial sensors: Accuracy of a novel sensor-to-segment calibration procedure based on pedaling motion. *Sensors* **2019**, *19*, 2474. [CrossRef] [PubMed]
39. Lugrís, U.; Carlín, J.; Pàmies-Vilà, R.; Font-Llagunes, J.M.; Cuadrado, J. Solution methods for the double-support indeterminacy in human gait. *Multibody Syst. Dyn.* **2013**, *30*, 247–263. [CrossRef]
40. Delp, S.L.; Anderson, F.C.; Arnold, A.S.; Loan, P.; Habib, A.; John, C.T.; Guendelman, E.; Thelen, D.G. OpenSim: Open-Source Software to Create and Analyze Dynamic Simulations of Movement. *IEEE Trans. Biomed. Eng.* **2007**, *54*, 1940–1950. [CrossRef] [PubMed]
41. Alonso, F.J.; Cuadrado, J.; Lugrís, U.; Pintado, P. A compact smoothing-differentiation and projection approach for the kinematic data consistency of biomechanical systems. *Multibody Syst. Dyn.* **2010**, *24*, 67–80. [CrossRef]
42. Bar-Shalom, Y.; Li, X.R.; Kirubarajan, T. *Estimation with Applications to Tracking and Navigation*; John Wiley & Sons, Inc.: New York, NY, USA, 2001.

43. Lutz, J.; Memmert, D.; Raabe, D.; Dornberger, R.; Donath, L. Wearables for integrative performance and tactic analyses: Opportunities, challenges, and future directions. *Int. J. Environ. Res. Public Health* **2020**, *17*, 59. [CrossRef]
44. Woodman, O.J. *An Introduction to Inertial Navigation*; UCAM-CL-TR-696; Computer Laboratory, University of Cambridge: Cambridge, UK, 2007.
45. Bartlett, R. *Introduction to Sports Biomechanics*; Routledge: London, UK, 2007.
46. Kristianslund, E.; Krosshaug, T.; van den Bogert, A.J. Effect of low pass filtering on joint moments from inverse dynamics: Implications for injury prevention. *J. Biomech.* **2012**, *45*, 666–671. [CrossRef]

Article

A Track Geometry Measuring System Based on Multibody Kinematics, Inertial Sensors and Computer Vision

José L. Escalona [1,*], Pedro Urda [1] and Sergio Muñoz [2]

[1] Department of Mechanical and Manufacturing Engineering, University of Seville, 41092 Seville, Spain; purda@us.es
[2] Department of Materials and Transportation Engineering, University of Seville, 41092 Seville, Spain; sergiomunoz@us.es
* Correspondence: escalona@us.es

Abstract: This paper describes the kinematics used for the calculation of track geometric irregularities of a new *Track Geometry Measuring System* (TGMS) to be installed in railway vehicles. The TGMS includes a computer for data acquisition and process, a set of sensors including an *inertial measuring unit* (IMU, 3D gyroscope and 3D accelerometer), two video cameras and an encoder. The kinematic description, that is borrowed from the multibody dynamics analysis of railway vehicles used in computer simulation codes, is used to calculate the relative motion between the vehicle and the track, and also for the computer vision system and its calibration. The multibody framework is thus used to find the formulas that are needed to calculate the track irregularities (gauge, cross-level, alignment and vertical profile) as a function of sensor data. The TGMS has been experimentally tested in a 1:10 scaled vehicle and track specifically designed for this investigation. The geometric irregularities of a 90 m-scale track have been measured with an alternative and accurate method and the results are compared with the results of the TGMS. Results show a good agreement between both methods of calculation of the geometric irregularities.

Keywords: rail vehicles; track irregularities; multibody dynamics; inertial sensors; computer vision

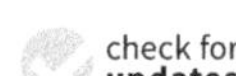

Citation: Escalona, J.L.; Urda, P.; Muñoz, S. A Track Geometry Measuring System Based on Multibody Kinematics, Inertial Sensors and Computer Vision. *Sensors* **2021**, *21*, 683. https://doi.org/10.3390/s21030683

Academic Editor: Felipe Jiménez
Received: 22 December 2020
Accepted: 15 January 2021
Published: 20 January 2021

Publisher's Note: MDPI stays neutral with regard to jurisdictional claims in published maps and institutional affiliations.

1. Introduction

Measurement of track geometry irregularities is a fundamental task in railway track maintenance with immense economic significance worldwide. Geometric irregularity is the most important track feature affecting safety and comfort of the rail transport. There are two different equipment that rail administrations and infrastructure managers use for track irregularity measurement: man-driven *rail track trolleys* (RTT) and *track recording vehicles* (TRV), also called *laboratory vehicles* or *inspection vehicles*. There are many companies that manufacture both types of equipment. The technology behind the RTT is simple and accurate. The relative irregularities, this is, gauge and cross-level are measured with a distance sensor, like an LVDT, and an inclinometer, respectively. The absolute irregularities, alignment and longitudinal profile require and absolute positioning system, like a total station or a very accurate GNSS. The technology used in TRV is more varied. Essentially, there are two technologies [1]: versine measurement systems (VMS), also called chord-based measurement systems and inertial measurement systems (IMS) or optical-inertial measurement systems. In addition to the sensors used, the main difference between these two methods is on the reference kinematics. The VMS use a physical reference, the chord, that can be a solid bar or a laser beam, while the IMS use an inertial reference frame or, in other words, the frame does not physically exist. The existence of the physical reference is an important advantage of the VMS that improves its accuracy. The main drawback of the VMS is that only irregularities with wave-lengths shorter than the length of the physical reference can be measured. A purely inertial IMS, as those based on accelerometer measurements, is based on the premise that the measuring device follows a trajectory that is parallel to one of the rails or the track centerline. This is in practice very difficult to obtain, or even impossible, if one wants to get the four irregularities with a single device. However, the most serious problem of the IMS is not that but the need to integrate the sensor signals in time to get the irregularities. The resulting accuracy depends very much on the filtering and other signal processing techniques that are needed to avoid the signal drift due to the integration.

The scientific foundation of the VMS is based on simple kinematics that only requires the measurements of distance sensors. The main challenge is to find the transfer function between the chord measurements and the vertical and lateral track irregularities [1,2]. The kinematics of the IMS is much more involved, due to the use of the inertial frame, as mentioned previously, and the use of different sensors (distance sensors, optical and inertial) requiring sensor fusion algorithms. To the authors' best knowledge, the description of the TRV used in the early eighties in British Rail by Lewis [3] remains the most complete description of the kinematics and signal processing behind the track geometry measurement with IMS. This paper tries to fill this gap. In a paper by Escalona [4], the calculation of the track vertical geometry using inertial sensors mounted on an un-suspended bogie was fully described, including analytical relationships, transfer functions, sensor fusion algorithms and signal processing. However, the method was valid only for the vertical geometry and the results were not experimentally validated.

Despite the existence of different measuring equipment, there is a need for improvement of the technology in track geometry inspection. RRT are much cheaper than TRV (small rail administrations cannot afford TRV) but much slower also. As a result, the frequency of track geometry inspection is not as high as desired. Due to this reason, there is an increasing tendency in the railway industry and research teams to developed track geometry measuring systems (TGMS) to be installed in in-service vehicles. That way, the frequency of the inspection is increased, the vehicle response to the irregularity can be measured simultaneously, and the time-evolution of the irregularities can be followed closely. However, what the sensors can get in an in-service vehicle, in general, is not a measurement of the irregularity but an estimation. There are two approaches: purely kinematic or model-based methods. Weston et al. published a set of papers [5–7] showing the measurement of vertical and lateral track irregularities using accelerometers and gyroscopes mounted in the bogie frame and a very simple kinematic model of the vehicle motion. Although these papers do not show results in frequency domain, the measured irregularities could only be relatively accurate for low frequencies, because the bogie frame does not follow the high frequency irregularities. Lee et al. [8] presented a Kalman filter data fusion approach based on accelerometer signals mounted on the body frame and the axle-box. Tsai et al. [9,10] presented a fast inspection technique based on the Hilbert–Huang Transform, that is a very useful time–frequency analysis technique, applied to the signal of an accelerometer mounted on the axle-box of an in-service car. Both methods that are based on axle-box-mounted accelerometers are only valid to measure vertical track irregularities. Tsumashima et al. [11] estimated the vertical track geometry using the accelerations measured in the car-body of the Japanese Shinkansen using a Kalman filter based on a very basic car model. In a more recent work, Tsumashima [12] has develop a method to detect and isolate track faults that are later classified using machine learning techniques. Tsumashima and Hirose [13] have applied the time–frequency Hilbert–Huang Transform method to identify track faults using the measured car-body acceleration as the input.

This paper describes a TGMS based on inertial sensors and computer vision that can be installed in in-service vehicles. The 3D kinematics used for the geometry measurements considers a general track design geometry, and the assumed kinematic approximations and simplifications are fully described. As a result of the background of the authors in on the development of multibody dynamics models for railway simulations [14–16], the presented kinematic equations follow the multibody formalism. The kinematic equations used for the computer vision are adapted to this formalism. The sensor fusion algorithms and signal processing tools used to find the track irregularities are explained also. The presented

TGMS has been built, tested and validated in an experimental scale track facility. Compared with the existing IMS methods that are implemented in the industry, the contribution of the method presented in this paper lies in the description of the methodology and the experimental validation. To the authors' best knowledge, the algorithms used to turn the sensors data into measured irregularities have not been described in the scientific literature, or they are described using an over-simplified track geometry, for example, assuming that the track is straight. This information is not provided by the companies that work in the important business of track inspection. In addition, an experimental proof of the accuracy of the equipment used for this task in not generally provided by these companies. This paper provides a full description of the signal processing that can be used to turn the sensors data into track irregularities. As shown below, the kinematic analysis needed to this end is not straightforward. This paper also provides an experimental validation of the measured irregularities. Authors do not claim that the presented method is more accurate or more efficient than existing methods, simply because the data are not accessible to compare.

This paper is organized as follows: Section 2 describes the sensors needed and the main features of the TGMS. Section 3 explains the kinematics of the irregular track and the rail vehicles based on the multibody dynamics formalism. Section 4 applies this formalism to the kinematics of the computer vision with the pin-hole camera model. Section 5 explains the method that is used to find the relative position of the rail cross-sections with respect to the moving vehicle using the computer vision data. Section 6 shows the kinematic relationships that are used to find the four track irregularities: gauge variation, cross-level, alignment and vertical profile, as a function of the sensors data. Section 7 describes a Kalman filter that is used to calculate the TGMS relative orientation with respect to the track. Section 8 describes an odometry algorithm that is used to locate at any instant the vehicle along the track. Section 9 explains the method that is used to find the relative trajectory of the TGMS with respect to the track centerline. The content of Sections 5–9 includes all the calculations needed to obtain the track irregularities out of the sensors data. Finally, Section 10 shows the experimental setup that has been used to build and validate the TGMS system and Section 11 compares the measured irregularities with accurate reference values. Summary and conclusions are given in Section 12.

2. Description of the TGMS

The TGMS sketched in Figure 1 (only right-side equipment in shown here) comprises:

1. Two video cameras.
2. Two laser line-projectors.
3. An IMU.
4. A rotary encoder.

It is important that the cameras, lasers and IMU are installed in a solid that can be considered as a rigid body when moving with the vehicle. The lasers and cameras must be equipped with orientation mechanisms that can be fully locked when the TGMS is working. The laser projectors draw red lines (when using a red laser) in the rail-heads (one on the left, one on the right) that are filmed by the video cameras. The information provided by the position and orientation of the red lines in the rails, together with the acceleration and angular velocity acquired with the IMU, are used to find the track geometry irregularities.

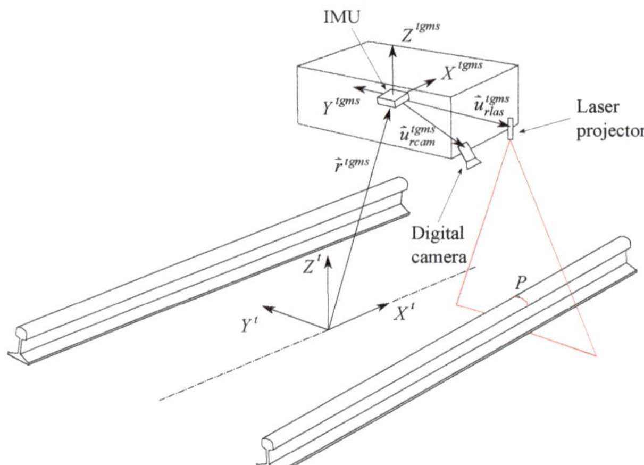

Figure 1. Kinematics of the Track Geometry Measuring System (TGMS) installed in a vehicle moving along the track.

The main features of the proposed system are:

1. It is capable to measure track alignment, vertical profile, cross-level, gauge, twist and rail-head profile using non-contact technology.
2. It can be installed in in-service vehicles. It is compact and low cost. Provided that the equipment sees the rail heads when the vehicle is moving, it can be installed in any body of the vehicle: at the wheelsets level, above primary suspension (bogie frame) or above the secondary suspension (car body).

Measuring devices must be selected taking into account the dusty environment under the train, affecting mainly the computer vision system, and the high level of accelerations that they will be subjected to. Protecting the video cameras and laser projectors may be solved, in part, locating them in a closed box with a polycarbonate window that must be cleaned frequently. Regarding the high level of accelerations, their expected value depends on the solid where the equipment is installed. The most critical body would be the axle-box (wheelset level) where the vertical accelerations can reach 100 g in high velocity trains. Reducing the possibility of damaging the measuring devices due to excessive accelerations is achieved if the equipment is installed at the car-body level.

3. Kinematics of the Irregular Track and the Railway Vehicle

This section includes the kinematic description of the rail geometry as a combination of a design geometry and the irregularities, and the kinematic description of an arbitrary body moving along the track, like the TGMS. Before presenting the kinematics, the different frames that are used and the nomenclature used to describe vectors, matrices and their components are described.

3.1. Frames of Reference

As shown in Figures 2 and 3, four different frames are used in railroad kinematics:

1. The inertial and global frame (GF) $< X, Y, Z >$. It is a frame fixed in space.
2. The track frame (TF) $< X^t, Y^t, Z^t >$. It is not a single frame but a field defined for each value of the arc-length coordinate along the track s. The position $\mathbf{R}^t(s)$ and orientation matrix $\mathbf{A}^t(s)$ of the TF with respect to the GF are functions of an arc-length coordinate s along the center line of the design track (without irregularities). These functions are implemented computationally in a track preprocessor.

3. The body frame (BF) $< X^i, Y^i, Z^i >$ of each body i. It is a frame rigidly attached to the body. In this document, the body i is the TGMS. The body frame of the TGMS is denoted as $< X^{tgms}, Y^{tgms}, Z^{tgms} >$

4. The rail profile frames. Left rail-profile frame (LRP), $< X^{lrp}, Y^{lrp}, Z^{lrp} >$, and right-profile frame (RRP), $< X^{rrp}, Y^{rrp}, Z^{rrp} >$. These frames are not unique frames but fields defined for each value of the arc-length coordinate along the track s. These frames are rigidly attached to the rail-heads.

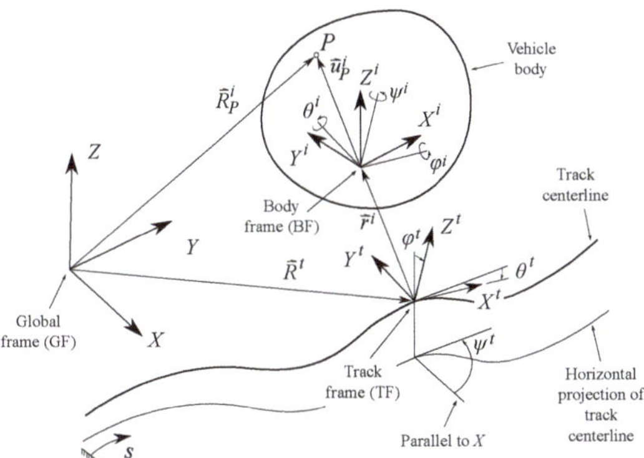

Figure 2. Kinematics of a body moving along the track.

The definition of the TF is such that the X^t axis is tangent to the track design centerline, the Y^t axis is perpendicular to X^t and connects the origin O^{lrp} of the LRP and the origin O^{rrp} of the RRP in the design track geometry (with no track irregularities) and the Z^t axis is perpendicular to both X^t and Y^t. Therefore, the TF is not the Frenet frame of the design track centerline. Each body i moving along the track has an associated TF at each instant of time. Its position and orientation can be obtained substituting the position of the body along the track, $s^i(t)$, in the functions $\mathbf{R}^t(s)$ and $\mathbf{A}^t(s)$ that are explained in next section. More details about the rail kinematics and the nomenclature used in this work for vectors and rotation matrices can be found in [17].

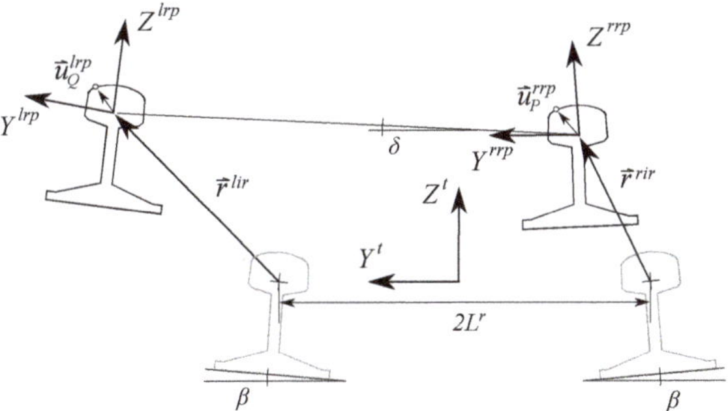

Figure 3. Kinematics of the irregular track cross-section.

3.2. Kinematics of the Design Track Centerline

Track geometry is the superposition of the design geometry and the irregularities. The components of the absolute position vector of an arbitrary point on the design track centerline with respect to an inertial and global frame is a function of the arc-length s, as follows:

$$\mathbf{R}^t(s) = \begin{bmatrix} R_x^t(s) \\ R_y^t(s) \\ R_z^t(s) \end{bmatrix} \tag{1}$$

where $\mathbf{R}^t$ contains the components of vector $\vec{R}^t$ shown in Figure 2. The geometry of the track centerline 3D-curve is defined by the horizontal profile and the vertical profile. Both profiles are defined in the rail industry using sections of variable length. Points between two sections are called vertices. Horizontal profile vertices do not necessary coincide with vertical profile vertices. Horizontal profile includes three types of sections: tangent (straight), curve (circular) and transitions (clothoid). Vertical profile includes two types of sections: constant-slope (straight) and transitions (cubic).

At each track section, the track centerline geometry is characterized by the following geometric values:

- Horizontal curvature: ρ_h.
- Vertical curvature: ρ_v.
- Twist curvature: ρ_{tw}.
- Spatial-derivative of horizontal curvature: $\rho_h{}'$.
- Vertical slope: α_v.

Figure 2 shows the TF $< X^t, Y^t, Z^t >$ associated with the track centerline at each value of s. The orientation of the TF with respect to a GF can be measured with the Euler angles ψ^t (*azimuth* or *heading* angle), θ^t (vertical slope, positive when downwards in the forward direction) and φ^t (*cant* or *superelevation angle*). The rotation matrix from the TF to the GF is given by:

$$\mathbf{A}^t(s) = \begin{bmatrix} c\theta^t c\psi^t & s\varphi^t s\theta^t c\psi^t - c\varphi^t s\psi^t & s\varphi^t s\psi^t + c\varphi^t s\theta^t c\psi^t \\ c\theta^t s\psi^t & c\varphi^t c\psi^t + s\varphi^t s\theta^t s\psi^t & c\varphi^t s\theta^t s\psi^t - s\varphi^t c\psi^t \\ -s\theta^t & s\varphi^t c\theta^t & c\varphi^t c\theta^t \end{bmatrix} \tag{2}$$

The azimuth ψ^t can have an arbitrary value, however, the slope θ^t and cant φ^t angles can be considered as small angles, such that the rotation matrix from the TF to the GF can be approximated to:

$$\mathbf{A}^t(s) \simeq \begin{bmatrix} c\psi^t & -s\psi^t & \varphi^t s\psi^t + \theta^t c\psi^t \\ s\psi^t & c\psi^t & \theta^t s\psi^t - \varphi^t c\psi^t \\ -\theta^t & \varphi^t & 1 \end{bmatrix} \tag{3}$$

An ideal body that moves along the track, taking the same orientation as the track frame with a forward velocity V and a forward acceleration $\dot{V}$, has the following absolute velocity and acceleration:

$$\dot{\bar{R}}^t = \begin{bmatrix} V \\ 0 \\ 0 \end{bmatrix}, \quad \ddot{\bar{R}}^t = \begin{bmatrix} \dot{V} \\ \rho_h V^2 \\ -\rho_v V^2 \end{bmatrix} \tag{4}$$

Similarly, the absolute angular velocity and the absolute angular acceleration of that body are given by:

$$\bar{\omega}^t = \begin{bmatrix} \rho_{tw} V \\ \rho_v V \\ \rho_h V \end{bmatrix}, \quad \bar{\alpha}^t = \begin{bmatrix} \rho_{tw} \dot{V} \\ \rho_v \dot{V} \\ \rho_h \dot{V} + \rho'_h V^2 \end{bmatrix} \tag{5}$$

3.3. Kinematics of the Irregular Track

Figure 3 shows the displacement of the rail heads due to irregularity in a cross-section of the track ($Y^t - Z^t$ plane). The irregularity vectors $\vec{r}^{lir}$ (*lir*, left rail *ir*regularity) and $\vec{r}^{rir}$

(*rir*, right rail *irregularity*) describe the displacement of the rail centerlines with respect to their design positions. The components of these vectors in the TF are functions of s, given by:

$$\bar{r}^{lir} = \begin{bmatrix} 0 \\ y^{lir} \\ z^{lir} \end{bmatrix}, \quad \bar{r}^{rir} = \begin{bmatrix} 0 \\ y^{rir} \\ z^{rir} \end{bmatrix} \tag{6}$$

In the railway industry, the following four combinations of the rail head centerlines irregularities are measured:

Alignment: $\quad \zeta_{al} = \left(y^{lir} + y^{rir} \right)/2$

Vertical profile: $\quad \zeta_{vp} = \left(z^{lir} + z^{rir} \right)/2$

Gauge variation: $\quad \zeta_{gv} = y^{lir} - y^{rir}$

Cross level: $\quad \zeta_{cl} = z^{lir} - z^{rir}$

The orientation of the rail head frames with respect to the TF is given by the following rotation matrices:

$$\mathbf{A}^{t,lrp} = \begin{bmatrix} 1 & 0 & 0 \\ 0 & \cos(\beta + \delta) & -\sin(\beta + \delta) \\ 0 & \sin(\beta + \delta) & \cos(\beta + \delta) \end{bmatrix},$$

$$\mathbf{A}^{t,rrp} = \begin{bmatrix} 1 & 0 & 0 \\ 0 & \cos(-\beta + \delta) & -\sin(-\beta + \delta) \\ 0 & \sin(-\beta + \delta) & \cos(-\beta + \delta) \end{bmatrix}, \tag{7}$$

where β is the orientation angle of the rail profiles, $\delta = \left(z^{lir} - z^{rir} \right)/2L_r$ is the linearized rotation angle due to the irregularity, and L_r is the distance from the track centerline to the rail profile frames. These angles and distance can be observed in Figure 3.

The absolute position vectors of two points, P and Q, defined in the right and left rail heads, respectively, are given by:

$$\mathbf{R}_P^{rrp} = \mathbf{R}^t + \mathbf{A}^t \left(\bar{r}^{rrp} + \bar{r}^{rir} + \mathbf{A}^{t,rrp} \hat{\mathbf{u}}_P^{rrp} \right)$$

$$\mathbf{R}_Q^{lrp} = \mathbf{R}^t + \mathbf{A}^t \left(\bar{r}^{lrp} + \bar{r}^{lir} + \mathbf{A}^{t,lrp} \hat{\mathbf{u}}_Q^{lrp} \right) \tag{8}$$

where $\hat{\mathbf{u}}_P^{rrp}$ and $\hat{\mathbf{u}}_Q^{lrp}$ contain the components of the position vector of points P and Q in the rail head profiles as shown in Figure 3. These vectors are parameterized following the rail head profile geometry:

$$\hat{\mathbf{u}}_P^{rrp} = \begin{bmatrix} 0 \\ s_2^{rr} \\ h^r\left(s_2^{rr} \right) \end{bmatrix}, \quad \hat{\mathbf{u}}_Q^{lrp} = \begin{bmatrix} 0 \\ s_2^{lr} \\ h^r\left(s_2^{lr} \right) \end{bmatrix} \tag{9}$$

where lr and rr stand for "left rail" and "right rail", and h^r is the function that defines the rail head profile.

3.4. Kinematics of a Body Moving along the Track

The coordinates used to describe the position and orientation of an arbitrary body i, as shown in Figure 2, or the TGMS shown in Figure 1, moving along the track are:

$$\mathbf{q}^i = \begin{bmatrix} s^i & r_y^i & r_z^i & \varphi^i & \theta^i & \psi^i \end{bmatrix}^T \tag{10}$$

where s^i is the arc-length along the track of the position of the body, r_y^i and r_z^i are the non-zero components of the position vector $\bar{r}^i$ of the BF with respect to the TF, this is

$\bar{\mathbf{r}}^i = \begin{bmatrix} 0 & r^i_y & r^i_z \end{bmatrix}^T$, and φ^i, θ^i and ψ^i are three Euler angles (roll, pitch and yaw, respectively) that define the orientation of the BF with respect to the TF. Note that, out of the six coordinates used for the kinematic description of a body, only the arc-length coordinate s^i is an absolute coordinate, being the other 5 track-relative coordinates.

The three Euler angles are assumed to be small, such that the following kinematic linearization is used:

$$\mathbf{A}^{t,i} \simeq \begin{bmatrix} 1 & -\psi^i & \theta^i \\ \psi^i & 1 & -\varphi^i \\ -\theta^i & \varphi^i & 1 \end{bmatrix} \tag{11}$$

The absolute position vector of point P that belongs to body i, as shown in Figure 2, is given by:

$$\mathbf{R}^i_P = \mathbf{R}^t + \mathbf{A}^t\left(\bar{\mathbf{r}}^i + \mathbf{A}^{t,i}\hat{\mathbf{u}}^i_P\right) \tag{12}$$

Using basic rigid body kinematics (see details in [17]), the absolute velocity and acceleration of point P are given by:

$$\dot{\mathbf{R}}^i_P = \dot{\mathbf{R}}^t + \dot{\bar{\mathbf{r}}}^i + \tilde{\bar{\omega}}^t\bar{\mathbf{r}}^i + \mathbf{A}^{t,i}\left(\tilde{\bar{\omega}}^i\hat{\mathbf{u}}^i_P\right) \tag{13}$$

$$\ddot{\mathbf{R}}^i_P = \ddot{\mathbf{R}}^t + \ddot{\bar{\mathbf{r}}}^i + \left(\tilde{\bar{\alpha}}^t + \tilde{\bar{\omega}}^t\tilde{\bar{\omega}}^t\right)\bar{\mathbf{r}}^i + 2\tilde{\bar{\omega}}^t\dot{\bar{\mathbf{r}}}^i + \mathbf{A}^{t,i}\left(\tilde{\bar{\alpha}}^i + \tilde{\bar{\omega}}^i\tilde{\bar{\omega}}^i\right)\hat{\mathbf{u}}^i_P, \tag{14}$$

where these vectors are projected to the track frame. In order to compute Equations (13) and (14), the orientation matrices, angular velocities and angular accelerations of the different frames need to be computed as a function of the generalized coordinates and velocities. The angular velocity $\bar{\omega}^t$ and acceleration $\bar{\alpha}^t$ vectors of the TF are given in Equation (5). The absolute angular velocity of body i is obtained as:

$$\hat{\omega}^i = \hat{\omega}^t + \hat{\omega}^{t,i} = \left(\mathbf{A}^{t,i}\right)^T\bar{\omega}^t + \hat{\omega}^{t,i} \tag{15}$$

The relative angular velocity of body i with respect to the TF, under the small-angles assumption, is given by:

$$\hat{\omega}^{t,i} = \begin{bmatrix} \dot{\varphi}^i \\ \dot{\theta}^i \\ \dot{\psi}^i \end{bmatrix}, \quad \bar{\omega}^{t,i} = \mathbf{A}^{t,i}\hat{\omega}^{t,i} = \begin{bmatrix} 1 & \psi^i & -\theta^i \\ -\psi^i & 1 & \varphi^i \\ \theta^i & -\varphi^i & 1 \end{bmatrix}\begin{bmatrix} \dot{\varphi}^i \\ \dot{\theta}^i \\ \dot{\psi}^i \end{bmatrix} \tag{16}$$

The absolute angular acceleration of body i, $\hat{\alpha}^t$ is simply calculated as the time-derivative of Equation (15).

4. Kinematics of the Computer Vision

Using the pin-hole model of the camera, Figure 4 shows the relation between the position vector $\vec{n}^{im}_{P'}$ of an arbitrary point P in the camera frame $< X^{cam}, Y^{cam}, Z^{cam} >$ (in our problem it can be left cam *lcam* or right cam *rcam*) and the position vector of the recorded point P' in the image plane $< X^{im}, Y^{im} >$.

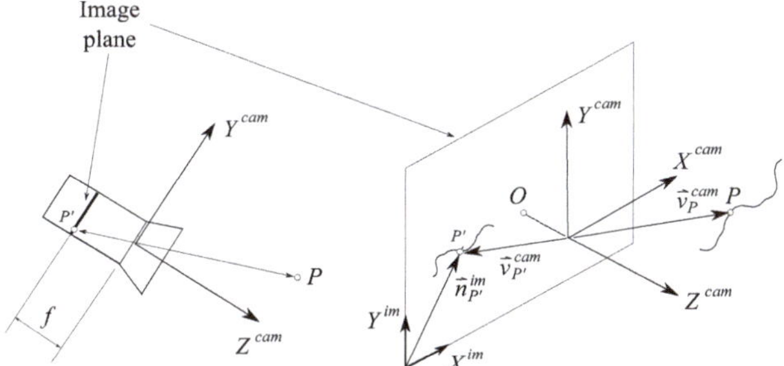

Figure 4. Kinematics of the computer vision.

Figure 5 shows the location of the camera in the TGMS and the relation between the position vector $\vec{v}_P^{cam}$ of the arbitrary point P in the camera frame and its position vector $\vec{u}_P^{tgms}$ in the TGMS frame.

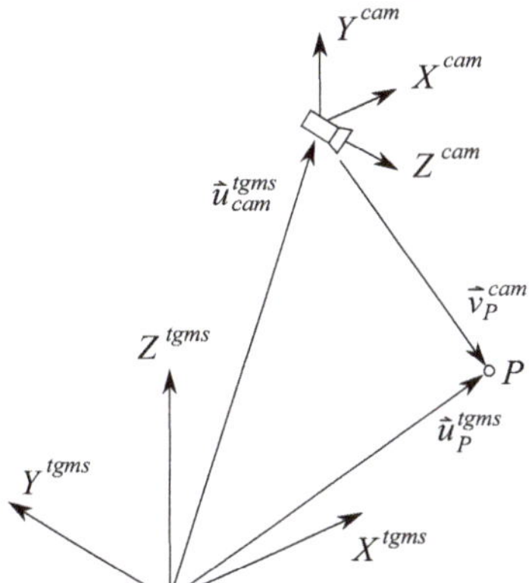

Figure 5. TGMS frame and camera frame.

The components of vectors $\vec{n}_{P'}^{im}$ and $\vec{u}_P^{tgms}$ are related through the equation [18,19]:

$$c \begin{bmatrix} \mathbf{n}_{P'}^{im} \\ 1 \end{bmatrix} = \mathbf{M}^{int}\mathbf{M}^{ext} \begin{bmatrix} \hat{\mathbf{u}}_P^{tgms} \\ 1 \end{bmatrix} \tag{17}$$

where c is an unknown *scale factor*. The matrix product on the right-hand side of this equation is called projection matrix $\mathbf{P} = \mathbf{M}^{int}\mathbf{M}^{ext}$. The column matrix $\mathbf{n}_{P'}^{im}$ is 2×1 (image is planar), and its components are given in pixel units (dimensionless) while the column matrix $\hat{\mathbf{u}}_P^{tgms}$ is 3×1, and its components are given in meters. Due to these dimensions, the projection matrix $\mathbf{P}$ is 3×4. Matrix $\mathbf{M}^{int}$ is 3×3 and it is called matrix of intrinsic parameters of the camera, and matrix $\mathbf{M}^{ext}$ is 3×4, it is called matrix of extrinsic parameters of the camera, and it is given by:

$$\mathbf{M}^{ext} = \begin{bmatrix} \left(\mathbf{A}^{tgms,cam}\right)^T & -\left(\mathbf{A}^{tgms,cam}\right)^T \hat{\mathbf{u}}_{cam}^{tgms} \end{bmatrix} \tag{18}$$

Matrices of intrinsic and extrinsic parameters can be experimentally obtained, for example, using the Zhang calibration method [18]. In this investigation, the calibration method is based on Zhang method but adapted to the application at hand. This method is fully detailed in [19].

Just using the 3 scalar equations in Equation (17), the position vector $\hat{\mathbf{u}}_P^{tgms}$ of the point P cannot be obtained using the values of $\mathbf{n}_{P'}^{im}$ because there are 4 unknowns (three components of $\hat{\mathbf{u}}_P^{tgms}$ and the scale factor c). One exception occurs when the point P moves on a surface whose equation is known in the TGMS frame. This is the case at hand if P belongs to the plane highlighted by the laser projector. In this case, the following system of equations can be solved to find $\hat{\mathbf{u}}_P^{tgms}$:

$$\begin{cases} c\begin{bmatrix} \mathbf{n}_{P'}^{im} \\ 1 \end{bmatrix} = \mathbf{M}^{int}\mathbf{M}^{ext}\begin{bmatrix} \hat{\mathbf{u}}_P^{tgms} \\ 1 \end{bmatrix} \\ A^{las}\left[u_P^{tgms}\right]_x + B^{las}\left[u_P^{tgms}\right]_y + C^{las}\left[u_P^{tgms}\right]_z + D^{las} = 0 \end{cases} \tag{19}$$

where A^{las}, B^{las}, C^{las} and D^{las} are the constants that define the laser plane (left laser, *llas*, or right laser, *rlas*, in our case), and $\left[u_P^{tgms}\right]_k$ $k = x, y, z$ means the component k of the vector $\vec{u}_P^{tgms}$ in the TGMS frame. The constants that define the laser planes have to be experimentally obtained in the TGMS computer vision-calibration process [19]. Equation (19) is a system of 4 algebraic equations with 4 unknowns that can be used to find vector components $\hat{\mathbf{u}}_P^{tgms}$ using as input data the vector components $\mathbf{n}_{P'}^{im}$ and the parameters of the cameras and the lasers.

5. Detecting the Rail Cross-Section from a Camera Frame

As a result of the solution of Equation (19) for all highlighted pixels in the image frames, a cloud of points P_i in the right rail and a cloud of points Q_i in the left rail, with position vectors $\hat{\mathbf{u}}_P^{tgms}$ and $\hat{\mathbf{u}}_Q^{tgms}$, respectively, that belong to the rails cross-sections can be identified. In fact, the points do not really belong to cross-sections, just to sections, because the laser planes are not necessarily perpendicular to the rails center line. However, because the relative angles of the TGMS with respect to the TF are very small, the irregularities are also small, and the lasers are set to project the light plane at right angles with respect to the rails, it will be assumed in this paper that the highlighted sections are actually cross-sections.

Figure 6a shows a sketch of the cloud of points and, in dashed line, the theoretical rail-head profile. The theoretical rail-head profiles, when they are new, not worn, have a known geometry that is made of circular and straight segments. An example is the UIC 54 E1 rail-head profile shown in Figure 6b. Detecting the rail cross-section from a camera frame can be solved with the well-developed computer vision algorithms of feature detection or feature tracking [20]. However, in this investigation, because the feature to be detected can be represented analytically, an optimization approach has been developed. The optimization problem consists of finding the position $\hat{\mathbf{u}}_{Orp}^{tgms}$ and angle $\varphi^{tgms,rp}$ of the rail profile that better fits the cloud of points. This will be the assumed position and orientation of the rail head profile (*lrp* or *rrp*) in the TGMS frame when the vehicle is moving. The objective function to minimize is the sum of the squares of the distance of the points in the cloud to the theoretical profile curve shown in Figure 6b.

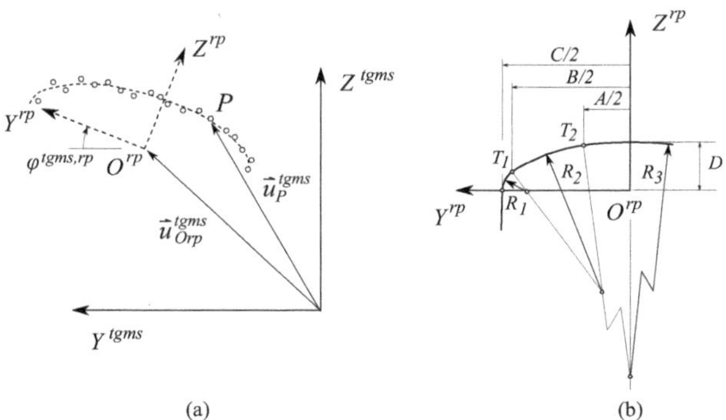

Figure 6. Cloud of points detected with computer vision (**a**). UIC 54 E1 rail-head profile (**b**).

6. Equations for Geometry Measurement

The equations that can be used to measure the track irregularities are easily deduced with the help of Figure 7. In this figure, vectors $\hat{\mathbf{u}}_{Olrp}^{tgms}$ and $\hat{\mathbf{u}}_{Orrp}^{tgms}$ are output data of the computer vision algorithm explained in previous section. The following equalities can be easily identified with the help of the figure:

$$\vec{r}^{tgms} + \vec{u}_{Olrp}^{tgms} = \vec{r}^{lrp} + \vec{r}^{lir}$$
$$\vec{r}^{tgms} + \vec{u}_{Orrp}^{tgms} = \vec{r}^{rrp} + \vec{r}^{rir} \tag{20}$$

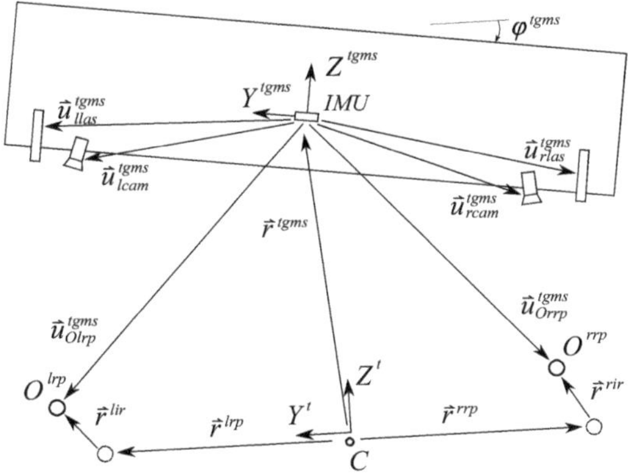

Figure 7. Planar view of the TGMS.

Subtracting these vector equations, one gets:

$$\vec{u}_{Olrp}^{tgms} - \vec{u}_{Orrp}^{tgms} = \vec{r}^{lrp} + \vec{r}^{lir} - \left(\vec{r}^{rrp} + \vec{r}^{rir}\right) \tag{21}$$

In this equation, the position vector $\vec{r}^{tgms}$ of the TGMS does not appear. This vector equation can be projected in the TF, as follows:

$$\mathbf{A}^{t,tgms}\left(\hat{\mathbf{u}}_{Olrp}^{tgms} - \hat{\mathbf{u}}_{Orrp}^{tgms}\right) = \bar{\mathbf{r}}^{lrp} - \bar{\mathbf{r}}^{rrp} + \bar{\mathbf{r}}^{lir} - \bar{\mathbf{r}}^{rir} \tag{22}$$

Using again the small-angles assumption, the Y, Z components of this equation are given by:

$$\begin{bmatrix} 1 & -\varphi^{tgms} \\ \varphi^{tgms} & 1 \end{bmatrix} \begin{bmatrix} \left[\hat{u}^{tgms}_{Olrp}\right]_y - \left[\hat{u}^{tgms}_{Orrp}\right]_y \\ \left[\hat{u}^{tgms}_{Olrp}\right]_z - \left[\hat{u}^{tgms}_{Orrp}\right]_z \end{bmatrix} = \begin{bmatrix} 2L^r \\ 0 \end{bmatrix} + \begin{bmatrix} r^{lir}_y - r^{rir}_y \\ r^{lir}_z - r^{rir}_z \end{bmatrix} \tag{23}$$

where L^r is half the distance between the rail-head profiles without irregularities. In this equation, the result $\bar{\mathbf{r}}^{lrp} - \bar{\mathbf{r}}^{rrp} = \begin{bmatrix} 2L^r & 0 \end{bmatrix}^T$ has been used. According to the definition given in Section 3.3, the components of the last column matrix of Equation (23) are the gauge variation (gv) and the cross-level (cl). Therefore, rearranging Equation (23) yields:

$$\begin{aligned} \xi_{gv} &= \left(\left[\hat{u}^{tgms}_{Olrp}\right]_y - \left[\hat{u}^{tgms}_{Orrp}\right]_y\right) - \varphi^{tgms}\left(\left[\hat{u}^{tgms}_{Olrp}\right]_z - \left[\hat{u}^{tgms}_{Orrp}\right]_z\right) - 2L^r \\ \xi_{cl} &= \varphi^{tgms}\left(\left[\hat{u}^{tgms}_{Olrp}\right]_y - \left[\hat{u}^{tgms}_{Orrp}\right]_y\right) + \left(\left[\hat{u}^{tgms}_{Olrp}\right]_z - \left[\hat{u}^{tgms}_{Orrp}\right]_z\right) \end{aligned} \tag{24}$$

Adding the vector Equation (20), one gets:

$$2\vec{r}^{tgms} + \vec{u}^{tgms}_Q + \vec{u}^{tgms}_P = \vec{r}^{lir} + \vec{r}^{rir} \tag{25}$$

where the fact that $\vec{r}^{lrp} + \vec{r}^{rrp} = \vec{0}$ has been used. Using again the small-angles assumption, the Y, Z components of this equation are given by:

$$2\begin{bmatrix} r^{tgms}_y \\ r^{tgms}_z \end{bmatrix} + \begin{bmatrix} 1 & -\varphi^{tgms} \\ \varphi^{tgms} & 1 \end{bmatrix} \begin{bmatrix} \left[\hat{u}^{tgms}_{Olrp}\right]_y + \left[\hat{u}^{tgms}_{Orrp}\right]_y \\ \left[\hat{u}^{tgms}_{Olrp}\right]_z + \left[\hat{u}^{tgms}_{Orrp}\right]_z \end{bmatrix} = \begin{bmatrix} r^{lir}_y + r^{rir}_y \\ r^{lir}_z + r^{rir}_z \end{bmatrix} \tag{26}$$

According to the definition given in Section 3.3, the components of the last column matrix of Equation (26) are twice the alignment irregularity (ξ_{al}) and twice the vertical profile (ξ_{vp}). Therefore, rearranging Equation (26) yields:

$$\begin{aligned} \xi_{al} &= \tfrac{1}{2}\left(\left[\hat{u}^{tgms}_{Olrp}\right]_y + \left[\hat{u}^{tgms}_{Orrp}\right]_y\right) - \tfrac{\varphi^{tgms}}{2}\left(\left[\hat{u}^{tgms}_{Olrp}\right]_z + \left[\hat{u}^{tgms}_{Orrp}\right]_z\right) + r^{tgms}_y \\ \xi_{vp} &= \tfrac{\varphi^{tgms}}{2}\left(\left[\hat{u}^{tgms}_{Olrp}\right]_y + \left[\hat{u}^{tgms}_{Orrp}\right]_y\right) + \tfrac{1}{2}\left(\left[\hat{u}^{tgms}_{Olrp}\right]_z + \left[\hat{u}^{tgms}_{Orrp}\right]_z\right) + r^{tgms}_z \end{aligned} \tag{27}$$

Therefore, Equations (24) and (27) can be used to find all track irregularities. The following conclusions are highlighted:

1. The calculation of the relative track irregularities (ξ_{gv} and ξ_{cl}), as shown in Equation (24), needs as an input the output of the computer vision $\hat{\mathbf{u}}^{tgms}_{Olrp}$ and $\hat{\mathbf{u}}^{tgms}_{Orrp}$ and the roll angle of the TGMS with respect to the track φ^{tgms}.
2. The calculation of the absolute track irregularities (ξ_{al} and ξ_{vp}), as shown in Equation (27), needs, in addition, the relative trajectory $\bar{\mathbf{r}}^{tgms}$ of the TGMS with respect to the TF.

The next section is devoted to the calculation of the TGMS to TF relative orientation, that includes the calculation of φ^{tgms}. Section 9 is devoted to the calculation of the relative trajectory $\bar{\mathbf{r}}^{tgms}$ of the TGMS with respect to the TF.

7. Measurement of TGMS to TF Relative Orientation

The calculation of the relative rotation of the TGMS with respect to the TF requires the calculation of the absolute orientation of the TGMS and the absolute orientation of the TF. As given in Equations (2) and (3), finding the orientation of the TF just requires the instantaneous value of the arc-length s, because the track design geometry is known.

The value of s has to be known accurately. The odometry algorithm explained in the next section has been developed to this end.

The calculation of the absolute orientation of the TGMS is obtained with the IMU sensor. Most IMU sensors come with their own internal algorithm to calculate the orientation angles or quaternions. However, these algorithms are not accurate in the application at hand. They are based on the sensor fusion of the data of the gyroscope, the accelerometer and the magnetometer, as follows:

1. The gyroscope provides three signals that are proportional to the components of the angular velocity vector in the sensor frame (the sensor frame is assumed to be parallel to the TGMS frame), as follows:

$$\omega^{imu} = \hat{\omega}^{tgms}_{abs} \tag{28}$$

These components are non-linearly related to the coordinates that define the TGMS orientation and their time-derivatives, as shown later.

2. The accelerometer signals are proportional to the components of the absolute acceleration in the local frame plus the absolute gravity vector field, as follows:

$$\mathbf{a}^{imu} = \mathbf{\hat{R}}^{tgms} + \left(\mathbf{A}^{tgms}\right)^{T} \begin{bmatrix} 0 & 0 & g \end{bmatrix}^{T} \tag{29}$$

This is the absolute acceleration in the sensor frame, plus the gravitational constant g, that is assumed to act in the absolute Z direction.

3. The magnetometer signals are proportional to the components of the Earth's magnetic field in the local frame. This information can be used to find the direction of the Earth's magnetic north.

In the algorithm developed in this investigation to find the TGMS absolute orientation, the magnetometer's signals are not used. This is mainly because out of the three Euler angles needed to define the orientation, the yaw angle lacks interest in our application. It is mainly the yaw angle which can be identified with the help of the magnetometer.

Following an Euler's angle sequence yaw-pitch-roll, as done with the TF to get the rotation matrix given in Equation (2), the absolute angular velocity of the TGMS that appear in Equation (28) can be obtained as follows:

$$
\begin{aligned}
\hat{\omega}^{tgms}_{abs} &=
\begin{bmatrix}
-\sin\theta^{tgms}_{abs} \\
\cos\theta^{tgms}_{abs}\sin\varphi^{tgms}_{abs} \\
\cos\theta^{tgms}_{abs}\cos\varphi^{tgms}_{abs}
\end{bmatrix}
\dot{\psi}^{tgms}_{abs}
+
\begin{bmatrix}
0 \\
\cos\varphi^{tgms}_{abs} \\
-\sin\varphi^{tgms}_{abs}
\end{bmatrix}
\dot{\theta}^{tgms}_{abs}
+
\begin{bmatrix}
1 \\
0 \\
0
\end{bmatrix}
\dot{\varphi}^{tgms}_{abs} = \\
&=
\begin{bmatrix}
1 & 0 & -\sin\theta^{tgms}_{abs} \\
0 & \cos\varphi^{tgms}_{abs} & \cos\theta^{tgms}_{abs}\sin\varphi^{tgms}_{abs} \\
0 & -\sin\varphi^{tgms}_{abs} & \cos\theta^{tgms}_{abs}\cos\varphi^{tgms}_{abs}
\end{bmatrix}
\begin{bmatrix}
\dot{\varphi}^{tgms}_{abs} \\
\dot{\theta}^{tgms}_{abs} \\
\dot{\psi}^{tgms}_{abs}
\end{bmatrix}
= \mathbf{\hat{G}}^{tgms}\mathbf{\dot{\Phi}}^{tgms}_{abs},
\end{aligned}
\tag{30}
$$

where $\mathbf{\Phi}^{tgms}_{abs} = \begin{bmatrix} \varphi^{tgms}_{abs} & \theta^{tgms}_{abs} & \psi^{tgms}_{abs} \end{bmatrix}^{T}$ is the set of absolute angles (with respect to the global frame) of the TGMS. This expression is non-linear in terms of the TGMS absolute angles, but linear in term of their time-derivative. This equation can be inverted to isolate the time-derivatives of the angles, as follows:

$$
\begin{bmatrix}
\dot{\varphi} \\
\dot{\theta} \\
\dot{\psi}
\end{bmatrix}
= \left(\mathbf{\hat{G}}\right)^{-1}\hat{\omega} =
\begin{bmatrix}
1 & \dfrac{\sin\varphi\sin\theta}{\cos\theta} & \dfrac{\cos\varphi\sin\theta}{\cos\theta} \\
0 & \cos\varphi & -\sin\varphi \\
0 & \dfrac{\sin\varphi}{\cos\theta} & \dfrac{\cos\varphi}{\cos\theta}
\end{bmatrix}
\begin{bmatrix}
\hat{\omega}_x \\
\hat{\omega}_y \\
\hat{\omega}_z
\end{bmatrix}
\tag{31}
$$

where, as done in the rest of this section, the superscript *tgms* and subscript *abs* have been eliminated in the symbols for simplicity. Using again the small-angles assumption of the roll and pitch angles, the following linearized formulas result:

$$
\begin{aligned}
\dot{\varphi} &= \hat{\omega}_x + \frac{\sin\varphi\sin\theta}{\cos\theta}\hat{\omega}_y + \frac{\cos\varphi\sin\theta}{\cos\theta}\hat{\omega}_z \simeq \hat{\omega}_x + \theta\hat{\omega}_z \\
\dot{\theta} &= \cos\varphi\,\hat{\omega}_y - \sin\varphi\,\hat{\omega}_z \simeq \hat{\omega}_y - \varphi\hat{\omega}_z \\
\dot{\psi} &= \frac{\sin\varphi}{\cos\theta}\hat{\omega}_y + \frac{\cos\varphi}{\cos\theta}\hat{\omega}_z \simeq \varphi\hat{\omega}_y + \hat{\omega}_z
\end{aligned}
\tag{32}
$$

For the accelerometer signals, Equation (29) can be linearized as follows:

$$
\mathbf{a}^{imu} \simeq \hat{\mathbf{R}}^{tgms} + g\begin{bmatrix} -\theta \\ \varphi \\ 1 \end{bmatrix}
\tag{33}
$$

where:

$$
\hat{\mathbf{R}}^{tgms} = \left[\mathbf{A}^{t,tgms}\right]^{\mathsf{T}}\ddot{\mathbf{R}}^{tgms}, \quad
\mathbf{A}^{t,tgms} \simeq \begin{bmatrix}
1 & -\psi^{tgms} & \theta^{tgms} \\
\psi^{tgms} & 1 & -\varphi^{tgms} \\
-\theta^{tgms} & \varphi^{tgms} & 1
\end{bmatrix}
\tag{34}
$$

Again, because of the small-angles assumption and for the sake of simplicity, the following approximation is used:

$$
\hat{\mathbf{R}}^{tgms} \simeq \ddot{\mathbf{R}}^{tgms}
\tag{35}
$$

Using the result of Equation (14), the absolute acceleration of the TGMS is given by:

$$
\ddot{\mathbf{R}}^{tgms} = \ddot{\mathbf{R}}^{t} + \ddot{\mathbf{r}}^{tgms} + \left(\tilde{\dot{\alpha}}^{t} + \tilde{\omega}^{t}\tilde{\omega}^{t}\right)\mathbf{r}^{tgms} + 2\tilde{\omega}^{t}\dot{\mathbf{r}}^{tgms}
\tag{36}
$$

In this equation, the last three terms on the right-hand side are unknown. These are the TGMS to TF relative acceleration, the tangential and centripetal relative accelerations and the Coriolis acceleration, respectively. In order to calculate these terms, the value of the relative position vector $\mathbf{r}^{tgms}$, velocity $\dot{\mathbf{r}}^{tgms}$ and acceleration $\ddot{\mathbf{r}}^{tgms}$ of the TGMS with respect to the TF are needed. As it can be observed in Figure 8, the trajectory followed by the TGMS when the vehicle is moving is a 3D curve that slightly differs with respect to the track centerline. In fact, the difference between these 3D curves is needed to measure the absolute track irregularities. However, in order to find the TGMS to TF relative orientation, it is going to be assumed that the acceleration equals the one of a particle moving along the track centerline, this is:

$$
\ddot{\mathbf{R}}^{tgms} \simeq \ddot{\mathbf{R}}^{t} = \begin{bmatrix} \dot{V} \\ \rho_h V^2 \\ -\rho_v V^2 \end{bmatrix}
\tag{37}
$$

Therefore, the influence of the last three terms in Equation (36) has been neglected.

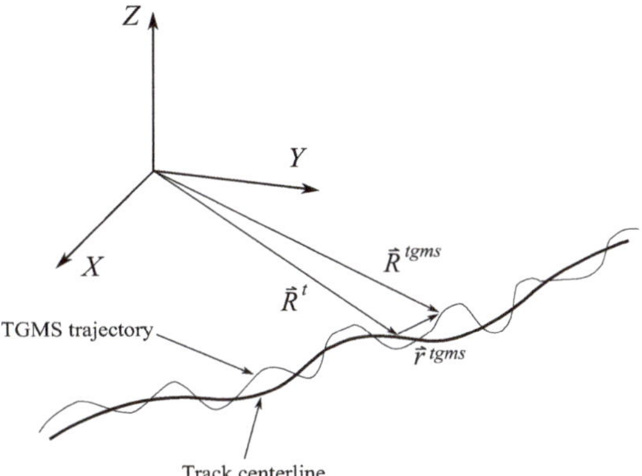

Figure 8. TGMS trajectory.

The sequence of approximations that has been assumed to calculate the acceleration of the TGMS may look rough for some readers. However, it has to be taken into account that in most algorithms used to find orientation from IMU's signals [21,22] the true acceleration measured by the sensor is neglected in comparison with the gravity term in Equation (33). In the application at hand, this approximation is not accurate and the "compensation" introduced with Equation (36) is much better than nothing.

Finally, the following measurement equation is used to account for the accelerometer signals:

$$\mathbf{a}_{corr}^{imu} = \mathbf{a}_{filt}^{imu} - \begin{bmatrix} \dot{V} \\ \rho_h V^2 \\ -\rho_v V^2 \end{bmatrix} \approx g \begin{bmatrix} -\theta \\ \varphi \\ 1 \end{bmatrix} \tag{38}$$

where $\mathbf{a}_{corr}^{imu}$ is the "corrected" accelerometer signal and $\mathbf{a}_{filt}^{imu}$ is the low-pass filtered acceleration signal. The reason for low-pass filtering the acceleration signals is that the effect of the TGMS to TF relative accelerations that have been neglected in Equation (36) contribute to relatively high-frequencies to the signals. This way their contribution is, in part, filtered out of the signals.

Using the two first gyroscope equations given in Equation (32) and the last two accelerometer equations given in Equation (38), the following linear dynamic system can be formulated in state-space form:

$$\dot{\mathbf{x}} = \begin{bmatrix} \dot{\varphi} \\ \dot{\theta} \end{bmatrix} = \begin{bmatrix} 0 & \hat{\omega}_z \\ -\hat{\omega}_z & 0 \end{bmatrix} \begin{bmatrix} \varphi \\ \theta \end{bmatrix} + \begin{bmatrix} \hat{\omega}_x \\ \hat{\omega}_y \end{bmatrix} = \mathbf{Fx} + \mathbf{u}$$

$$\mathbf{z} = \begin{bmatrix} \left(a_{corr}^{imu}\right)_x \\ \left(a_{corr}^{imu}\right)_y \end{bmatrix} = \begin{bmatrix} 0 & -g \\ g & 0 \end{bmatrix} \begin{bmatrix} \varphi \\ \theta \end{bmatrix} = \mathbf{Hx} \tag{39}$$

where the state vector $\mathbf{x} = \begin{bmatrix} \varphi & \theta \end{bmatrix}^T$ just includes the roll and pitch angles, the state transition matrix $\mathbf{F}$ is not constant but depends on the measured vertical angular velocity, the measurement matrix $\mathbf{H}$ is constant, the measurement vector $\mathbf{z}$ includes the x and y measured-corrected accelerations and the input vector $\mathbf{u}$ includes the x and y measured angular velocities. The linear system shown in (39) is a very simple set of equations that can be used with a standard Kalman filter algorithm to find the TGMS to TF relative angles φ and θ.

8. Odometry Algorithm

The odometry algorithm presented here can be used when the TGMS has no access to the data of a precise odometer of the vehicle and/or a GNSS cannot be used, for example, as it happens in underground trains. Underground trains use to be metropolitan. Being metropolitan, there used to be many narrow curved sections. Curved sections facilitate the method presented next.

As shown in Figure 9, the design geometry of a railway track (horizontal profile, as explained in Section 3.2) is a succession of segments of three types: straight (*s* in the figure) with zero curvature, circular (*c* in the figure) with constant curvature and transitions (*t* in the figure) with linearly varying curvature. The curvature function can be decomposed into a set of zero segments (straight segments) plus a set of curvature functions that have trapezoidal shape (normal curve) or double-trapezoidal shape (*S-curve*). The location of the curvature functions (start and end points) is exactly identified along the track using the design geometry provided by the track preprocessor.

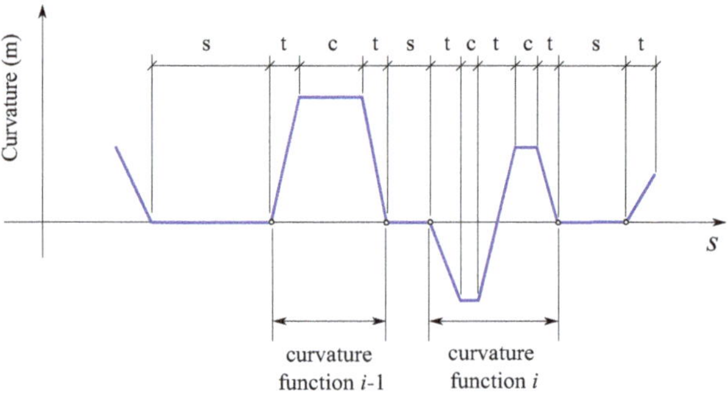

Figure 9. Design horizontal curvature of a railway track.

The curvature of the track can be experimentally approximated in the TGMS with the installed sensors. The curvature of the trajectory followed by the TGMS can be obtained as:

$$\rho_h^{exp} \simeq \frac{\hat{\omega}_z^{tgms}}{V} \tag{40}$$

Of course, this approximate measure is a noisy version of the track horizontal curvature. However, experimental measures show that the overall shape of the curvature functions can be clearly obtained with this approximation.

The concept of the odometry algorithm is to monitor the experimental curvature during the ride of the train using Equation (40) and to store the data together with the approximate coordinate s_{app}^{tgms} obtained with the help of the installed encoder and the assumption of rolling-without-slipping of the wheel. The plot of the experimental curvature looks like the plot at the top of Figure 10. Using the track preprocessor, the design value of the curvature of the track ρ_h^{design} in the area where the train is located, may look like the lower plot in Figure 10. As shown in the figure, this information can be used to correct the value of s_{app}^{tgms} at points 1, 2, 3 and 4 located at the entry or exit of the curves. Measures of s_{app}^{tgms} between these corrected points are also corrected using a linear mapping, as shown in Figure 11. Where, black arrows mean odometry corrections due to the algorithm output. Coloured arrows mean odometry correction using linear interpolation of the algorithm output. Different colours are just used to distinguish the curvature function sections

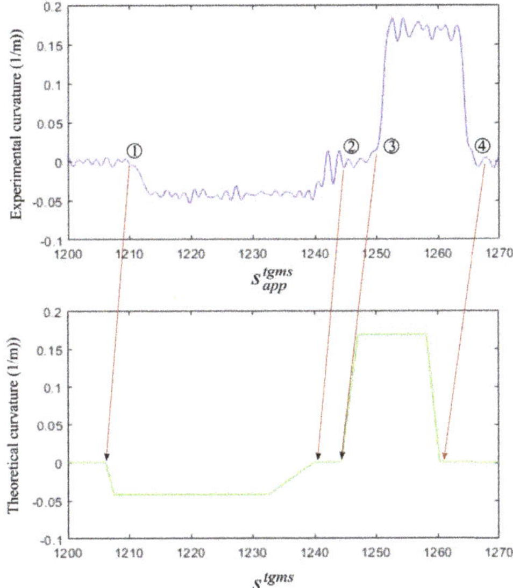

Figure 10. Odometry algorithm.

The problem is how to detect the entry and exit of the curves using the functions $\rho_h^{\exp}\left(s_{app}^{tgms}\right)$ and $\rho_h^{design}\left(s^{tgms}\right)$. In fact, it is the exit of the curves that is detected first. Once the TGMS leaves a curve, the shape of the curvature function that the TGMS has ahead is known. Therefore, when the measured curvature $\rho_h^{\exp}\left(s_{app}^{tgms}\right)$ "looks similar" to the expected curvature function $\rho_h^{design}\left(s^{tgms}\right)$, the exit of the curve has been reached. This similarity is computed by calculating at each instant the squared-error of the experimentally measured curvature and the expected curvature function, as follows:

$$e2(s) = \int_{\bar{s}=s-\Delta s}^{\bar{s}=s} \left[\rho_h^{\exp}(\bar{s}) - \rho_h^{design}(\bar{s} - s + s_{exit})\right]^2 d\bar{s} \tag{41}$$

where $e2(s)$ is the squared error (s substitute s_{app}^{tgms} for simplicity in the formula), Δs is the length of the expected curvature function and s_{exit} is the location of the exit of the curve in the design geometry. For a better accuracy, the value of the squared error is normalized for each curvature function using the following factor:

$$I2 = \int_{\bar{s}=0}^{\bar{s}=\Delta s} \left[\rho_h^{design}(\bar{s})\right]^2 d\bar{s} \tag{42}$$

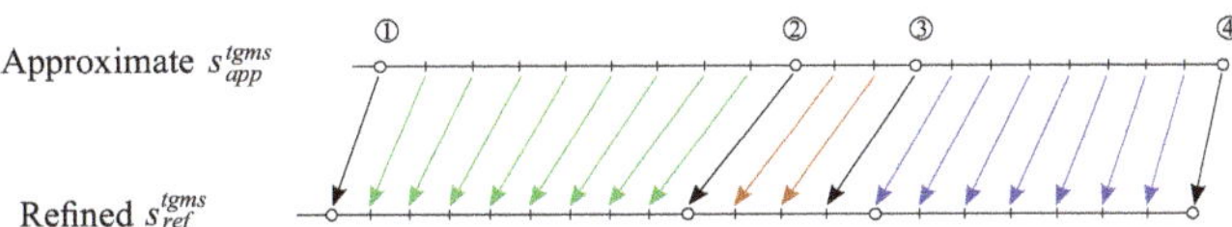

Figure 11. Correction of ^{tgms}.

The normalization factors, that are different for each curvature function, are of course computed in a preprocesing stage. The normalized squared-error is given by:

$$ne2(s) = \frac{e2(s)}{I2} \tag{43}$$

Thanks to the normalization, the value of $ne2$ varies between approximately 1, in straight track sections, and 0 when there is a perfect matching between $\rho_h^{\exp}\left(s_{app}^{tgms}\right)$ and $\rho_h^{design}\left(s^{tgms}\right)$. A typical plot of the function is observed in Figure 12. This function used to be smooth, such that detecting the local minimum that indicates the detection of the exit of the curve is a very easy task. Once the exit of the curve is detected, the expected curvature function is substituted by the next curve ahead along the track. It can be shown that this method can be run in real-time. The main computational cost is the one associated with the calculation of the integral given in Equation (41).

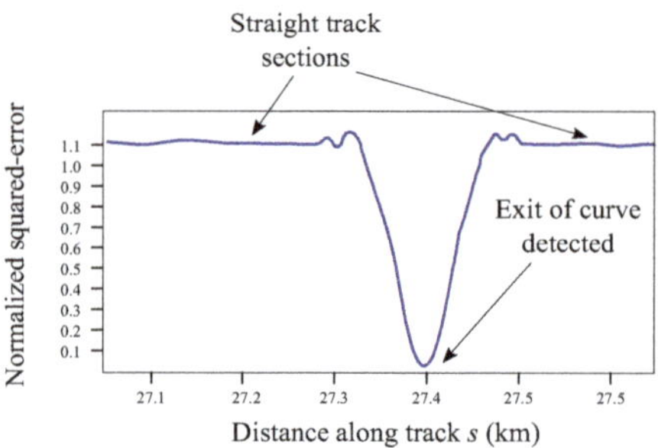

Figure 12. Normalized squared-error.

9. Measurement of TGMS to TF Relative Motion

The final and most difficult step needed to find the absolute irregularities of the track using Equation (27) is to obtain the relative trajectory $\bar{r}^{tgms}$ of the TGMS with respect to the TF shown in Figure 8. Equation (36) can be treated as a second order differential equation in terms of $\bar{r}^{tgms}$. This equation has the following scalar components:

$$\ddot{\bar{R}}^{tgms} = \begin{bmatrix} \dot{V} \\ \rho_h V^2 \\ -\rho_v V^2 \end{bmatrix} + \begin{bmatrix} 0 \\ \ddot{r}_y^{tgms} \\ \ddot{r}_z^{tgms} \end{bmatrix} +$$

$$\begin{bmatrix} -r_y^{tgms}\left(\dot{V}\rho_h + V^2(\rho'_h - \rho_{tw}\rho_v)\right) + r_z^{tgms}\left(V^2\rho_{tw}\rho_h + \dot{V}\rho_v\right) \\ -r_y^{tgms}\left(V^2(\rho_{tw}^2 + \rho_h^2)\right) - r_z^{tgms}\left(-V^2\rho_v\rho_h + \dot{V}\rho_{tw}\right) \\ r_y^{tgms}\left(V^2\rho_v\rho_h + \dot{V}\rho_{tw}\right) - r_z^{tgms}\left(V^2(\rho_{tw}^2 + \rho_v^2)\right) \end{bmatrix} +$$

$$+2\begin{bmatrix} V\dot{r}_z^{tgms}\rho_v - V\dot{r}_y^{tgms}\rho_h \\ -V\dot{r}_z^{tgms}\rho_{tw} \\ V\dot{r}_y^{tgms}\rho_{tw} \end{bmatrix} \tag{44}$$

Alternatively, the absolute acceleration of the TGMS can be obtained using the measurement of the accelerometer given in Equation (29). If the vectors in Equation (29) are projected to the TF, it yields:

$$\ddot{\bar{R}}^{tgms} = \mathbf{A}^{t,tgms}\mathbf{a}^{imu} - \left(\mathbf{A}^t\right)^T\begin{bmatrix} 0 & 0 & g \end{bmatrix}^T =$$

$$\begin{bmatrix} a_x^{imu} - a_y^{imu}\psi^{tgms} + a_z^{imu}\theta^{tgms} \\ a_y^{imu} + a_x^{imu}\psi^{tgms} - a_z^{imu}\varphi^{tgms} \\ a_z^{imu} - a_x^{imu}\theta^{tgms} + a_y^{imu}\varphi^{tgms} \end{bmatrix} + \begin{bmatrix} g\theta^t \\ -g\varphi^t \\ -g \end{bmatrix} \tag{45}$$

where the small-angles assumption has been used again. The left-hand sides of Equations (44) and (45) represent the same physical magnitudes, therefore, the right hand sides of these equations can be equated. Using just the second and third components of the right hand sides and rearranging yields:

$$\begin{bmatrix} \ddot{r}_y^{tgms} \\ \ddot{r}_z^{tgms} \end{bmatrix} + \begin{bmatrix} 0 & -2V\rho_{tw} \\ 2V\rho_{tw} & 0 \end{bmatrix} \begin{bmatrix} \dot{r}_y^{tgms} \\ \dot{r}_z^{tgms} \end{bmatrix} + \begin{bmatrix} -V^2\left(\rho_{tw}^2 + \rho_h^2\right) & V^2\rho_v\rho_h - \dot{V}\rho_{tw} \\ V^2\rho_v\rho_h + \dot{V}\rho_{tw} & -V^2\left(\rho_{tw}^2 + \rho_v^2\right) \end{bmatrix} \begin{bmatrix} r_y^{tgms} \\ r_z^{tgms} \end{bmatrix} = \begin{bmatrix} a_y^{imu} + a_x^{imu}\psi^{tgms} - a_z^{imu}\varphi^{tgms} - g\varphi^t - \rho_h V^2 \\ a_z^{imu} - a_x^{imu}\theta^{tgms} + a_y^{imu}\varphi^{tgms} - g + \rho_v V^2 \end{bmatrix}$$

$$(46)$$

This is a *2nd* order linear system of ordinary differential equations (ODE) with time-variant coefficients (linear time-varying system, LTV). This ODE has to be integrated forward in time to find the TGMS to TF relative trajectory ($r_y^{tgms}(t)$ and $r_z^{tgms}(t)$). The inputs of these equations are all calculated or measured so far. These inputs are:

1. The accelerometer data $\mathbf{a}^{imu}$.
2. The instantaneous forward velocity V and acceleration $\dot{V}$ of the vehicle. This is obtained from the encoder data.
3. The position s^{tgms} of the TGMS along the track. This is the output of the odometry algorithm explained in previous section. The position s^{tgms} is used as an entry to the track preprocessor to get the track design cant angle φ^t and the curvatures ρ_{tw}, ρ_v and ρ_{tw}.
4. The relative orientation of the TGMS with respect to the TF that is calculated in Section 7.

In the case of a tangent (straight) track, where all track curvatures are zero, Equation (46) reduces to:

$$\begin{bmatrix} \ddot{r}_y^{tgms} \\ \ddot{r}_z^{tgms} \end{bmatrix} = \begin{bmatrix} a_y^{imu} + a_x^{imu}\psi^{tgms} - a_z^{imu}\varphi^{tgms} \\ a_z^{imu} - a_x^{imu}\theta^{tgms} + a_y^{imu}\varphi^{tgms} - g \end{bmatrix}$$

$$(47)$$

Numerical integration of Equation (46) can be used to find the TGMS to TF relative motion. However, a close look to Equation (46) shows that these differential equations are equivalent to the equations of motion of a linear 2-degrees of freedom system with negative-definite stiffness and damping matrices. It is a highly unstable system. Therefore, serious problems of drift of the resulting displacements are expected. As a result that the track irregularity is a relatively-high frequency component of the track geometry (being the design geometry the low-frequency component), the drift of the solution can be solved using several methods, like these two:

1. Using a digital signal processing approach for the integration that combines double-integration and high-pass filtering of the signal [3].
2. Using a Kalman filter approach that adds to the system dynamic equations, Equation (46), a set of measurements. These measurements are virtual sensors that in practice always provide a zero value of the TGMS to TF relative motion. The assumed covariance of these measurements is the expected covariance of the track irregularities. This method has been successfully applied in [23] to eliminate the drift in the results while keeping a good accuracy.

The second method is used in this investigation to get the results presented in Section 11.

10. Experimental Setup

A scale track has been built at the rooftop of the School of Engineering at the University of Seville. Figure 13 on the left shows an aerial view of the 90 m-track. The scale is approximately 1:10 (5-inch gauge). The track includes a tangent section (next to end A), a 26 m-curved section and a 12 m-curved section, among other features. The track is supported on a set of mechanisms that can be observed on the right of the figure that are separated 10 cm. These mechanisms can be used to create any irregularity profile. Details of the track design can be found in [24].

<table>
<tr><td align="center">(a)</td><td align="center">(b)</td></tr>
</table>

Figure 13. Scaled track at the School of Engineering, University of Seville. (**a**) Aerial view, (**b**) detail of track supports.

Figure 14 shows the scale vehicle that has been used to install the TGMS. The vehicle has a classical architecture of 4 wheelsets, 2 bogies and 1 car-body. The TGMS is attached to the car-body. The right picture shows a detail of the vehicle where the video cameras and the laser beam can be observed. The vehicle drive is an electric motor with a chain transmission. Details of the vehicle design can be found in [24]. The TGMS is equipped with two video cameras Ximea MQ003CG-CM. The VGA cameras have a resolution of 648×488 pixels. They include a CMOS RGB bayer matrix sensor. Their maximum frame acquisition rate reaches 500 fps. The camera has a C-mount lens FUJINN 2/3″ 12.5 F1.2–F1.6 MI E1.5MP MV that allows the adjustment of focus and exposure. The projection lasers have a power of 10 mW and they illuminate in red. The IMU is a LORD Microstrain 3DM-GX5-25 AHRS, an triaxial accelerometer, gyroscope and magnetometer industrial sensor fully calibrated and temperature compensate. It has a maximum sampling rate of 1 kH. The accelerometer has two different measuring ranges ± 8 g ($25\ \mu g/\sqrt{\text{Hz}}$) or ± 20 g ($80\ \mu g/\sqrt{\text{Hz}}$). The gyroscope has an angle random walk of $0.3°/\sqrt{\text{h}}$. The traveled distance by the vehicle along the track is obtained using a precision digital encoder Kubler 2400 series. It is a two channel encoder with up to 1440 pulses per revolution.

In order to have a reference value of the track irregularities, an accurate track geometry measurement has been performed. The relative irregularities have been measured with the instrument that can be observed in the right of Figure 15. This instrument slides along the track. It includes a LVDT to measure the gauge variation and an inclinometer to measure the cross-level. In order to measure the absolute irregularities, alignment and longitudinal profile, the total station shown in the left hand side of the figure has been used. Details of the track geometry measurement can be found in [24].

(a)

(b)

Figure 14. Scale vehicle, (**a**) global view of instrumented scale vehicle, (**b**) detail showing video cameras and laser beam.

(a)

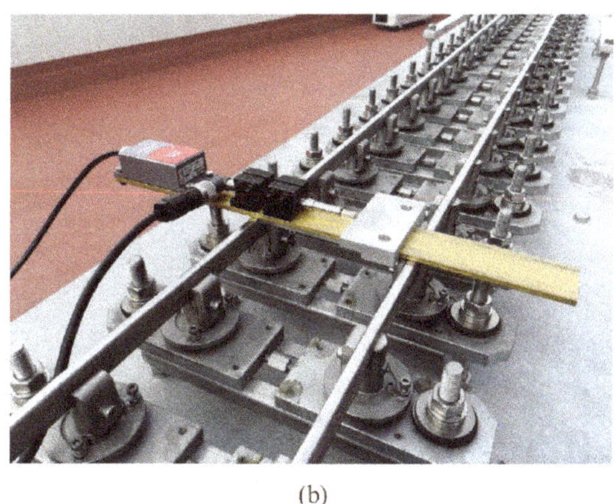
(b)

Figure 15. Irregularity measurement. (**a**) Robotic total station, (**b**) LVDT and inclinometer.

11. Computer Implementation and Comparison of TGMS Measurement with Reference Irregularity

The methods and algorithms described in this paper have been implemented in Matlab environment. However, only built in functions of the basic package have been used, avoiding the use of functions, for example, of the computer vision toolbox. The program that produces the track irregularity out of the sensor data includes the following modules:

1. Pre-process: Camera calibration module. It finds the cameras' intrinsic and extrinsic parameters using pictures of a checkerboard pattern [19].
2. Pre-process: Track pre-processor module. It is set to provide the design geometry of the track.
3. Process: Computer vision module. It finds the position and orientation of the rail cross-sections using the recorded frames and the method described in Section 5.
4. Process: Odometry module. As described in Section 8, it finds the position, velocity and acceleration of the TGMS every time instant.

5. Process: TGMS orientation module. It finds the orientation of the TGMS using the method described in Section 7.
6. Process: TGMS trajectory module. It finds the relative trajectory of the TGMS with respect to the TF using the method described in Section 9.
7. Process: Irregularity module. It calculates the track irregularities with the method of Section 6.

The results of this program are now compared with the reference value of the irregularities that were obtained with the method described in the previous section. Figures 16–19 show the comparison between the reference geometry measurement explained in the previous section and the measurement of the TGMS developed in this investigation. The experiment was done with the vehicle shown in Figure 14 at a forward velocity of 2.5 m/s, that, considering the scale, it is equivalent to 25 m/s = 90 km/h of a real train. This is approximately the maximum forward velocity of metropolitan trains. The figures show the irregularities in the first 56 m of the track. The complete length of the scale track could not be measured because the vehicle had difficulties to negotiate the 12 m-radius curve at that forward velocity. The irregularities shown in the figures are high-pass filtered with a cut-off wave length of 3 m (wavelengths above 3 m are filtered out). It is a common practice in the geometry measurement of metropolitan trains to measure just wave lengths below 30 m, what is equivalent to 3 m in our scaled track.

The cross-level shown in Figure 16 shows a good agreement between the reference geometry and the measure of the TGMS. The agreement in gauge variation shown in Figure 17 is even better. Both curves are almost identical. However, in both figures, some edge effect is observed at the beginning of the measurement.

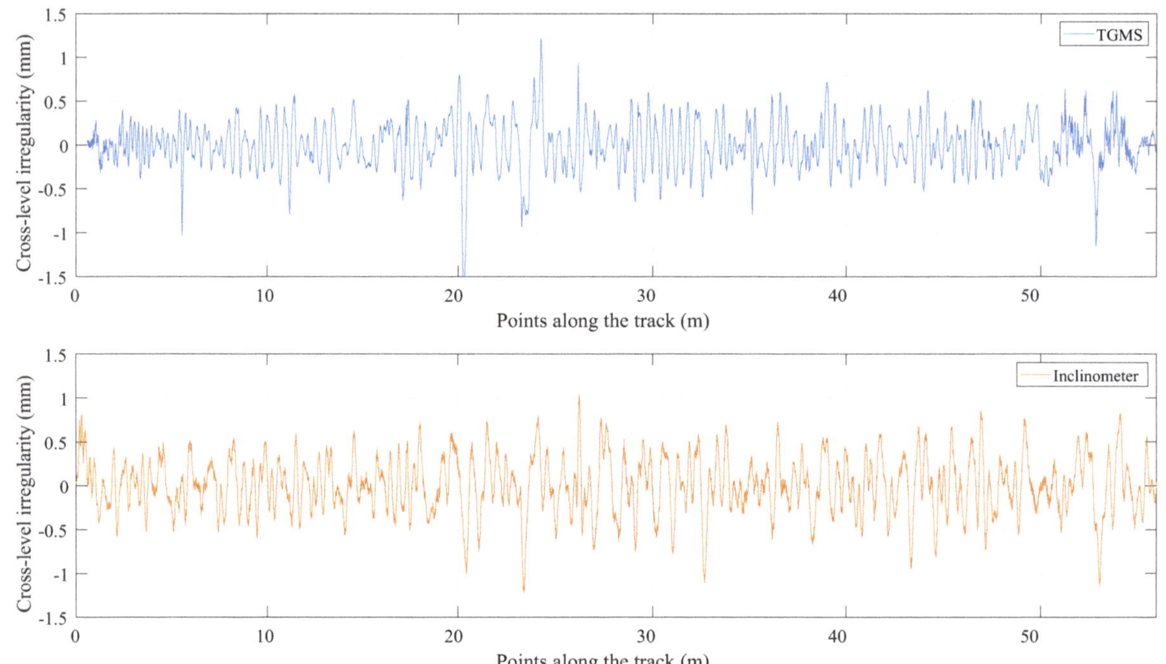

Figure 16. Measurements of cross level.

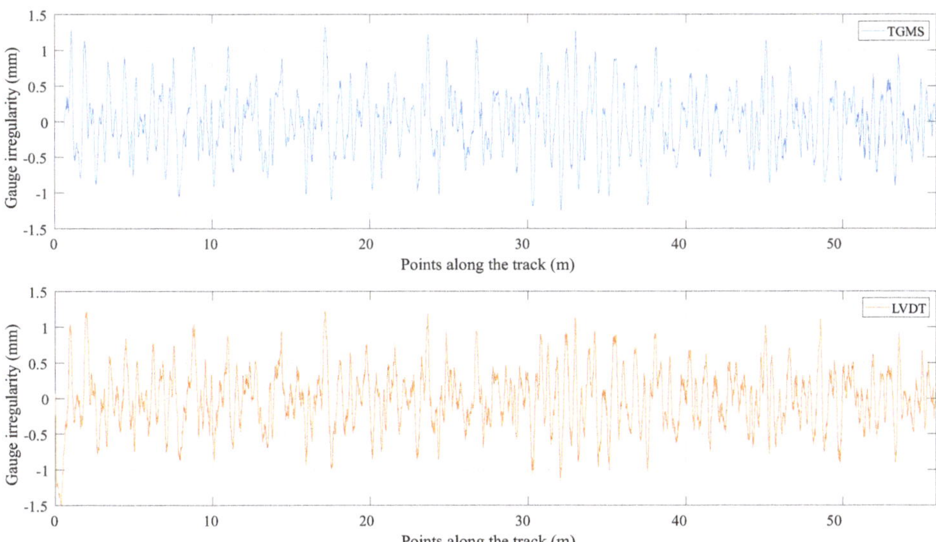

Figure 17. Measurements of gauge variation.

The absolute irregularities are compared in Figures 18 and 19. As it can be observed, the agreement is fine, but not as good as the agreement of the relative track irregularities. The reason is clearly the influence in the results of the approximations and noise associated with the calculation of the TGMS to TF relative trajectory explained in Section 9. Recall that this relative trajectory is not needed for the calculation of the relative irregularities. It can also be observed that the measurements of the designed TGMS shows higher spatial-frequencies than the measurement with the total station. However, these higher frequencies must not be considered as noise, because the measuring method used with the total station can not detect high-frequency (short wavelength) irregularities.

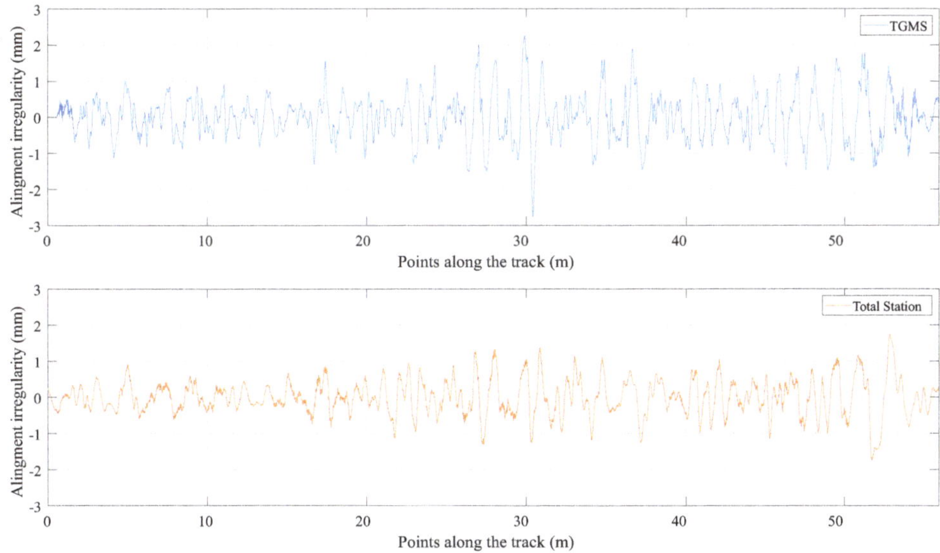

Figure 18. Measurements of alignment.

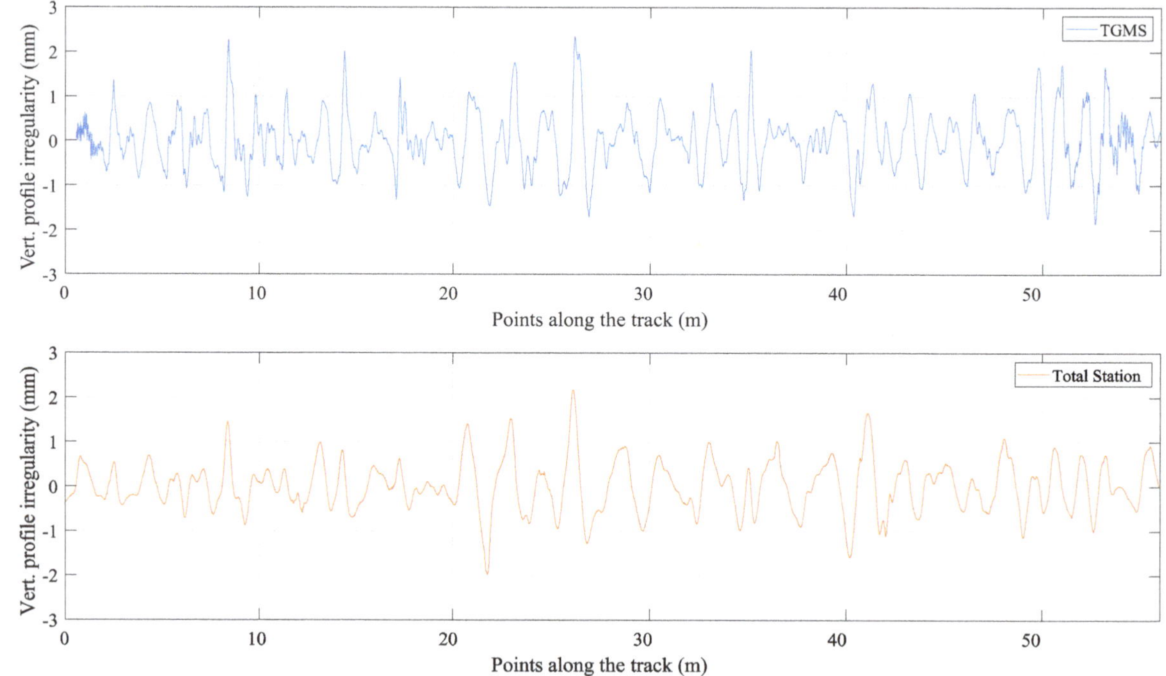

Figure 19. Measurements of vertical profile.

12. Summary and Conclusions

This paper explains the numeric algorithms used to calculate track geometry irregularities with a new TGMS. These algorithms are basically kinematic relationships that describe the motion of a body moving along a track. After the description of the system and the sensors needed, the multibody-based kinematics is described in detail as well as the approximations used due to the small angles assumption. In this kinematic description, the track geometry (design geometry plus irregularities) is assumed totally arbitrary. As a result that the TGMS uses a computer vision system, the kinematic equations needed to locate the recorded rail cross-sections in the TGMS frame have been deduced, as well as a method to find the relative coordinates of the rail cross-sections with respect to the TGMS frame.

Section 6 shows the equations that can be used to find the value of the four independent track irregularities: gauge and cross-level in Equation (24), and alignment and vertical profile in Equation (27), as a function of the output of the computer vision system (explained in Section 5), the relative TGMS to TF relative orientation and the relative TGMS to TF relative trajectory. However, there is a long way to go to apply these equations.

Section 7 explains the development of a linear dynamic model that can be used to find the TGMS to TF relative orientation using the installed inertial sensors, 3D accelerometer and 3D gyroscope. This model is the basis of a Kalman filter that provides as the output the required TGMS angles. The application of this algorithm requires a precise knowledge of the position of the TGMS along the track. This position together with the vehicle forward velocity are needed to calculate the accelerations due to the motion of the TGMS along the design geometry of the track. In turn, these accelerations are needed to calculate the relative orientation based on the accelerometer signals. Section 8 explains the odometry algorithm that is used in this investigation for the precise calculation of the value of the arc-length s of the TGMS along the track. This algorithm is based on the measurement of the track horizontal curvature and the knowledge of the track design geometry.

The most difficult step for the calculation of the track irregularities is to obtain the TGMS to TF relative trajectory. Fortunately, this trajectory is needed to find the absolute irregularities but not needed to find the relative irregularities. The calculation of the relative trajectory is explained in Section 9. After a set of approximations, Equation (46) shows a second-order linear system of ordinary differential equations that can be solved to find the TGMS to TF relative trajectory. However, these equations are intrinsically unstable, requiring special solution methods that need to be improved, as it can be deduced from the experimental results.

The calculation of the track irregularities has been experimentally validated in a scale track that is described in Section 10 as well as the alternative method used to measure the track irregularities. This alternative method is accurate and based on the sensors used in the commercial rail track trolleys, LVDT and inclinometer, and a total station. The comparison of the measurements of the TGMS presented in this paper and this reference measurement is shown in Section 11. The relative irregularities are measured with high accuracy. The measurement of the absolute irregularities can be considered as acceptable, but the accuracy is much less. It can be concluded that the method of calculation of the TGMS to TF relative trajectory, that is the additional input needed for the calculation of the absolute irregularities, needs to be improved.

Author Contributions: Conceptualization and algorithm implementation, J.L.E.; Investigation, J.L.E., P.U. and S.M.; Design of the experiments, J.L.E., P.U. and S.M.; Instrumentation of the vehicle and data acquisition, P.U.; Experimentation, P.U. and S.M.; Writing, J.L.E.; Supervision, J.L.E. All authors have read and agreed to the published version of the manuscript.

Funding: This research was supported by the *Consejería de Economía, Conocimiento, Empresas y Universidad de la Junta de Andalucía* under the project reference US-1257665. This support is gratefully acknowledged.

Data Availability Statement: The data presented in this study are available on request from the corresponding author.

Conflicts of Interest: The authors declare no conflict of interest.

Abbreviations

The following abbreviations are used in this manuscript:

TGMS	Track geometry measurement system.
RTT	Rail track trolleys.
TRV	Track recording vehicles.
VMS	Versine measurement systems.
IMS	Inertial measurement systems.
$<X,Y,Z>$	Global frame (GF).
$<X^t,Y^t,Z^t>$	Track frame (TF).
$<X^i,Y^i,Z^i>$	Body frame i (BF).
$<X^{tgms},Y^{tgms},Z^{tgms}>$	TGMS frame.
$<X^{lrp},Y^{lrp},Z^{lrp}>$	Left rail-profile frame (LRP).
$<X^{rrp},Y^{rrp},Z^{rrp}>$	Right rail-profile frame (RRP).
$\vec{R}$	Position vector with respect to the GF.
$\vec{R}^t$	Position vector of the TF with respect to the GF.
$\vec{r}$	Position vector with respect to the TF.
$\vec{u}$	Position vector with respect to the BF, LRP or RRP.
$\vec{u}^i_P$	Position vector of point P that belongs to body i.
$\mathbf{v}$	Components of a generic vector $\vec{v}$ in the GF.
$\bar{\mathbf{v}}$	Components of a generic vector $\vec{v}$ in the TF.
$\hat{\mathbf{v}}$	Components of a generic vector $\vec{v}$ in the BF, LRP or RRP.

$\mathbf{A}^t$	Rotation matrix of the TF with respect to the GF.
$\mathbf{A}^i$	Rotation matrix of the BF of body i with respect to the GF.
$\mathbf{A}^{t,i}$	Rotation matrix from the BF of body i to the TF.
$\rho_h, \rho_v, \rho_{tw}$	Horizontal, vertical and twist curvatures of the track centerline.
$\rho_h{'}$	Spatial-derivative of horizontal curvature.
α_v	Vertical slope of the track centerline.
$\psi^t, \theta^t, \varphi^t$	Euler angles that describe the orientation of the TF with respect to the GF.
$\psi^i, \theta^i, \varphi^i$	Euler angles that describe the orientation of the body i with respect to the TF.
$\psi^i_{abs}, \theta^i_{abs}, \varphi^i_{abs}$	Euler angles that describe the orientation of the body i with respect to the GF.
s	Arc-length coordinate.
$V, \dot{V}$	Forward velocity and acceleration.
$\vec{\omega}^t, \vec{\alpha}^t$	Absolute angular velocity and acceleration vectors of the TF.
$\vec{\omega}^i, \vec{\alpha}^i$	Absolute angular velocity and acceleration vectors of body i.
$\vec{\omega}^{t,i}, \vec{\alpha}^{t,i}$	Angular velocity and acceleration vectors of body i with respect to the TF.
$\vec{r}^{lir} = [\, 0, y^{lir}, z^{lir}\,]$	Irregularity vector of the left rail.
$\vec{r}^{rir} = [\, 0, y^{rir}, z^{rir}\,]$	Irregularity vector of the right rail.
$\xi_{al}, \xi_{vp}, \xi_{gv}, \xi_{cl}$	Alignment, vertical profile, gauge variation and cross level.

References

1. Grassie, S.L. Measurement of railhead longitudinal profiles: A comparison of different techniques. *Wear* **1996**, *191*, 245–251. [CrossRef]
2. Aknin, P.; Chollet, H. A new approach for the modelling of track geometry recording vehicles and the deconvolution of versine measurements. *Veh. Syst. Dyn.* **1999**, *33*, 59–70. [CrossRef]
3. Lewis, R.B. Track-recording techniques used on British Rail. *IEE Proc. B Electr. Power Appl.* **1984**, *131*, 73–81. [CrossRef]
4. Escalona, J.L. Vertical Track Geometry Monitoring Using Inertial Sensors and Complementary Filters. In *Proceedings of IDETC/CIE 2016*; ASME: Charlotte, NC, USA, 2016.
5. Weston, P.; Ling, C.S.; Roberts, C.; Goodman, C.J.; Li, P.; Goodall, R.M. Monitoring vertical track irregularity from in-service railway vehicles. *Proc. Inst. Mech. Eng. Part F: J. Rail Rapid Transit* **2007**, *221*, 75–88. [CrossRef]
6. Weston, P.; Ling, C.S.; Goodman, C.J.; Roberts, C.; Li, P.; Goodall, R.M. Monitoring lateral track irregularity from in-service railway vehicles. *Proc. Inst. Mech. Eng. Part F: J. Rail Rapid Transit* **2007**, *221*, 89–100. [CrossRef]
7. Weston, P.; Roberts, C.; Yeo, G.; Stewart, E. Perspectives on railway track geometry condition monitoring from in-service railway vehicles. *Veh. Syst. Dyn.* **2015**, *53*, 1063–1091. [CrossRef]
8. Seok Lee, J.; Choi, S.; Kim, S.S.; Park, C.; Guk Kim, Y. A Mixed Filtering Approach for Track Condition Monitoring Using Accelerometers on the Axle Box and Bogie. *IEEE T. Instrum. Meas.* **2012**, *61*, 749–758.
9. Tsai, H.C.; Wang, C.Y.; Huang, N. Fast Inspection and Identification Techniques for Track Irregularities Based on HHT Analysis. *Adv. Adapt. Data Anal.* **2012**, *4*. [CrossRef]
10. Tsai, H.C.; Wang, C.Y.; Huang, N.; Kuo, T.W.; Chieng, W.H. Railway track inspection based on the vibration response to a scheduled train and the Hilbert-Huang transform. *Proc. Inst. Mech. Eng. Part F: J. Rail Rapid Transit* **2014**, *229*. [CrossRef]
11. Tsunashima, H.; Naganuma, Y.; Kobayashi, T. Track geometry estimation from car-body vibration. *Veh. Syst. Dyn.* **2014**, *52*, 207–219. [CrossRef]
12. Tsunashima, H. Condition Monitoring of Railway Tracks from Car-Body Vibration Using a Machine Learning Technique. *Appl. Sci.* **2019**, *9*, 2734–2746. [CrossRef]
13. Tsunashima, H.; Hirose, R. Condition monitoring of railway track from car-body vibration using time-frequency analysis. *Veh. Syst. Dyn.* **2020**, 1–18. [CrossRef]
14. Muñoz, S.; Aceituno, J.F.; Urda, P.; Escalona, J.L. Multibody model of railway vehicles with weakly coupled vertical and lateral dynamics. *Mech. Syst. Signal Process.* **2019**, *115*, 570–592. [CrossRef]
15. Escalona, J.L.; Aceituno, J.F. Multibody simulation of railway vehicles with contact lookup tables. *Int. J. Mech. Sci.* **2019**, *155*, 571–582. [CrossRef]
16. Escalona, J.L.; Aceituno, J.F.; Urda, P.; Balling, O. Railway multibody simulation with the knife-edge-equivalent wheel-rail constraint equations. *Multibody Syst. Dyn.* **2020**, *48*, 373–402. [CrossRef]
17. Escalona, J.L. A methodology for the measurement of track geometry based on computer vision and inertial sensors. *arXiv* **2020**, arXiv:2008.03763v1.
18. Zhang, Z. *A Flexible New Technique for Camera Calibration*; Technical Report MSR-TR-98-71 for Microsoft Research; Microsoft Corporation: Redmond, WA, USA, 1998.
19. Escalona, J.L. Kinematics of motion tracking using computer vision. *arXiv* **2020**, arXiv:2008.00813v1.

20. Szelinski, R. *Computer Vision. Algorithms and Applications*; Springer: Berlin/Heidelberg, Germany, 2011.
21. Madgwick, S.; Harrison, A.; Vaidyanathan, A. Estimation of IMU and MARG orientation using a gradient descent algorithm. In Proceedings of the IEEE International Conference on Rehabilitation Robotics, Zurich, Switzerland, 29 June–1 July 2011.
22. Valenti, R.; Dryanovski, I.; Xiao, J. Keeping a good attitude: A quaternion-based orientation filter for IMUs and MARGs. *Sensors* **2015**, *15*, 19302–19330. [CrossRef] [PubMed]
23. Muñoz, S.; Ros, J.; Escalona, J.L. Estimation of lateral track irregularity through Kalman filtering techniques. *arXiv* **2020**, arXiv:2003.05222.
24. Urda, P.; Muñoz, S.F.; Aceituno, J.; Escalona, J. Application and Experimental Validation of a Multibody Model with Weakly Coupled Lateral and Vertical Dynamics to a Scaled Railway Vehicle. *Sensors* **2020**, *20*, 3700. [CrossRef] [PubMed]

sensors

MDPI

Article

Vision-Based Hybrid Controller to Release a 4-DOF Parallel Robot from a Type II Singularity

José L. Pulloquinga [1,*], **Rafael J. Escarabajal** [1], **Jesús Ferrándiz** [1], **Marina Vallés** [1], **Vicente Mata** [2] **and Mónica Urízar** [3]

[1] Departamento de Ingeniería de Sistemas y Automática, Instituto de Automática e Informática Industrial, Universitat Politècnica de València, 46022 Valencia, Spain; raessan2@etsii.upv.es (R.J.E.); jeferal@etsid.upv.es (J.F.); mvalles@isa.upv.es (M.V.)

[2] Departamento de Ingeniería Mecánica y de Materiales, Instituto Universitario de Investigación Concertado de Ingeniería Mecánica y Biomecánica, Universitat Politècnica de València, 46022 Valencia, Spain; vmata@mcm.upv.es

[3] Department of Mechanical Engineering, Faculty of Engineering in Bilbao, University of the Basque Country, 48013 Bilbao, Spain; monica.urizar@ehu.es

* Correspondence: jopulza@doctor.upv.es; Tel.: +34-96-387-70-00 (ext. 75783)

Abstract: The high accuracy and dynamic performance of parallel robots (PRs) make them suitable to ensure safe operation in human–robot interaction. However, these advantages come at the expense of a reduced workspace and the possible appearance of type II singularities. The latter is due to the loss of control of the PR and requires further analysis to keep the stiffness of the PR even after a singular configuration is reached. All or a subset of the limbs could be responsible for a type II singularity, and they can be detected by using the angle between two output twist screws (OTSs). However, this angle has not been applied in control because it requires an accurate measure of the pose of the PR. This paper proposes a new hybrid controller to release a 4-DOF PR from a type II singularity based on a real time vision system. The vision system data are used to automatically readapt the configuration of the PR by moving the limbs identified by the angle between two OTSs. This controller is intended for a knee rehabilitation PR, and the results show how this release is accomplished with smooth controlled movements where the patient's safety is not compromised.

Keywords: singular configuration; parallel robot; motion control; 3D tracking; screw theory

Citation: Pulloquinga, J.L.; Escarabajal, R.J.; Ferrándiz, J.; Vallés, M.; Mata, V.; Urízar, M. Vision-Based Hybrid Controller to Release a 4-DOF Parallel Robot from a Type II Singularity. *Sensors* **2021**, *21*, 4080. https://doi.org/10.3390/s21124080

Academic Editors: Javier Cuadrado and Miguel Ángel Naya Villaverde

Received: 7 May 2021
Accepted: 9 June 2021
Published: 13 June 2021

Publisher's Note: MDPI stays neutral with regard to jurisdictional claims in published maps and institutional affiliations.

1. Introduction

Parallel robots (PRs) are composed of two or more closed kinematic chains connecting a fixed and a mobile platform that defines the end-effector to be controlled [1]. As opposed to their serial counterpart, they benefit from greater accuracy, stiffness, and load capacity, making them suitable for a great variety of applications [2,3]. Human–robot interaction is one of the major applications, for instance, in the context of medical rehabilitation [4]. Within this field, lower limb rehabilitation [5–9] is an active research area. However, PRs also present several drawbacks regarding the size of their workspace and the presence of singularities within the workspace. The former can be addressed by means of a proper mechanical design of the PR to cover the workspace as required, while the latter requires further analysis.

Singularities in a PR were first analysed by Gosselin and Angeles [10], who established a classification of singular configurations according to the characteristics of the Jacobian matrices calculated from constraint equations. They defined a type I (or inverse kinematic) singularity to refer to the loss of at least one degree of freedom (DOF) due to a degeneracy of the inverse Jacobian matrix ($\|J_I\| = 0$) and a type II (or forward kinematic) singularity to indicate the gain of at least one DOF caused by the degeneracy of the forward Jacobian matrix ($\|J_D\| = 0$). Some other related classifications of singular configurations can be

Sensors **2021**, *21*, 4080. https://doi.org/10.3390/s21124080   https://www.mdpi.com/journal/sensors

found in [11,12]. type I singularities typically occur as the manipulator approaches the boundary of the workspace and are easy to detect and avoid, but type II singularities can arise within the workspace and are more difficult to treat [13].

Type II singularities prevent the mobile platform from bearing external forces despite having all the actuators locked, leading to an uncontrolled motion of the end-effector. The main goal of lower-limb rehabilitation is to perform specific movements that stimulate the motor plasticity of the patient to improve the motor recovery [5]. In conventional rehabilitation, the movements of the patient are controlled and monitored by a physiotherapist, while in robotic rehabilitation, the control task is performed by a PR. For this task, a PR must ensure stiff behaviour despite the presence of type II singularities to maintain control during the rehabilitation process.

Extensive research has been conducted to tackle type II singularities. The determinant of the forward Jacobian J_D gives no further information than the proximity to a singularity, as it lacks a physical meaning [14]. Based on screw theory [15], a transmission index (TI) was designed by Yuan et al. [16] to express the quality of force and motion transmission by using the transmission wrench screw (TWS) and output twist screw (OTS). Takeda and Funabashi [17] designed a TI that expresses how each actuator individually contributes to the motion of the mobile platform by leaving just one actuator active and the rest locked. Subsequently, Wang et al. [18] using the TI proposed by Takeda and Funabashi established that for a type II singularity, at least two OTSs are linearly dependent. Pulloquinga et al. [19] proposed the angle between two instantaneous screw axes from the OTSs (Ω) as a measure for the proximity detection of type II singularities, providing physical meaning and the capability to determine the chains producing the singular configuration.

The extensive analysis of type II singularities presented has been incorporated in motion/force performance evaluation [18,20], path planning, and the design of reconfigurable PRs [21,22]. These analyses have been developed offline, and very little has been found about including this information in the control unit of the PR [23,24]. Abgarwal et al. [24] designed a control scheme to avoid type II singularities of a planar PR by using artificial potential functions. The potential functions are activated near the singularity to alter the trajectory by means of repulsion forces. This setting prevents the PR from entering into a singular configuration by avoiding it. However, an evader controller cannot deal with the situation in which the robot is initially in a type II singularity. Such a task would require extra instrumentation, since solving the forward kinematic problem based on the joint variable measures does not have a single solution. The various possible positions of the mobile platform are due to the degeneracy of the forward Jacobian matrix.

One unambiguous solution to estimate the actual pose of the PR is by using a vision system [25]. Huynh et al. [26] implemented a vision/position hybrid control for a Hexa PR by defining a two-level closed-loop controller. Amarasinghe et al. [27] designed a vision-based hybrid control on a mobile robot. It could autonomously reach a docking station by using a finite-state machine and proportional control combined with image processing.

However, to the best of the authors' knowledge, no research has been published focusing on PR singularity releasing, i.e., letting the robot autonomously get out of a type II singularity. In this paper, a novel algorithm based on online readings from a vision system is proposed to release the PR from a type II singularity. The proposed algorithm is the first to use the angle Ω as an online detector for the proximity to singular configurations. This algorithm is integrated into a two-level closed-loop hybrid controller that results in more compliant manipulation when performing knee rehabilitation tasks. In the inner loop, there is an algebraic closed-loop controller. The outer loop implements a vision-based controller whose algorithm determines the two limbs that most affect the type II singularity by means of the angle Ω. Then, only the references of those two limbs are modified online to feed the inner loop of the controller.

Section 2 describes the 4-DOF PR for knee rehabilitation used to perform the simulations and experiments. Next, the mathematical foundations of type II singularities and the angle Ω are explained. Then, the 3D vision system that has been used to keep track of the pose of the PR is presented, together with a detailed description of the proposed vision-based hybrid controller. Section 3 begins with a description of the requirements for simulation and experimentation as well as the singular trajectories that were designed for this research. The main results are also presented in this section. Finally, the results are discussed in Section 4.

2. Materials and Methods

This section presents the mathematical foundation used in the development of the angle Ω that detects the proximity to a type II singularity in a knee rehabilitation PR. Subsequently, the 3D tracking system (3DTS) used to measure the actual pose of the mobile platform is described. Then the 3DTS and Ω are combined to develop a novel vision-based hybrid controller to release the actual PR under study from a type II singularity. This section also includes a detailed explanation of the algorithm corresponding to this hybrid controller, which detects and moves the actuators according to the angle Ω.

2.1. 3UPS+RPU Parallel Robot

After knee surgery, the diagnosis and rehabilitation tasks require two translational movements (x_m, z_m) in the tibiofemoral plane, one rotation (ψ) around the coronal plane and one rotation (θ) around the tibiofemoral plane [28]. These four DOFs are shown in Figure 1. In order to accomplish these requirements, a PR with 4-DOF has been designed, built [29], and optimized [30] at the Universitat Politècnica de València. The PR under study is named 3UPS+RPU due to its four-limb architecture. The external limbs or open kinematic chains have a UPS configuration, while the central one has an RPU configuration (see Figure 1). The letters R, U, S and P stand for revolute, universal, spherical and prismatic joints, respectively, and the actuated joints are indicated by the underlined format.

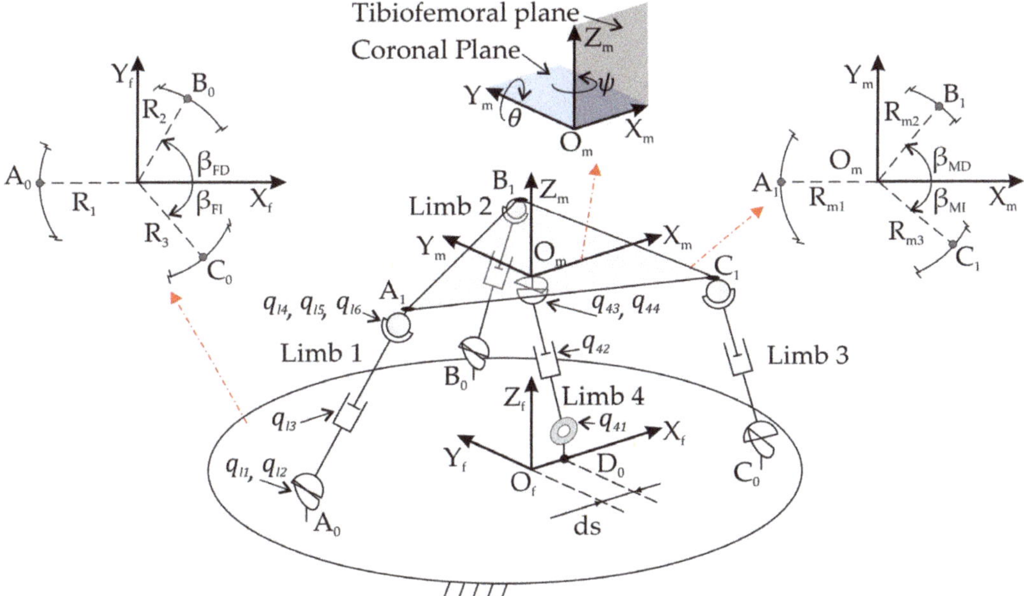

Figure 1. Mechanical configuration of the 3UPS+RPU PR.

The kinematic model of the 3UPS+RPU PR is established by 15 generalized coordinates as follows:

- The position (x_m, z_m) and the orientation (θ, ψ) of the mobile platform.
- The orientation of the four universal joints: q_{l1}, q_{l2} for limbs $l = 1 \ldots 3$ and q_{43}, q_{44} for limb 4.
- The length of the four linear actuators given by q_{l3} for limbs $l = 1 \ldots 3$ and q_{42} for limb 4.
- The orientation of the three spherical joints represented by q_{l4}, q_{l5}, q_{l6} for external limbs $l = 1 \ldots 3$.
- The orientation of the revolute joint q_{41}.

The variables q are measured with respect to a local reference system attached to each joint. The coordinates x_m, z_m, θ, and ψ are measured with respect to the reference system $\left\{ O_f - X_F Y_F Z_F \right\}$ attached to the centre of the fixed platform to reduce the complexity of the model.

The locations of A_0, B_0, C_0, and D_0 that link the four limbs to the fixed platform are defined by the geometric variables $R_1, R_2, R_3, \beta_{FD}, \beta_{FI}$, and d_s. The locations of A_1, B_1, C_1, and O_m that link the limbs to the mobile platform are defined by the geometric variables $R_{m1}, R_{m2}, R_{m3}, \beta_{MD}$, and β_{MI}. Figure 1 shows the location of A_0, B_0, and C_0 and A_1, B_1, and C_1 on the fixed and mobile platform. Table 1 shows the values of $R_1, R_2, R_3, \beta_{FD}, \beta_{FI}$, and d_s measured with respect to $\left\{ O_f - X_F Y_F Z_F \right\}$ and $R_{m1}, R_{m2}, R_{m3}, \beta_{MD}$, and β_{MI} with respect to $\{ O_m - X_M Y_M Z_M \}$. The mobile reference system $\{ O_m - X_M Y_M Z_M \}$ is attached to the centre of the mobile platform.

Table 1. Geometric parameters for the 3UPS+RPU PR.

R_1 (*m*)	R_2 (*m*)	R_3 (*m*)	β_{FD} (°)	β_{FI} (°)	d_s (*m*)
0.4	0.4	0.4	90	45	0.15

R_{m1} (*m*)	R_{m2} (*m*)	R_{m3} (*m*)	β_{MD} (°)	β_{MI} (°)	
0.3	0.3	0.3	50	90	

2.2. Type II Singularities

The velocity equations of a PR [10] are defined by the time derivatives of the geometrical constraint equations $(\vec{\Phi})$ as follows:

$$J_D \dot{\vec{X}} + J_I \dot{\vec{q}}_{ind} = \vec{0} \tag{1}$$

$\vec{X}$ is the set of outputs that represents the DOFs of the mobile platform; $\dot{\vec{q}}_{ind}$ is the set of inputs that corresponds to the active joint length; and J_D and J_I are the forward and inverse Jacobian matrices, respectively. J_D and J_I are $F x F$ square matrices for a non-redundant PR, where F is the number of DOF.

A type II singularity takes place when J_D is rank deficient; i.e., its determinant is zero ($\| J_D \| = 0$). In this configuration, if an external force is applied to the PR, the mobile platform may move ($\dot{\vec{X}} \neq 0$) despite the actuators being locked ($\dot{\vec{q}}_{ind} = 0$). For this reason, in a type II singularity, the control over the PR is lost, becoming potentially dangerous for the user or the robot itself. The PR under study must ensure a safe interaction with the knee of the patient, and, therefore, the treatment of type II singularities is an important problem to solve. A general method to detect the proximity to a type II singularity is by calculating the $\| J_D \|$.

In the 3UPS+RPU PR, J_D is defined as a 4×4 matrix,

$$J_D = \left[\frac{\partial \vec{\Phi}}{\partial x_m} \quad \frac{\partial \vec{\Phi}}{\partial z_m} \quad \frac{\partial \vec{\Phi}}{\partial \theta} \quad \frac{\partial \vec{\Phi}}{\partial \psi} \right] \tag{2}$$

with $\vec{X} = [x_m \ z_m \ \theta \ \psi]^T$ and $\vec{q}_{ind} = [q_{13} \ q_{23} \ q_{33} \ q_{42}]^T$.

The online calculation of J_D requires an accurate measure of $\vec{X}$. In a model-based controller, $\vec{X}$ is estimated by solving the forward kinematic problem using the sensors installed in the actuated joints. In a type II singularity, the forward kinematic problem presents several feasible solutions, and an unambiguous estimation of the actual $\vec{X}$ is not possible. The accurate measure of the actual $\vec{X}$ of the PR requires direct sensing of the mobile platform by means of extra instrumentation, such as a 3DTS.

2.3. Angle between Two Output Twist Screws

The motion of the mobile platform of a PR is produced by the combined action of the active joints, making it difficult to identify the contribution of each actuator. Takeda and Funabashi [17] divided the movement of the mobile platform ($) into F OTSs as follows:

$$\$ = \rho_1 \hat{\$}_{O1} + \rho_2 \hat{\$}_{O2} + \ldots + \rho_F \hat{\$}_{OF} \tag{3}$$

with

$$\hat{\$}_O = \left(\vec{\mu}_{\omega O}; \vec{\mu}_{v O}^{\,*} \right) \tag{4}$$

ρ_i is the amplitude for each OTS, and $\vec{\mu}_{\omega O}$ and $\vec{\mu}_{v O}^{\,*}$ are the instantaneous screw axis and the linear component of the normalized OTS ($\hat{\$}_O$), respectively. Each $\hat{\$}_O$ is determined by solving Equation (5) considering that $F - 1$ actuators are locked.

$$\hat{\$}_{Oi} \circ \hat{\$}_{Tj} = 0 \ (i,j = 1,2,\ldots,F, \ i \neq j) \tag{5}$$

where $\circ$ stands for the reciprocal product, and $\hat{\$}_T$ is the unitary TWS.

In [18], Wang et al. proved that for a singular configuration of a PR, at least two $\hat{\$}_O$s are linearly dependent. This means that in a type II singularity, both $\vec{\mu}_{\omega O}$ and $\vec{\mu}_{v O}^{\,*}$ are equal or parallel. Based on this feature, a novel type II singularity proximity index is defined by measuring the angle $\Omega_{i,j}$ between two $\vec{\mu}_{\omega O}$s and verifying the equality of their respective $\vec{\mu}_{v O}^{\,*}$. Grouping in pairs the F $\vec{\mu}_{\omega O}$, there are $\binom{F}{2}$ angles Ω, which are defined as:

$$\Omega_{i,j} = acos\left(\vec{\mu}_{\omega O_i} \cdot \vec{\mu}_{\omega O_j} \right) \ (i,j = 1,2,\ldots,F, \ i \neq j) \tag{6}$$

where i and j represent the selected $\vec{\mu}_{\omega O}$.

In contrast with the $\|J_D\|$, the index $\Omega_{i,j}$ provides a physical scale for the measure of the proximity to a type II singularity. When $\Omega_{i,j} = 0$ and $\vec{\mu}_{v O_i}^{\,*} = \vec{\mu}_{v O_j}^{\,*}$, $\hat{\$}_{Oi}$ and $\hat{\$}_{Oj}$ are linearly dependent in Equation (3), identifying the open chains (i, j) involved in the type II singularity. Considering the centre of the mobile platform of the 3UPS+RPU PR, six possible $\Omega_{i,j}$s are considered:

$$\begin{aligned}
\Omega_{1,2} &= arccos\left(\vec{\mu}_{\omega O_1} \cdot \vec{\mu}_{\omega O_2} \right) & \Omega_{2,3} &= arccos\left(\vec{\mu}_{\omega O_2} \cdot \vec{\mu}_{\omega O_3} \right) \\
\Omega_{1,3} &= arccos\left(\vec{\mu}_{\omega O_1} \cdot \vec{\mu}_{\omega O_3} \right) & \Omega_{2,4} &= arccos\left(\vec{\mu}_{\omega O_2} \cdot \vec{\mu}_{\omega O_4} \right) \\
\Omega_{1,4} &= arccos\left(\vec{\mu}_{\omega O_1} \cdot \vec{\mu}_{\omega O_4} \right) & \Omega_{3,4} &= arccos\left(\vec{\mu}_{\omega O_3} \cdot \vec{\mu}_{\omega O_4} \right)
\end{aligned} \tag{7}$$

The variables $\vec{\mu}_{\omega O_1} \ldots \vec{\mu}_{\omega O_4}$ are calculated by solving Equation (5) with the four $\hat{\$}_T$s of the linear actuators defined as

$$\hat{\$}_{T1} = \begin{bmatrix} \vec{z}_{12} \\ \vec{r}_{O_m A_1} \times \vec{z}_{12} \end{bmatrix}, \ \hat{\$}_{T2} = \begin{bmatrix} \vec{z}_{22} \\ \vec{r}_{O_m B_1} \times \vec{z}_{22} \end{bmatrix}, \hat{\$}_{T3} = \begin{bmatrix} \vec{z}_{32} \\ \vec{r}_{O_m C_1} \times \vec{z}_{32} \end{bmatrix}, \hat{\$}_{T4} = \begin{bmatrix} \vec{z}_{41} \\ \vec{0} \end{bmatrix} \tag{8}$$

where $\vec{z}$ is the unit vector of the forces applied by the actuators, and $\vec{r}$ is the position vector for the connection point of the limbs with the mobile platform measured from O_m; see Figure 2.

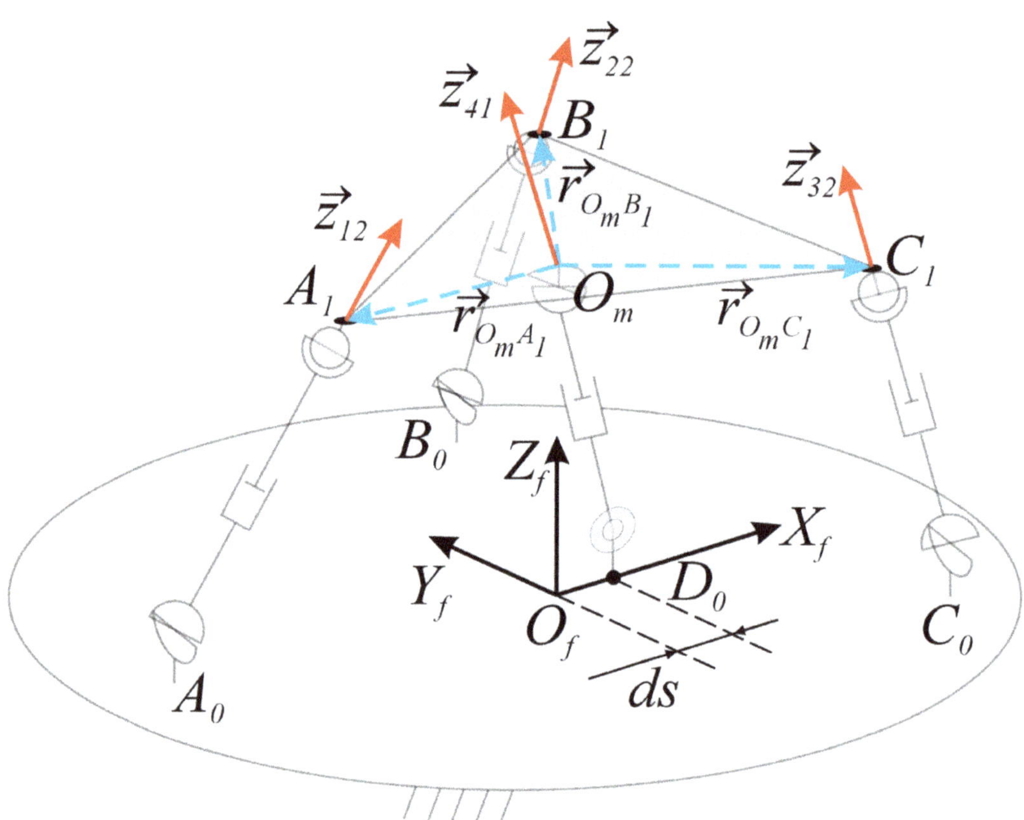

Figure 2. TWSs in the 3UPS+RPU PR.

The capability to detect the proximity to a type II singularity given by the six $\Omega_{i,j}$ indices defined in Equation (7) has been verified from an analytical and experimental perspective [19]. However, the capability to identify the pair of limbs responsible for the type II singularity has not been exploited. Therefore, this study proposes a novel hybrid controller that takes advantage of the index $\Omega_{i,j}$ to release the 3UPS+RPU PR from a type II singular configuration. The index $\Omega_{i,j}$ is defined by means of the position and orientation of the mobile platform. For this reason, an accurate measure of the actual $\vec{X}$ is essential for developing the hybrid controller proposed.

2.4. 3D Tracking System

To be able to capture the movements of the mobile platform of the PR, a 3D tracking system (3DTS) based on artificial vision was used. The system consists of 10 Flex13 cameras from the manufacturer OptiTrack (Corvallis, OR, USA). These cameras use the infrared emission principle to be able to capture and detect the reflection that it creates on markers made of reflective 3M material.

Figure 3 shows the Robotics Laboratory and some cameras of the 3DTS used in this work. The cameras have a 1.3 Megapixel resolution and a capture velocity of 120 Hz. They have a latency or frame delay of 8.3 ms. The set of 10 cameras and the use of high-quality 14 mm markers make it possible to obtain an accuracy of more than 0.1 mm.

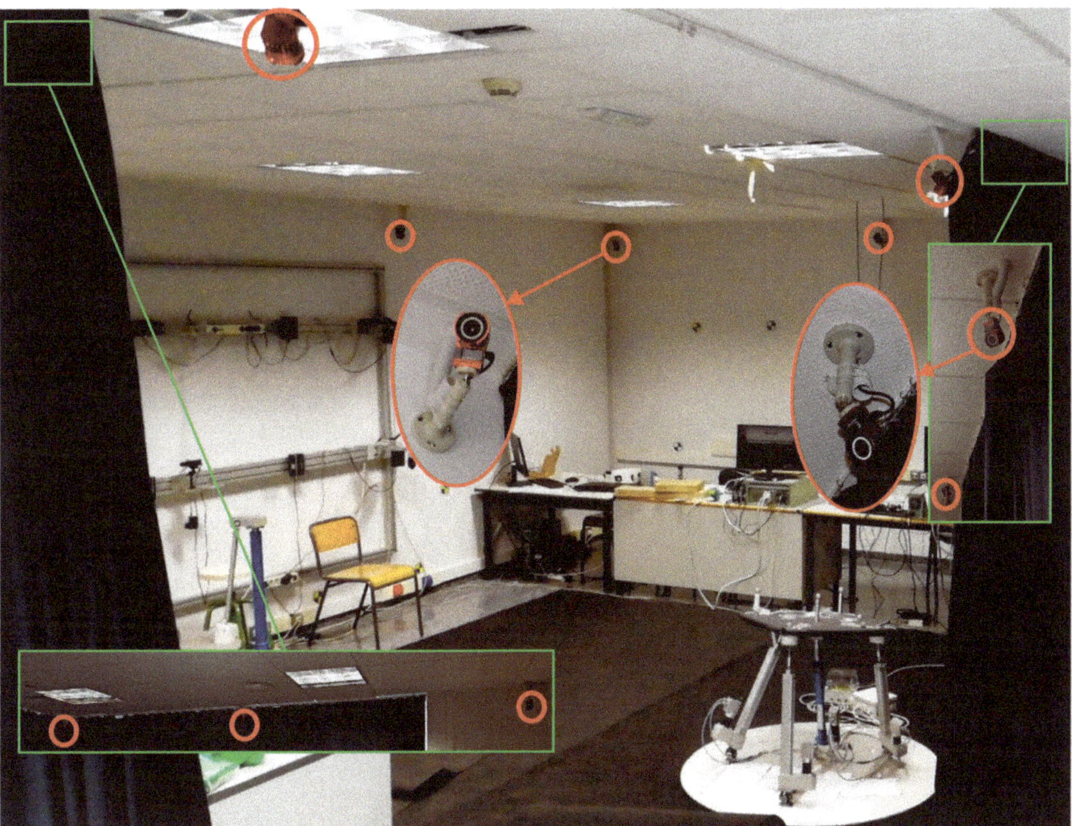

Figure 3. Robotics Laboratory equipped with the OptiTrack 3DTS.

The cameras are connected to two OptiHub2 devices. The OptiHub2 allows higher and more consistent power delivery to cameras for enhanced tracking range, simpler camera setup and cabling, and support for camera synchronization. The OptiHub2 devices are connected to high-speed USB ports in the camera control computer, and this computer communicates with the robot control computer using an Ethernet connection. The Figure 4 below shows the architecture of the OptiTrack 3DTS of the laboratory.

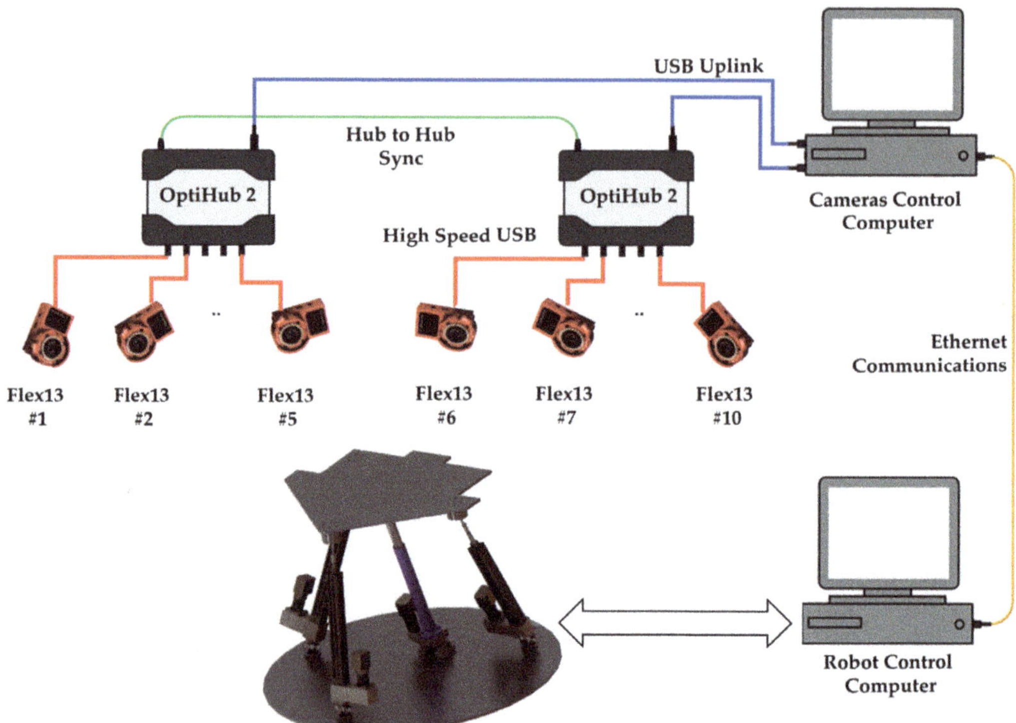

Figure 4. Laboratory OptiTrack 3DTS architecture.

The Motive Tracker software (Motive) from the same manufacturer, OptiTrack, is used on the camera control computer. This software is used to perform vision system calibration and obtain 6-DOF positioning results of objects within the tracking area. Motive uses high-level tracking filters and constraints to fine tune the performance of the high-speed object tracking. Motive associates a custom set of markers to a virtual element called rigid body and offers data access at any stage in the pipeline, i.e., 2D camera images, marker centroid data, labelled markers, and rigid bodies. In addition, it is possible to completely replace the Motive user interface and directly control the system operation in a new application with the NatNet SDK.

NatNet's client/server architecture allows client applications to run on the same system as the tracking software (Motive), on separate system(s), or both. The SDK integrates seamlessly with standard APIs (C/C++/.NET), tools (Microsoft Visual Studio), and protocols (UDP/Unicast/Multicast). Using the NatNet SDK, developers can quickly integrate OptiTrack motion tracking data into new and existing applications, including custom plugins for third-party applications and engines for real-time streaming. In addition, this SDK provides a .NET interface and sample programs that work directly with MATLAB core, requiring no additional MATLAB modules. Figure 5 summarizes the software architecture of the 3DTS used in this study.

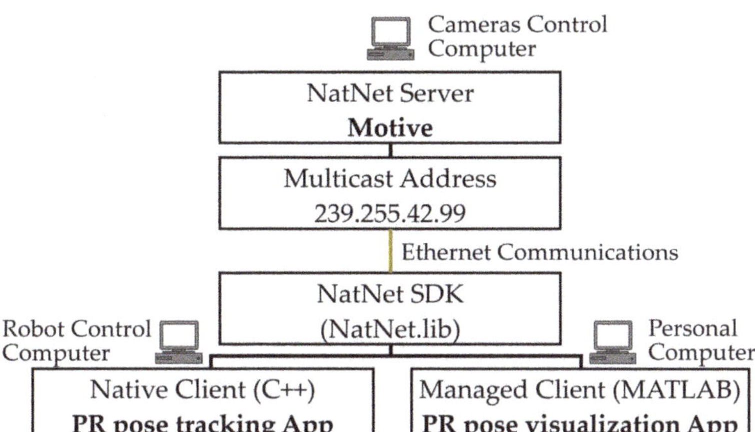

Figure 5. Software architecture of the OptiTrack 3DTS.

Regarding the experimental setup, each camera individually builds a 2D image based on the markers' location, so a calibration process is required prior to the experiments in order to ensure that the system correctly reconstructs the 3D position of every marker.

The first step involved in this process is the correct orientation of the cameras to aim them at the workspace and, specifically, at the tracking volume, which, in this case, is the 4-DOF PR. Since the robot always operates in the same location and its workspace is limited, no changes in the camera location or orientation are required, and, therefore, they remain in the same position from the moment they are installed.

Another aspect to control is the brightness and illumination of the scene, as this allows the markers to be visible for the cameras, and, as such, no other unwanted objects are detected. Since the lighting conditions are the same for all experiments, some configuration parameters of the cameras, such as the exposure time, the gain, and the threshold, are set at constant values for all cameras using the software. If any intrusive markers are detected, they can either be manually covered by a cloth or masked in the software before performing the calibration.

After configuring the cameras, the calibration process starts with an empty scene where no markers should be detected, except for those attached to the calibration wand.

By moving the calibration wand, which is provided by OptiTrack, around the workspace, the cameras provide successive 2D projections of the markers. The 2D projections are used to compute the relative position of the cameras. The software shows the increasing precision of this estimation as the process progresses (Figure 6), and when a high enough quality is achieved, the process is manually stopped.

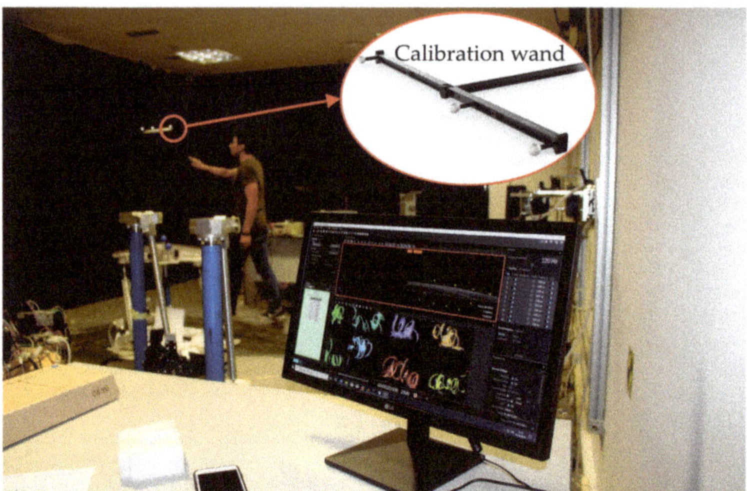

Figure 6. Calibration wand and experiment to determine the location of the markers.

The second tool, which concludes the calibration process, is the calibration square shown in Figure 7. This object includes three markers in right angle that define the origin and axes of the world coordinate system (also called ground plane by Motive). The ground plane is placed on the floor within the workspace area in such a way that its markers can be visible by as many cameras as possible. This tool incorporates a level to ensure its horizontal position.

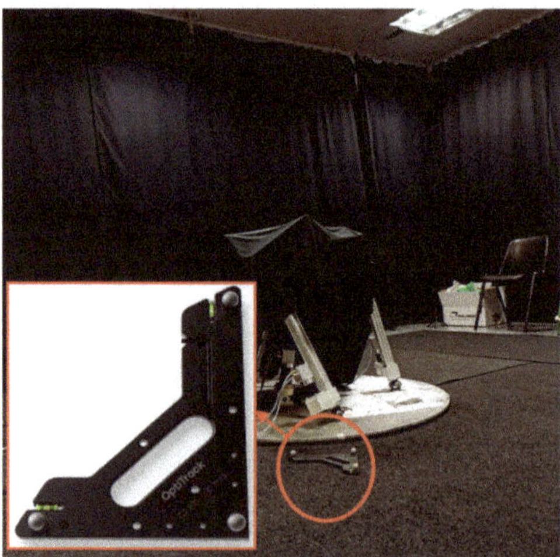

Figure 7. Calibration square.

Although all the cameras remain in the same position, minor movements of any of the cameras between experiments (for example, due to vibrations) can lead to poor tracking performance. For that reason, the calibration process must be performed once a day to ensure reliable 3D tracking. The calibration steps take no more than five minutes. After calibrating the cameras, Motive starts streaming data from all rigid bodies within the workspace. Rigid bodies are a set of 3 or more (maximum 20) markers whose relative distances remain constant. In this research, there are 2 rigid bodies represented by the fixed and mobile platform, respectively, and a set of 4 markers was attached to them. Three of the markers describe the cartesian coordinate frame of both platforms, and the fourth is added in a random (but known) position. If one of the markers is missed by the software during an experiment, the other three make it possible to reconstruct its position and keep streaming enough accurate data.

In the PR pose tracking App presented in this paper, the NatNet SDK provides a client class to communicate with the Motive server. A data handler is attached to this client, which works as a call-back that is executed every time there is a new frame of data available from the server. This handler has been used for retrieval of the x, y, z position of three markers placed on the fixed platform and another three placed on the mobile platform of the PR. Given the coordinates of the six markers, the actual position and orientation of the mobile platform ($\vec{X}_c$) are calculated with respect to the $\left\{ O_f - X_F Y_F Z_F \right\}$ coordinate frame. Finally, $\vec{X}_c$ is sent through ROS2 messages to feed the control system. A MATLAB program has also been designed to provide an online view of $\vec{X}_c$ and calculate the actual actuator's length by solving the inverse kinematics. Figure 8 presents the graphic user interface (GUI) for online measures of $\vec{X}_c$. It is important to note that this program is independent of the control system (and therefore runs in a personal computer) and simply offers a viewing tool.

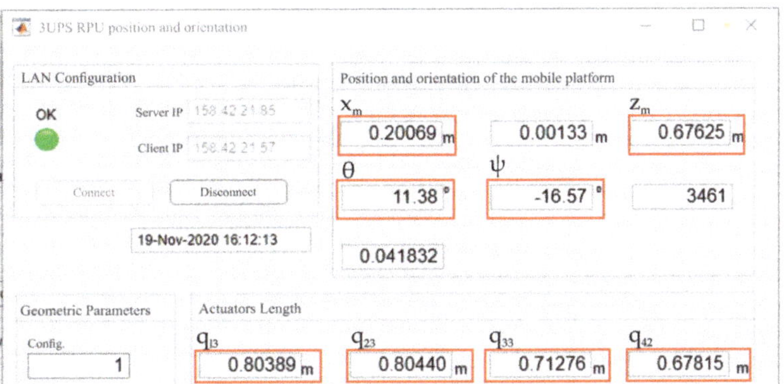

Figure 8. GUI for position and orientation tracking designed in MATLAB.

2.5. Hybrid Controller Description

If a PR reaches a type II singularity, a controller must move the actuators to release the PR from the singularity, maintaining a minimum deviation from the original configuration. Therefore, a method to identify the best set of actuators to be moved is needed. The index $\Omega_{i,j}$, using the position and orientation of the mobile platform, is able to identify the actuators involved in the type II singularity. However, in a type II singularity, the measurement of the actual position and orientation of the PR require an external sensor, such as a 3DTS. For this reason, a novel controller able to release the PR under study from a type II singularity using the index $\Omega_{i,j}$ and a 3DTS is proposed. It is important to note that this is the first time that the index $\Omega_{i,j}$ is employed as an online proximity detector to a type II singularity.

The novel vision-based hybrid controller to release the 3UPS+RPU PR from a type II singularity is shown in Figure 9. The hybrid controller combines two-level closed loops: an algebraic algorithm (inner loop) and a type II singularity releaser (outer loop). The type II singularity releaser calculates the $\Omega_{i,j}$ indices using the position and orientation of the PR provides by the OptiTrack 3DTS.

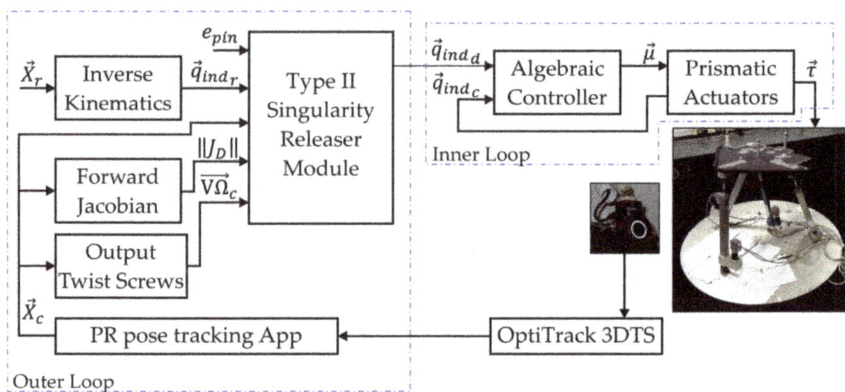

Figure 9. Hybrid controller architecture.

In the inner loop, the control signals ($\vec{\mu}$) to track the desired actuator's location $\vec{q}_{ind_d}$ are calculated by an algebraic algorithm based on the measured location of the actuators $\vec{q}_{ind_c}$. The $\vec{\mu}$ is proportional to the forces ($\vec{\tau}$) applied by the linear actuators to move the mobile platform.

In the outer loop, the reference location of the actuators ($\vec{q}_{ind_r}$) is obtained by solving the inverse kinematics for a knee rehabilitation trajectory ($\vec{X}_r$), and $\vec{X}_r$ is designed for the 4-DOF of the 3UPS+RPU PR.

Based on the $\vec{X}_c$ measured by the 3DTS, the proximity to a type II singularity is detected by $\vec{V\Omega}_c$ and $||J_D||_c$ at every time step. $\vec{V\Omega}_c$ stores the six $\Omega_{i,j}$ indices as $\begin{bmatrix} \Omega_{1,2} & \Omega_{1,3} & \Omega_{1,4} & \Omega_{2,3} & \Omega_{2,4} & \Omega_{3,4} \end{bmatrix}^T$. If the 3UPS+RPU PR gets close to a type II singularity, $\vec{q}_{ind_r}$ is modified to define $\vec{q}_{ind_d}$. In Figure 9, the type II singularity releaser module (SRM) calculates the desired location of the actuator as follows:

$$\vec{q}_{ind_d} = \vec{q}_{ind_r} + v_d \cdot t_s \cdot \vec{\Delta i} \tag{9}$$

where v_d is the releasing velocity module for each actuator, t_s stands for the controller sample time, and $\vec{\Delta i}$ represents an integer vector that counts the deviation required in the F actuators to release the PR from a type II singularity.

The SRM calculates $\vec{q}_{ind_d}$ at every time step, although $\vec{\Delta i}$ is modified only if an enable pin (e_{pin}) is activated. Two versions of the algorithms have been proposed to contrast the results when (i) moving the actuators that cause the singularity and (ii) moving the actuators that do not cause the singularity according to $\vec{V\Omega}_c$.

The first version (SRM-V1) releases a PR from a type II singularity by moving the limbs identified by $\min\Omega_c$, which represents the minimum value of $\vec{V\Omega}_c$. If $\min\Omega_c$ or $||J_D||_c$ is lower than a certain limit, Ω_{lim} and $||J_D||_{lim}$, respectively, the two rows of $\vec{\Delta i}$ that have to change are identified by $\vec{i}_{ch}$. The possible change combinations for the two rows of $\vec{\Delta i}$ are defined by the columns of M_{inc} as follows:

$$M_{inc} = \begin{bmatrix} 1 & -1 & 1 & -1 & 1 & -1 & 0 & 0 \\ 1 & -1 & -1 & 1 & 0 & 0 & 1 & -1 \end{bmatrix} \tag{10}$$

where 0, 1, and −1 correspond to the stop, unit forward motion, and unit backward motion commands for an actuator, respectively.

For each column of M_{inc}, an auxiliary variable $\vec{q}_{ch}$ is initialized as $\vec{q}_{ind_d}$, and then its elements indexed by $\vec{i}_{ch}$ are modified using the current M_{inc} column. Then, it is checked that this position is confined within the geometrical limits. If $\vec{q}_{ch}$ is inside the actuators' displacement range, the forward kinematic problem is solved ($\vec{X}_{ch}$). Next, the angles reached by the spherical joints ($\vec{\alpha}_{ch}$) are calculated. If $\vec{\alpha}_{ch}$ is within the working range, a new $\Omega_{i,j}$ is calculated for the limbs identified by $\vec{i}_{ch}$, and it is added to $\vec{V\Omega}_{ch}$. Then, $\vec{\Delta i}$ takes the value of the column of M_{inc} that produces the maximum element of $\vec{V\Omega}_{ch}$ ($\max\Omega_{ch}$), as that combination contributes the most to releasing the singularity without exceeding any range limit. Finally, $\vec{q}_{ind_d}$ is updated using the new $\vec{\Delta i}$.

An alternative algorithm called SRM-V2 has been proposed to test the behaviour when moving the wrong limbs. It modifies the rows of $\vec{\Delta i}$ that are not related to $\min\Omega_c$ ($\vec{i}_{nc}$) to release the PR from the type II singularity caused by the actuators $\vec{i}_{ch}$. SRM-V2 is designed to verify that moving the actuators identified by $\min\Omega_c$ is the best way to release the 3UPS+RPU PR from a type II singularity.

The complete process performed by SRM-V1 is described in the pseudocode shown in Algorithm 1, where SRM-V2 is obtained by adding and replacing, the lines marked with $*$ and $**$, respectively. A description of the variables used in Algorithm 1 is presented in Table 2.

Table 2. Description of parameters, inputs, and outputs of SRM-V1 and SRM-V2.

Parameters		
Variable	**Description**	**Default**
v_d	releasing velocity module in m/s	0.01
t_s	controller sample time in s	0.01
$\|J_D\|_{lim}$	experimental limit for $\|J_D\|$	0.015
Ω_{lim}	experimental limit for $\Omega_{i,j}$	1.8°
$\vec{max}q_{ind}$	maximum feasible values for the actuators' length in m, 4x1 vector	$\begin{bmatrix} 0.93 & 0.92 & 0.93 & 0.82 \end{bmatrix}^T$
$\vec{min}q_{ind}$	minimum feasible values for the actuators' length in m, 4x1 vector	$\begin{bmatrix} 0.65 & 0.64 & 0.65 & 0.54 \end{bmatrix}^T$
$\vec{\alpha}_{lim}$	experimental limits for the spherical joints, 3x1 vector	38°
M_{inc}	possible increments/decrements for $\vec{\Delta i}$	See equation (10)
$\vec{\Delta i}$	column vector 4x1, persistent variable	-
Inputs		
Variable	**Description**	**Default**
e_{pin}	enable pin	-
$\|J_D\|_c$	determinant of the forward Jacobian matrix, feedback signal	-
$\vec{V\Omega}_c$	column vector with the six $\Omega_{i,j}$ indices, feedback signals	-
$\vec{X}_c$	position and orientation of the mobile platform, feedback signal	-
$\vec{q}_{ind_r}$	trajectory for the actuators, reference signal	-
Outputs		
Variable	**Description**	**Default**
$\vec{q}_{ind_d}$	trajectory for the actuators, desired signal	-

---

Algorithm 1. Initialization 3

INITIALIZATION

$$\vec{\Delta i} = \vec{0}$$

N_{ch} = number of columns of M_{inc}.
BEGIN
$\vec{q}_{ind_d} = \vec{q}_{ind_r} + v_d \cdot t_s \cdot \vec{\Delta i}$
IF $e_{pin} == true$
    $\min\Omega_c$ = minimum element in $\vec{V\Omega}_c$
    IF $\min\Omega_c < \Omega_{lim}$ **OR** $\|J_D\|_c < \|J_D\|_{lim}$
        IF $\min\Omega_c == \vec{V\Omega}_c(1)$
        $\vec{i}_{ch} = \begin{bmatrix} 1 & 2 \end{bmatrix}$
        $*\ \vec{i}_{nc} = \begin{bmatrix} 3 & 4 \end{bmatrix}$
        ELSEIF $\min\Omega_c == \vec{V\Omega}_c(2)$
        $\vec{i}_{ch} = \begin{bmatrix} 1 & 3 \end{bmatrix}$
        $*\ \vec{i}_{nc} = \begin{bmatrix} 2 & 4 \end{bmatrix}$
        ELSEIF $\min\Omega_c == \vec{V\Omega}_c(3)$
        $\vec{i}_{ch} = \begin{bmatrix} 1 & 4 \end{bmatrix}$
        $*\ \vec{i}_{nc} = \begin{bmatrix} 2 & 3 \end{bmatrix}$
        ELSEIF $\min\Omega_c == \vec{V\Omega}_c(4)$
        $\vec{i}_{ch} = \begin{bmatrix} 2 & 3 \end{bmatrix}$
        $*\ \vec{i}_{nc} = \begin{bmatrix} 1 & 4 \end{bmatrix}$
        ELSEIF $\min\Omega_c == \vec{V\Omega}_c(5)$
        $\vec{i}_{ch} = \begin{bmatrix} 2 & 4 \end{bmatrix}$
        $*\ \vec{i}_{nc} = \begin{bmatrix} 1 & 3 \end{bmatrix}$
        ELSE $\min\Omega_c == \vec{V\Omega}_c(6)$
        $\vec{i}_{ch} = \begin{bmatrix} 3 & 4 \end{bmatrix}$
        $*\ \vec{i}_{nc} = \begin{bmatrix} 1 & 2 \end{bmatrix}$
    ENDIF
    $\vec{V\Omega}_{ch}$ = column vector of N_{ch} zeros
    FOR $c_1 = 1 : N_{ch}$
    $\vec{q}_{ch} = \vec{q}_{ind_d}$
    $\vec{q}_{ch}\left(\vec{i}_{ch}\right) = \vec{q}_{ch}\left(\vec{i}_{ch}\right) + v_d \cdot t_s \cdot M_{inc}(:, c_1)$
    $*\!*\ \vec{q}_{ch}\left(\vec{i}_{nc}\right) = \vec{q}_{ch}\left(\vec{i}_{nc}\right) + v_d \cdot t_s \cdot M_{inc}(:, c_1)$
        IF $\vec{minq}_{ind} < \vec{q}_{ch} < \vec{maxq}_{ind}$ (element-wise comparison)
            $\vec{X}_{ch}$ = Solve the Forward Kinematics for $\vec{q}_{ch}$, using $\vec{X}_c$ as initial condition
            $\vec{\alpha}_{ch}$ = Angle of spherical joints for $\vec{X}_{ch}$
            IF $\vec{\alpha}_{ch} < \vec{\alpha}_{lim}$ (element-wise comparison)
                $\vec{V\Omega}_{ch}(c_1)$ = Calculate the index $\Omega_{i,j}$ for $\vec{X}_{ch}$ with $i, j = \vec{i}_{ch}$
            ENDIF
        ENDIF
    ENDFOR
    $c_1 = argmax\left(\vec{V\Omega}_{ch}\right)$
    $\vec{\Delta i}\left(\vec{i}_{ch}\right) = \vec{\Delta i}\left(\vec{i}_{ch}\right) + M_{inc}(:, c_1)$
    $\vec{q}_{ind_d} = \vec{q}_{ind_r} + v_d \cdot t_s \cdot \vec{\Delta i}$
    ENDIF
ENDIF
END

Due to the properties of the index $\Omega_{i,j}$, the SRM algorithm has the advantage of moving a pair of F actuators simultaneously in each time step of the controller. For this reason, the SRM reduces the consumption of computing resources and the difference between $\vec{q}_{ind_d}$ and the original $\vec{q}_{ind_r}$.

3. Results

This section begins with a detailed description of the simulation setup, including the singular trajectories to be tested with SRM-V1 and SRM-V2 versions of the hybrid controller. Next, the performance of the hybrid controller in simulation is evaluated, where SRM-V1 appears to be better than SRM-V2. Subsequently, the experimental setup and the features of the actual 3UPS+RPU PR are detailed. Finally, the main experimental results show the effectiveness of the hybrid controller using SRM-V1 to release the PR under study from a type II singularity.

3.1. Simulation of the Vision-Based Hybrid Controller

Prior to implementing the algorithm on the actual PR, some simulations are performed on a kinematic and dynamic model of the 3UPS+RPU PR designed in MATLAB/Simulink. In both simulation and experimentation, the PR is moved from the initial position to a singular test configuration without activating the releaser. Then, it remains in the singular configuration for 15 s, after which the loop of the SRM is activated via e_{pin}. In that moment, one of the SRMs in Section 2.5 is launched based on the assumption that it will help release the robot from the type II singularity. The SRM launched has a lapse of 15s, allowing it to move the PR under study to a non-singular configuration.

Due to the lack of a simulated model of the 3DTS (see Figure 9) for MATLAB/Simulink, $\vec{X}_c$ is calculated directly by solving the forward kinematic problem. The main objective of the simulation is to test that the novel hybrid controller increases the values of $\|J_D\|$ and $\Omega_{i,j}$ in the vicinity of a type II singular configuration; i.e., it is able to release the PR under study from the type II singularity.

Since the 3UPS+RPU PR was built to interact with human knees, it is used to execute three rehabilitation movements: flexion of the hip, flexion–extension of the knee, and internal–external rotation of the knee [19]. This study, combining these three fundamental movements for simulation and experimentation, performs five knee rehabilitation trajectories ending in a type II singular configuration (see Table 3). The singular configurations of these five trajectories have $\|J_D\|$ and $\Omega_{i,j}$ close to zero but not exactly zero, avoiding several forward kinematic solutions in the simulation. All five knee trajectories are designed with constant velocity; in this case, the translational DOFs move at 0.02 m/s and the rotational ones at 0.03 rad/s.

Table 3. Description of the trajectories with a type II singularity at the end.

Trajectory	Description	Type II Singularity			
		x_m (m)	z_m (m)	θ (rad)	ψ (rad)
1	Hip flexion	0.01	0.70	0.15	0.31
2	Partial internal–external knee rotation	0.01	0.70	−0.02	0.14
3	Flexion–extension of the knee combined with ankle and knee rotations	0.05	0.72	−0.01	0.15
4	Flexion–extension of the knee combined with hip flexion	0.12	0.77	−0.06	0.11
5	Complete internal–external knee rotation	−0.05	0.73	0.10	0.33

The simulation of the five knee rehabilitation trajectories verifies that SRM-V1 and SRM-V2 release the 3UPS+RPU PR from a singular configuration. Figure 10 shows how the type II singularity indices $\|J_D\|_c$ and $\min\Omega_c$ increase when e_{pin} is activated for trajectory

1. These results verify from an analytical perspective that the hybrid controller proposed releases the 3UPS+RPU PR from a type II singularity.

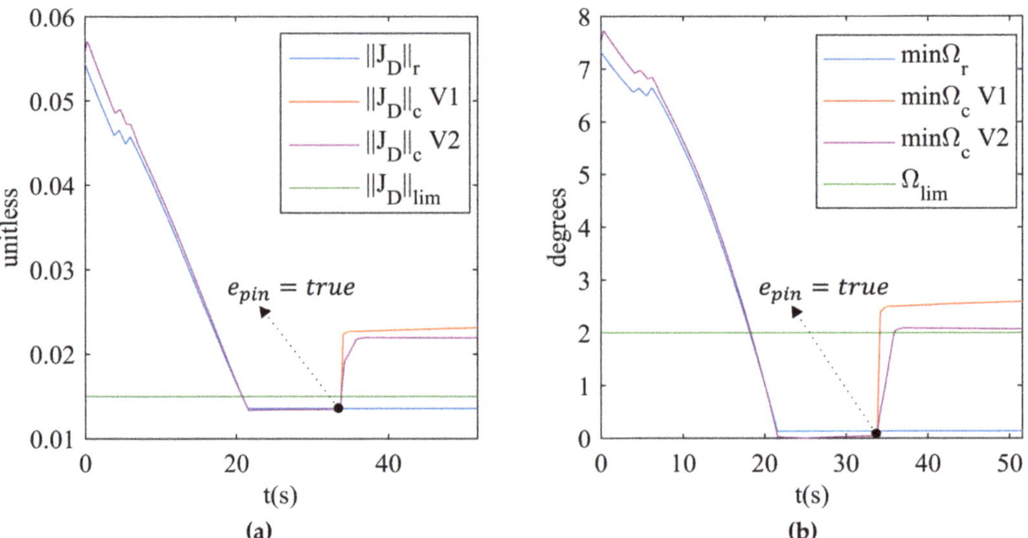

Figure 10. (a) $\|J_D\|$ (b) $\min\Omega$ for trajectory 1 in the simulation.

The performance of the proposed hybrid controller in tracking $\vec{q}_{indr}$ is evaluated by three overall measures:

- The mean absolute error (MAE)

$$\text{MAE} = \frac{1}{F} \sum_{i=1}^{F} \left(\frac{1}{n} \sum_{j=1}^{n} |q_{indr}(i,j) - q_{indc}(i,j)| \right) \tag{11}$$

- The mean absolute percentage error (MAPE)

$$\text{MAPE} = \frac{100}{F} \sum_{i=1}^{F} \left(\frac{1}{n} \sum_{j=1}^{n} \left| \frac{q_{indr}(i,j) - q_{indc}(i,j)}{q_{indr}(i,j)} \right| \right) \tag{12}$$

- The mean distance travelled for type II singularity release (MDSR)

$$\text{MDSR} = \frac{1}{F} \sum_{i=1}^{F} \left(\sum_{j=1}^{n} |q_{indr}(i,k) - q_{indc}(i,j)| \right) \tag{13}$$

where n is the number of samples taken after the activation of e_{pin} at instant k, and i and j are the actuator and the time instant, respectively.

Table 4 shows the MAE, MAPE, and MDSR results for the simulation of the hybrid controller with SRM-V1 and SRM-V2. In this table, the MAE and the MAPE show that SRM-V1 has less error in position tracking than that of SRM-V2 during release from the type II singularity. In addition, the MDSR shows that SRM-V1 needs fewer movements of the actuators than SRM-V2 to release the PR from a singular configuration. These results show that moving the pair of actuators identified by the index $\Omega_{i,j}$ (SRM-V1) is the best option to release a PR from a type II singularity.

Table 4. Performance of the hybrid controller using SRM-V1 and SRM-V2 in the simulation.

Trajectory	MAE (mm)		MAPE (%)		MDSR (mm)	
	SRM-V1	**SRM-V2**	**SRM-V1**	**SRM-V2**	**SRM-V1**	**SRM-V2**
1	3.87	10.74	0.53	1.40	7.01	18.18
2	1.09	2.04	0.14	0.28	5.05	2.92
3	1.77	6.15	0.24	0.82	4.78	6.74
4	3.00	10.24	0.38	1.25	7.48	10.81
5	10.74	10.44	1.43	1.37	15.47	35.23
MEAN	**4.09**	**7.92**	**0.54**	**1.02**	**7.95**	**14.77**

3.2. Experimentation of the Vision-Based Hybrid Controller

After testing the novel vision-based controller in simulation, the next step is implementing the hybrid controller on the real robot according to the diagram shown in Figure 9. Although both the simulation and experimentation have the same procedure, the experimentation presents two notable differences:

- $\vec{X}_c$ is provided by processing the data stream from the 3DTS in real time.
- During the 15 s before the SRM is activated, an external perturbation is applied to the PR. Since in a type II Singularity the PR can vary its position and orientation without moving any actuators, the researcher can apply some forces to the PR by hand to check whether the mobile platform experiences uncontrolled motion.

In the experimental context, the type II singularity release can be tested by trying to move the PR by hand before (when the PR is expected to move) and after the SRM is activated. After the activation of the SRM, the 3UPS+RPU PR will regain the stiffness required to ensure safe interaction with a patient.

Regarding the actual robot, the external limbs are driven by Festo DNCE 32-BS10 prismatic actuators, and the central limb is driven by a NIASA M100-F16 prismatic actuator. All the actuators are attached to Maxon 148867 150 W DC motors commanded by ESCON 50/5 servo controllers, which control the current by means of pulse width modulation (PWM). The current is proportional to the applied voltage (which comes from the control actions), and the torque is in turn proportional to the current. The DC motors are equipped with incremental encoders with a resolution of 500 counts per turn.

The control unit is connected to an industrial computer using acquisition cards. A PCI 1784 Advantech card is used to read the position from the encoders, having four 32-bit quadruple AB phase encoder counters. On the other hand, a 12-bit, 4-channel PCI 1720 Advantech card is used to send the control actions $\vec{\mu}$.

The proposed vision-based hybrid controller runs on the Robot Operating System 2 (ROS2) [31,32]. The two levels of the hybrid controller and the processing of the data stream from the 3DTS are implemented in a modular way using the C++ and Python programming languages. The controller receives the set of references $\vec{q}_{ind_r}$ from the solution of the inverse kinematics given the Cartesian references for the end-effector. The $\vec{q}_{ind_r}$ is sampled at a rate of 100 Hz, and the desired releasing velocity v_d is set to 0.01 $\frac{m}{s}$. These parameters are suitable for knee rehabilitation requirements.

For the actual PR, a fourth performance index is added to evaluate the smoothness of the movements performed by the controller, which is measured with the absolute variation rate (AVR) of the control actions as follows:

$$\text{AVR} = \frac{1}{F} \sum_{i=1}^{F} \left(\sum_{j=2}^{n} |\tau(i,j) - \tau(i,j-1)| \right) \tag{14}$$

During the first run of trajectory 1 using the hybrid controller with SRM-V2, the actual 3UPS+RPU PR reaches an AVR of 8N, which is too high for knee rehabilitation. For this reason, the experiment on the actual PR under study only focuses on the hybrid controller with SRM-V1. This decision is also supported by the better performance shown in the simulation (see Table 4).

Table 5 shows the results of performance tracking of $\vec{q}_{ind_r}$ of the hybrid controller with SRM-V1 implemented on the 3UPS+RPU PR. The MAE and MAPE for experimentation are similar to the simulation results, with a low AVR ensuring smooth movements of the mobile platform. In contrast, the actual MDSR is lower than the values calculated in the simulation. The reduction in MDSR is due to the accurate measure of $\vec{X}_c$ provided by the 3DTS, which is fundamental for a proper measure of the proximity to a type II singularity.

Table 5. Performance of the hybrid controller using SRM-V1 in the experimentation.

Trajectory	MAE (mm)	MAPE (%)	MDSR (mm)	AVR (N)
1	3.26	0.45	3.64	0.22
2	3.02	0.41	7.61	0.52
3	2.05	0.27	1.60	0.17
4	2.14	0.27	1.90	0.46
5	10.66	1.42	11.82	1.44
MEAN	**4.22**	**0.56**	**5.31**	**0.56**

Figure 11 shows the measures of the two indices ($\|J_D\|_c$ and $\min\Omega_c$) when the actual PR is released from a singular configuration, corresponding to trajectory 1 with $\Omega_{3,4}$ as $\min\Omega_c$. The variation of $\|J_D\|_c$ and $\min\Omega_c$ before SRM-V1 is activated is due to the external force applied to the actual PR. It is important to mention that the actual PR recovers its stiffness at the end of all experiments. To the best of the author's knowledge, this is the first time that an actual PR has been driven to a type II singularity and successfully released from it by using the index $\Omega_{i,j}$. The results can be seen in Video 1 and Video 2 provided as Supplementary Materials of this research.

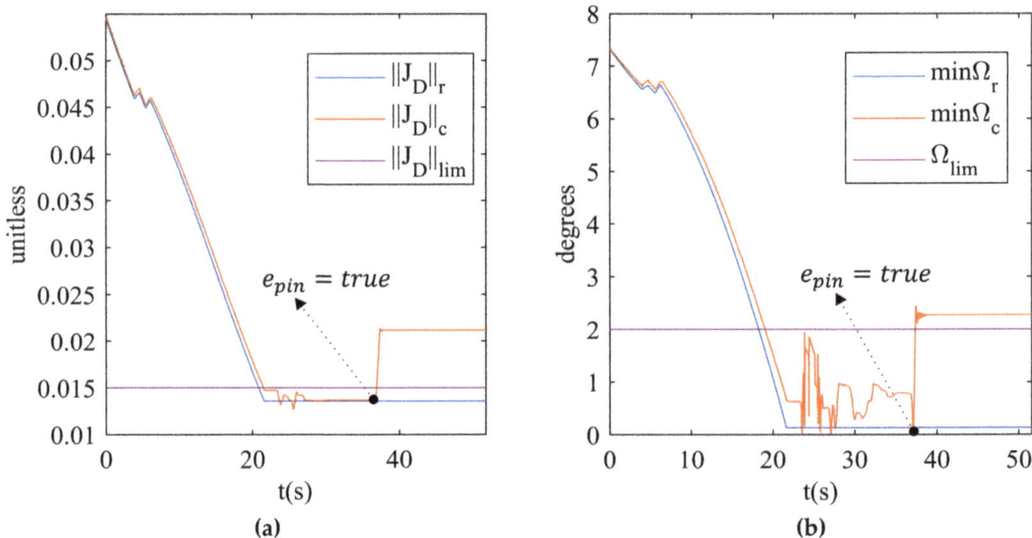

Figure 11. (a) $\|J_D\|$ (b) $\min\Omega$ for trajectory 1 in the experimentation.

Figure 12 shows the reference (*r*) trajectory for x_m in contrast to its estimation ($\hat{c}$) by using the forward kinematic model and the experimental measures (*c*) based on data streaming from the 3DTS. Despite both estimated and experimental measures being calculated online, only the experimental measure detects the movement produced by the external force applied to the PR. This verifies that when the 3UPS+RPU PR is in a type II singularity, the actual x_m cannot be determined by solving the forward kinematic.

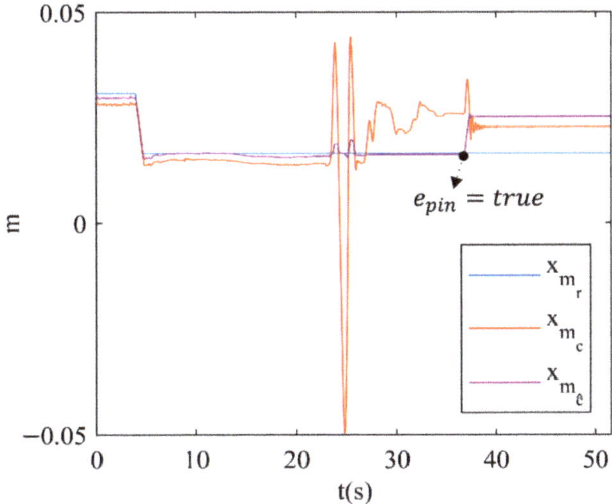

Figure 12. x_m position for trajectory 1.

Figure 13a shows the position for limb 3, which is one of the two limbs involved in the type II singularity in trajectory 1. In this figure, the measured position (*c*) accurately tracks the desired position (*d*), which differs from the reference (*r*) only after SRM-V1 activation. Furthermore, Figure 13a clearly shows that the desired position is modified by a few millimetres from the reference to release the actual PR from the type II singularity. Finally, Figure 13b shows the smooth control actions calculated by the hybrid controller implemented on the actual PR using SMR-V1. Video 1 provides an interactive view of the results presented in Figures 12 and 13 and can be found in the Supplementary Materials Section.

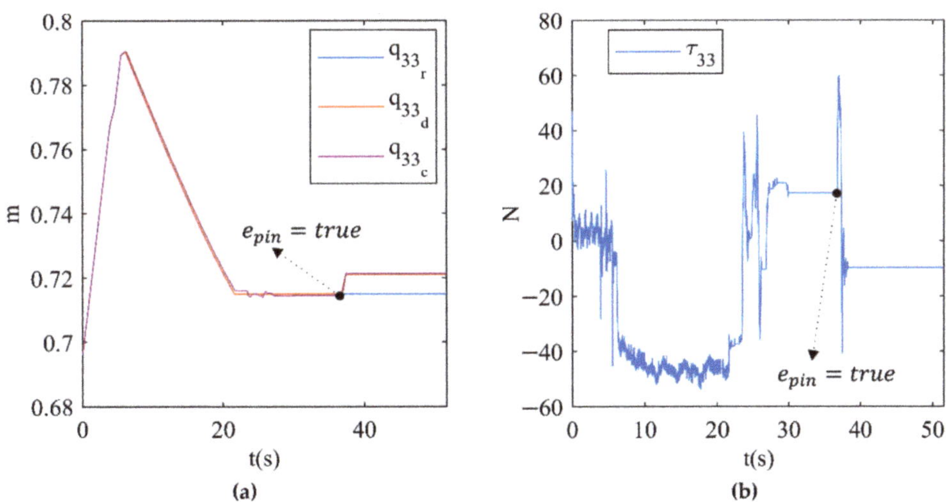

Figure 13. (a) q_{ind} (b) τ on limb 3 for trajectory 1.

The experimental results conclude that the vision-based hybrid controller with SMR-V1 releases an actual PR from a type II singularity with minimum deviation from the original reference. In addition, the OptiTrack 3DTS allows the hybrid controller with SMR-V1 to take advantage of the features of the index $\Omega_{i,j}$.

4. Discussion

This study has addressed the novel task of releasing a 4-DOF PR from type II singular configuration using the index $\Omega_{i,j}$ to identify the limbs involved in the singularity. The hybrid controller proposed combines an algebraic controller with an external computational loop that modifies the joint references only for the limbs that are causing the singularity. This mechanism can be activated whenever the robot enters into a type II singularity by measuring the $\|J_D\|_c$ and $\min\Omega_c$. Both $\|J_D\|_c$ and $\min\Omega_c$ are measured based on the actual position and orientation of the mobile platform that is provided online by a OptiTrack 3DTS. The embedded sensorization includes a set of encoders attached to the motors to ascertain the joint positions.

To show the effectiveness of the designed method, several experiments have been conducted with trajectories that leave the robot in distinct singular configurations, where the releasing algorithm is activated. This scheme has been implemented in both simulation and actual settings to compare the differences in performance when moving the limbs involved (SRM-V1) or not involved (SRM-V2) in the type II singularity. The algorithm for SRM-V1 and SRM-V2 defined the movement of the actuators based on the results of $\min\Omega_c$.

The results of the simulation in Section 3.1 clearly show that SRM-V1 makes the robot behave better in terms of all the performance measures with respect to SRM-V2. According to Table 4, SRM-V1 presents a 0.54% (4.09 mm) mean error in tracking the original reference with a mean distance travelled of 7.95 mm for releasing the PR from a type II singularity. These errors are approximately half of those obtained with SRM-V2, thus verifying that moving the actuators identified by $\min\Omega_c$ is the best option to release the 3UPS+RPU PR from a type II singularity. In fact, no trajectories were performed with the actual robot using SRM-V2, as a first experiment using this algorithm showed that the robot was struggling to get out of the singular configuration, with sharper control actions than those obtained in simulation.

Section 3.2 shows that by using knowledge of the true position and orientation of the mobile platform, the hybrid controller with SRM-V1 can successfully release the actual PR from a type II singularity. All singular trajectories were overcome, even in the cases where the mobile platform was manipulated to change its position during the standby time. The results show how the simulated and real experiments are alike, as all of the indicators for SMR-V1 are somewhat similar. These errors are proven to be dependent on the starting singular configuration, since trajectory 5 is harder for the PR to overcome.

Based on the results of simulation and experimentation, this is the first use of a vision-based hybrid controller capable of releasing a 4-DOF PR from a singular configuration. It is also notable that the effectiveness of the release from a type II singularity with a minimum deviation of the original reference depends on $\min\Omega_c$. The smoother response of the vision-based hybrid controller is achieved because of the accurate measures of the 3DTS, making it a fundamental element of the hybrid controller. It is important to highlight that before this research, the $\Omega_{i,j}$ had not been used as an online detector of the proximity to type II singularities for controlling purposes.

The proposed vision-base hybrid controller compensates a main drawback of PRs, and it represents a step forward towards compliant manipulation of PRs. This system improves the performance of knee rehabilitation tasks by ensuring the safety of the patient during human–robot interaction, even if the PRs arise a type II singularity.

In future research, SRM-V1 can be extended for its use in the task of type II singularity avoidance, i.e., preventing the PR from entering into a singular configuration. Although little literature exists regarding this field, the SRM-V1 algorithm offers valuable insight into the limbs that are expected to lead the robot to a type II singularity. After adding the possibility of returning to the original reference to SRM-V1, the avoidance of type II singularities could be achieved in a more reliable way than using other methods such as artificial potential fields.

Supplementary Materials: The following videos are available online at https://imbio3r.ai2.upv.es/videos/TypeII_singularities: Video 1: 4-DOF parallel robot: vision-based hybrid controller to release from a type II singularity. Trajectory 1; Video 2: 4-DOF parallel robot: vision-based hybrid controller to release from a type II singularity. Trajectory 5.

Author Contributions: Conceptualization, V.M., M.V., and M.U.; methodology, V.M., M.V., and J.L.P.; software, R.J.E. and J.F.; validation, V.M. and M.U.; formal analysis, V.M.; investigation, J.L.P.; resources, V.M. and M.V.; data curation, R.J.E. and J.F.; writing—original draft preparation, M.V.; writing—review and editing, J.L.P.; visualization, M.V.; supervision, V.M.; project administration, V.M. and M.V.; funding acquisition, V.M. and M.V. All authors have read and agreed to the published version of the manuscript.

Funding: This research was funded by the FEDER-CICYT project with reference PID2020-119522RB-I00 (ROBOTS PARALELOS DE REHABILITACION: DETECCION Y CONTROL DE SINGULARIDADES EN PRESENCIA DE ERRORES DE MANUFACTURA), Spain.

Institutional Review Board Statement: Not applicable.

Informed Consent Statement: Not applicable.

Data Availability Statement: Not applicable.

Conflicts of Interest: The authors declare no conflict of interest.

References

1. Briot, S.; Khalil, W. *Dynamics of Parallel Robots—From Rigid Links to Flexible Elements*; Springer: Berlin/Heidelberg, Germany, 2015; ISBN 978-3-319-19787-6.
2. Patel, Y.D.; George, P.M. Parallel Manipulators Applications—A Survey. *Mod. Mech. Eng.* **2012**, *2*, 57–64. [CrossRef]
3. Aliakbari, M.; Mahboubkhah, M. An adaptive computer-aided path planning to eliminate errors of contact probes on free-form surfaces using a 4-DOF parallel robot CMM and a turn-table. *Meas. J. Int. Meas. Confed.* **2020**, *166*, 108216. [CrossRef]
4. Xie, S. *Advanced Robotics for Medical Rehabilitation: Current State of the Art and Recent Advances*; Springer: Cham, Switzerland, 2016; ISBN 978-3-319-19895-8.
5. Díaz, I.; Gil, J.J.; Sánchez, E. Lower-Limb Robotic Rehabilitation: Literature Review and Challenges. *J. Robot.* **2011**, *2011*, 759764. [CrossRef]
6. Sui, P.; Yao, L.; Lin, Z.; Yan, H.; Dai, J.S. Analysis and synthesis of ankle motion and rehabilitation robots. In Proceedings of the 2009 IEEE International Conference on Robotics and Biomimetics (ROBIO), Guilin, China, 19–23 December 2009; pp. 2533–2538. [CrossRef]
7. Ai, Q.; Zhu, C.; Zuo, J.; Meng, W.; Liu, Q.; Xie, S.; Yang, M. Disturbance-Estimated Adaptive Backstepping Sliding Mode Control of a Pneumatic Muscles-Driven Ankle Rehabilitation Robot. *Sensors* **2017**, *18*, 66. [CrossRef]
8. Atashzar, S.F.; Shahbazi, M.; Patel, R.V. Haptics-enabled Interactive NeuroRehabilitation Mechatronics: Classification, Functionality, Challenges and Ongoing Research. *Mechatronics* **2019**, *57*, 1–19. [CrossRef]
9. Kataoka, Y.; Takeda, R.; Tadano, S.; Ishida, T.; Saito, Y.; Osuka, S.; Samukawa, M.; Tohyama, H. Analysis of 3-d kinematics using h-gait system during walking on a lower body positive pressure treadmill. *Sensors* **2021**, *21*, 2619. [CrossRef]
10. Gosselin, C.; Angeles, J. Singularity Analysis of Closed-Loop Kinematic Chains. *IEEE Trans. Robot. Autom.* **1990**, *6*, 281–290. [CrossRef]
11. Park, F.C.; Kim, J.W. Singularity Analysis of Closed Kinematic Chains. *J. Mech. Des.* **1999**, *121*, 32–38. [CrossRef]
12. di Gregorio, R.; Parenti-Castelli, V. Mobility analysis of the 3-UPU parallel mechanism assembled for a pure translational motion. In *Proceedings of the 1999 IEEE/ASME International Conference on Advanced Intelligent Mechatronics (Cat. No.99TH8399), Atlanta, GA, USA, 19–23 September 1999*; IEEE: New York, NY, USA, 1999; pp. 520–525.
13. Slavutin, M.; Shai, O.; Sheffer, A.; Reich, Y. A novel criterion for singularity analysis of parallel mechanisms. *Mech. Mach. Theory* **2019**, *137*, 459–475. [CrossRef]
14. Voglewede, P.A.; Ebert-Uphoff, I. Measuring "closeness" to singularities for parallel manipulators. In Proceedings of the IEEE International Conference on Robotics and Automation, New Orleans, LA, USA, 26 April–1 May 2004; IEEE: New York, NY, USA, 2004; Volume 5, pp. 4539–4544.
15. Davidson, J.K.; Hunt, K.H.; Pennock, G.R. Robots and Screw Theory: Applications of Kinematics and Statics to Robotics. *J. Mech. Des.* **2004**, *126*, 763. [CrossRef]

16. Yuan, M.S.C.; Freudenstein, F.; Woo, L.S. Kinematic Analysis of Spatial Mechanisms by Means of Screw Coordinates. Part 2—Analysis of Spatial Mechanisms. *J. Eng. Ind.* **1971**, *93*, 67. [CrossRef]
17. Takeda, Y.; Funabashi, H. Motion Transmissibility of In-Parallel Actuated Manipulators. *JSME Int. J. Ser. C Dyn. Control Robot. Des. Manuf.* **1995**, *38*, 749–755. [CrossRef]
18. Wang, J.; Wu, C.; Liu, X.-J. Performance evaluation of parallel manipulators: Motion/force transmissibility and its index. *Mech. Mach. Theory* **2010**, *45*, 1462–1476. [CrossRef]
19. Pulloquinga, J.L.; Mata, V.; Valera, Á.; Zamora-Ortiz, P.; Díaz-Rodríguez, M.; Zambrano, I. Experimental analysis of Type II singularities and assembly change points in a 3UPS+RPU parallel robot. *Mech. Mach. Theory* **2021**, *158*, 104242. [CrossRef]
20. Scalera, L.; Carabin, G.; Vidoni, R.; Wongratanaphisan, T. Energy efficiency in a 4-dof parallel robot featuring compliant elements. *Int. J. Mech. Control* **2019**, *20*, 1–9.
21. Marchi, T.; Mottola, G.; Porta, J.M.; Thomas, F.; Carricato, M. Position and singularity analysis of a class of planar parallel manipulators with a reconfigurable end-effector. *Machines* **2021**, *9*, 7. [CrossRef]
22. Llopis-Albert, C.; Rubio, F.; Valero, F. Optimization approaches for robot trajectory planning. *Multidiscip. J. Educ. Soc. Technol. Sci.* **2018**, *5*, 1–16. [CrossRef]
23. Bordalba, R.; Porta, J.M.; Ros, L. A Singularity-Robust LQR Controller for Parallel Robots. In Proceedings of the IEEE International Conference on Intelligent Robots and Systems, Madrid, Spain, 1–5 October 2018; Institute of Electrical and Electronics Engineers Inc.: New York, NY, USA, 2018; pp. 5048–5054.
24. Agarwal, A.; Nasa, C.; Bandyopadhyay, S. Dynamic singularity avoidance for parallel manipulators using a task-priority based control scheme. *Mech. Mach. Theory* **2016**, *96*, 107–126. [CrossRef]
25. Lee, L.W.; Chiang, H.H.; Li, I.H. Development and control of a pneumatic-actuator 3-dof translational parallel manipulator with robot vision. *Sensors* **2019**, *19*, 1459. [CrossRef]
26. Huynh, B.-P.; Su, S.-F.; Kuo, Y.-L. Vision/Position Hybrid Control for a Hexa Robot Using Bacterial Foraging Optimization in Real-time Pose Adjustment. *Symmetry* **2020**, *12*, 564. [CrossRef]
27. Amarasinghe, D.; Mann GK, I.; Gosine, R.G. *Vision-Based Hybrid Control Strategy for Autonomous Docking of a Mobile Robot*; Institute of Electrical and Electronics Engineers (IEEE): New York, NY, USA, 2005; pp. 1600–1605.
28. Araujo-Gómez, P.; Mata, V.; Díaz-Rodríguez, M.; Valera, A.; Page, A. Design and Kinematic Analysis of a Novel 3UPS/RPU Parallel Kinematic Mechanism With 2T2R Motion for Knee Diagnosis and Rehabilitation Tasks. *J. Mech. Robot.* **2017**, *9*, 061004. [CrossRef]
29. Vallés, M.; Araujo-Gómez, P.; Mata, V.; Valera, A.; Díaz-Rodríguez, M.; Page, Á.; Farhat, N.M. Mechatronic design, experimental setup, and control architecture design of a novel 4 DoF parallel manipulator. *Mech. Based Des. Struct. Mach.* **2018**, *46*, 425–439. [CrossRef]
30. Valero, F.; Díaz-Rodríguez, M.; Vallés, M.; Besa, A.; Bernabéu, E.; Valera, Á. Reconfiguration of a parallel kinematic manipulator with 2T2R motions for avoiding singularities through minimizing actuator forces. *Mechatronics* **2020**, *69*, 102382. [CrossRef]
31. Maruyama, Y.; Kato, S.; Azumi, T. Exploring the performance of ROS2. In Proceedings of the Proceedings of the 13th International Conference on Embedded Software, EMSOFT 2016, Pittsburgh, PA, USA, 1–7 October 2016; Association for Computing Machinery, Inc.: New York, NY, USA, 2016; pp. 1–10.
32. Jiang, Z.; Gong, Y.; Zhai, J.; Wang, Y.P.; Liu, W.; Wu, H.; Jin, J. Message Passing Optimization in Robot Operating System. *Int. J. Parallel Program.* **2020**, *48*, 119–136. [CrossRef]

sensors

Article

A Discrete-Time Extended Kalman Filter Approach Tailored for Multibody Models: State-Input Estimation

Rocco Adduci [1,2,*], Martijn Vermaut [1,2], Frank Naets [1,2], Jan Croes [1,2] and Wim Desmet [1,2]

[1] LMSD Research Group, Mechanical Engineering Department, KU Leuven University, 3000 Leuven, Belgium; martijn.vermaut@kuleuven.be (M.V.); frank.naets@kuleuven.be (F.N.); jan.croes@kuleuven.be (J.C.); wim.desmet@kuleuven.be (W.D.)

[2] DMMS Core Labs, Flanders Make, 3001 Heverlee, Belgium

* Correspondence: rocco.adduci@kuleuven.be

Abstract: Model-based force estimation is an emerging methodology in the mechatronic community given the possibility to exploit physically inspired high-fidelity models in tandem with ready-to-use cheap sensors. In this work, an inverse input load identification methodology is presented combining high-fidelity multibody models with a Kalman filter-based estimator and providing the means for an accurate and computationally efficient state-input estimation strategy. A particular challenge addressed in this work is the handling of the redundant state-description encountered in common multibody model descriptions. A novel linearization framework is proposed on the time-discretized equations in order to extract the required system model matrices for the Kalman filter. The presented framework is experimentally validated on a slider-crank mechanism. The nonlinear kinematics and dynamics are well represented through a rigid multibody model with lumped flexibilities to account for localized interaction phenomena among bodies. The proposed methodology is validated estimating the input torque delivered by a driver electro-motor together with the system states and comparing the experimental data with the estimated quantities. The results show the stability and accuracy of the estimation framework by only employing the angular motor velocity, measured by the motor encoder sensor and available in most of the commercial electro-motors.

Keywords: multibody dynamics; Kalman filtering; coupled states-inputs estimation; virtual sensors; slider-crank mechanism

Citation: Adduci, R.; Vermaut, M.; Naets, F.; Croes, J.; Desmet, W. A Discrete-Time Extended Kalman Filter Approach Tailored for Multibody Models: State-Input Estimation. *Sensors* **2021**, *21*, 4495. https://doi.org/10.3390/s21134495

Academic Editor: Giuseppe Ferri

Received: 21 May 2021
Accepted: 28 June 2021
Published: 30 June 2021

Publisher's Note: MDPI stays neutral with regard to jurisdictional claims in published maps and institutional affiliations.

1. Introduction

In mechatronic systems, operational forces and moments are essential quantities in the different stages of the development cycle and strongly impact the design, durability, diagnostic, prognostic, maintenance, and advanced control strategies [1]. However, forces and moments are also difficult, even impossible, quantities to measure. This is due to high force sensor costs and the geometrical constraints (space limitations) that would make the sensor integration impossible without influencing the overall system design and behavior.

In past decades, different test-driven and model-based inverse force methods have been presented in the literature to overcome these limitations. Initially, the challenge of inverse load identification was tackled in offline test-based strategies. One of the most commonly used technique for mechanical applications relates to the field of Transfer Path Analysis (TPA) [2].

TPA summarizes the family of test-based methodologies to study the vibration transmission in mechanical systems where the Matrix Inversion and Mount Stiffness approaches are the most commonly used to estimate inputs and transmitted forces respectively. Despite the wide variety of methods and extensive use in the industrial world, TPA still remains quite expensive from an experimental point of view in terms of preparation and execution time.

The growing computational power of modern computers opened new opportunities to exploit numerical model-based methods. These models can be exploited in virtual sensing approaches [3] which enable the exploitation of low cost, accessible and non-collocated measurements together with first-principle/physically inspired models to obtain state and input estimates.

State estimation techniques such as the Kalman Filter (KF) methods allow the joint estimation of unknown inputs and model states [4] in an efficient manner. By regularly feeding back the measurements on a physical asset, KF techniques enable the compensation of drift in the model while reducing the noise from the direct measurements.

Multibody (MB) modeling approaches are regularly used in the literature and industry [5–7] for full scale rigid and flexible mechanical systems where conventional Finite Element (FE)-based methods would be unnecessarily expensive. The MB methods establish a good trade-off between model fidelity and computational cost. Moreover, MB models disclose 3D system-level information, enabling dynamic interaction phenomena among bodies due to distributed and/or localized flexibilities.

However, the link between MB models and estimation algorithms is nontrivial since most estimators require an ordinary differential state-space representation of the system dynamics. Instead, the MB model dynamics, depicted by the Equations of Motion (EOMs), are generally described by a set of Implicit-Differential Algebraic Equations (I-DAEs) that makes the state-space representation difficult to be met. On the other hand, Explicit-Ordinary Differential Equations (E-ODEs) are well suited for a state-space representation but specific MB formulations should be employed to obtain this structure (e.g., [8]), otherwise, dedicated manipulations of the EOMs are demanded to achieve an ODE form.

In [9] the Matrix-R method was proposed to eliminate the constraint equations of the MB model reducing the initial EOMs to an ODE form in independent coordinates. The aim of this work was to combine an extended KF estimator with detailed MB models to obtain an automotive real-time observer. Despite the high accuracy of the estimated quantities, the real-time target was not achieved due to the costly solver iterations.

Similarly in [10–12] the Matrix-R method was used to deal with the DAE structure of the EOMs. Here, different KF estimators are compared in terms of accuracy and performance on a rigid 4 and 5-bar linkage mechanisms.

Alternatively, in [13] a kinematic state observer is presented. It combines the constrained kinematic MB equations with nonlinear estimators. Here, the dynamic equations of the MB system are not considered therefore leading to an estimation which is less sensitive to input and mechanical (properties) uncertainties. Moreover the system accelerations are treated as random walk models and augmented to the kinematic discrete state vector that imply the use of a more extensive number of measurement sensors.

In [14,15] an interesting approach based on the combination of deep learning and MB dynamics information was proposed to achieve this transformation. It allows reducing a generic MB model to minimal coordinates allowing the description of the EOMs through E-ODEs while not requiring a specific formulation or access to the constraint equations. However, the methodology depends on a reference numerical simulation as training data which must cover the mechanism workspace; moreover only rigid MB systems can be tackled by the technique.

In [16] an Augmented Discrete Extended Kalman filter (ADE-KF) approach tailored for flexible MB models to construct a state-input estimator is presented.The methodology demonstrates the advantages of using analytical expressions to cover the necessary linearized and explicit EOMs. However, this approach relies on the use of a penalty constraint formulation to achieve E-ODE type of equations. This leads to a relatively poorly conditioned problem and introducing additional model parameters, namely the penalty factors, in comparison to a Lagrange-multiplier approach.

In this work, a generalization of the methodology described in [16] is presented which is compatible with a Lagrange multiplier approach for the constraint equations.

The proposed methodology starts from a novel linearization approach of the EOMs that includes the algebraic variables (Lagrange multipliers) to the system states. Consequently, the resulting unconstrained discrete-time state-space model is employed in a constraint KF scheme where the kinematic constraints are enforced trough the augmented measurement equations, therefore eliminating the effort of selecting effective penalty factors.

The scientific contribution is structured as follows: in Section 2 a general overview of the governing EOMs of the MB system dynamics is given; in Section 3 the implicit constrained EOMs are linearized and made explicit through a first order Taylor expansion; in Section 4 the system and measurement Kalman filter equations are introduced. Here, the system and measurement matrices are analytically assembled thanks to the use of the in-house Multibody Research Code (MBRC) [17]; finally, in Section 5 the methodology is validated on an industrial relevant application comparing the estimated quantities with the experimentally measured one.

2. Multi-Body Model and Time-Disretization

This section summarizes the derivation of the EOMs of flexible multibody systems, as they will be employed in this work.

2.1. The Multibody Equations of Motion

The mathematical description of the system dynamics can be derived by means of the Lagrange's equations for constrained mechanical systems [18]:

$$\frac{d}{dt}\left(\frac{\partial \mathcal{L}(\dot{q},q,\lambda)}{\partial[\dot{q},\dot{\lambda}]}\right) - \frac{\partial \mathcal{L}(\dot{q},q,\lambda)}{\partial[q,\lambda]} = u^e(q,u), \tag{1}$$

with the Lagrangian defined as:

$$\mathcal{L} = \mathcal{T} - \mathcal{V} - \phi(q)^T\lambda. \tag{2}$$

$\mathcal{L}$ represents the Lagrangian functional, $\mathcal{T}$ the kinetic energy, $\mathcal{V}$ the potential energy, $\phi(q)^T\lambda$ the constraint contribution with the Lagrange multipliers λ, and u^e is the vector of the external actions. MB models describe the dynamics of several rigid and/or flexible interacting bodies linked together through the definition of kinematic joints which are mathematically represented by the constraint equations $\phi(q)$ while $J = \partial\phi(q)/\partial q$ represent the Jacobian of the constraint equations. $q \in \mathbb{R}^{n_q}$ is the generalized coordinates vector, $\lambda \in \mathbb{R}^{n_\lambda}$ is known as Lagrange multipliers and $u \in \mathbb{R}^{n_u}$ is the input vector. Through the definition of the assembled body coordinates and the motion parametrization Equation (1) can be written in a residual form as a fully implicit real-valued non-linear function:

$$g(\ddot{q},\dot{q},q,\lambda,u) = 0. \tag{3}$$

2.2. The Differential-Algebraic form of the EOMs

A set of natural coordinates $q_n \in \mathbb{R}^{n_{q_n}}$ was proposed in [18], where redundant degrees of freedom are employed to define the system coordinates of the assembled bodies. Moreover, including the motion parameterization employed in the Flexible Natural Coordinate Formulation (FNCF) [19] allows deriving a constant singular mass matrix $M \in \mathbb{R}^{n_{q_n} \times n_{q_n}}$. Assuming this formulation, Equation (3) can be written in the so-called index-3 form:

$$\begin{cases} g_1 = M_n\ddot{q}_n + f_{nl}(\dot{q}_n,q_n,u) + J^T\lambda = 0_q \\ g_2 = \phi(q_n) = 0_\lambda \end{cases}. \tag{4}$$

Here, $f_{nl} \in \mathbb{R}^{n_{q_n}}$ is the non-linear generalized force vector expressed as:

$$f_{nl}(\dot{q}_n,q_n,u) = f_v(\dot{q}_n,q_n) + f_{int}(\dot{q}_n,q_n) + f_{ext}(\dot{q}_n,q_n,u). \tag{5}$$

f_v represents the quadratic velocity vector related to the gyroscopic forces of the bodies, which is zero for FNCF formulation. f_{int} is the internal force vectors which accounts for the elastic energy stored by deformable bodies and if rigid bodies are assumed f_{int} vanishes; f_{ext} is the external force vector and can be spilt in the sum of two contributions, the interaction forces among bodies f_b (i.e., contact and friction forces) and the input forces f_u. They can be summarized as follows:

$$f_{ext}(\dot{q}_n, q_n, u) = f_b(\dot{q}_n, q_n) + f_u(q_n, u). \tag{6}$$

Here, f_u can be written as $f_u(q_n, u) = U_t(q_n)u$, where U_t is tangent input matrix defined as:

$$U_t = \frac{\partial f_u}{\partial u}. \tag{7}$$

Due to the structure of the EOM, Equation (4), for the FNCF formulation, derivatives can be more readily obtained than for may alternative flexible multibody formulations. Therefore, the above mentioned coordinates definition and motion parameterization will be considered in this work. For the sake of brevity, we omit the subscript n referred to the natural coordinate formulation for the remainder of this manuscript.

Despite the computational advantages of the above mentioned MB approach, the methodologies that will be introduced in the next sections can be easily extended to alternative MB formulation, such as the floating-frame of reference component mode synthesis approach or the generalized component mode synthesis [5,20].

The I-DAEs form of Equation (4) are generally not suitable for estimation algorithms such as the Kalman Filter family, since these have been designed to handle E-ODEs type of equations.

In the next section, we present a new methodology to directly linearize the I-DAEs starting from its discrete form but without employing any explicit constraint elimination technique.

2.3. EOMs: The Discrete Index-3 Form

To transform the second order differential equation into first-order differential equations, it is common to introduce the velocity variable $\dot{q} = v$, allowing to write Equation (4) in a first-order form as:

$$g(\dot{v}, v, \dot{q}, q, \lambda, u) = \begin{cases} g_1 = v - \dot{q} = 0_v \\ g_2 = M(q)\dot{v} + f_{nl}(v, q, u) + J^T \lambda = 0_q \\ g_3 = \phi(q) = 0_\lambda \end{cases} . \tag{8}$$

These equations represent a system of constrained-DAEs of index 3 [21]. Numerical differentiation is generally employed [22] to convert Equation (8) into a discrete form.

In this work, a first order, constant time-step Backward Differentiation Formula (BDF-1), also known as Backward Euler, is employed. This choice does not imply that other differentiation schemes cannot be applied to discretize the EOMs in time. However, the time-discretization must be consistent with the defined estimation sampling and particular attention should be paid to choosing the differentiation schemes (e.g., forward or backward) because it influences the achievement of the discrete-time E-ODE form of the EOMs, as will be discussed in Section 3.

The single step method BDF-1 can be written as

$$\begin{cases} \dot{v}_{k+1} = \frac{1}{h}(v_{k+1} - v_k) \\ v_{k+1} = \frac{1}{h}(q_{k+1} - q_k) \end{cases} , \tag{9}$$

where h represents the constant time step size and $k \in \mathbb{Z}$ the $k - th$ time instance. By substituting Equation (9) into Equation (8) the discrete-time EOMs g_d are obtained:

$$\begin{cases} g_{d1} = v_{k+1} - \frac{1}{h}(q_{k+1} - q_k) = 0_v \\ g_{d2} = M(q_{k+1})\frac{v_{k+1}-v_k}{h} + f_{nl}(q_{k+1}, q_k, u_{k+1}) + J^T(q_{k+1})\lambda_{k+1} = 0_q \\ g_{d3} = \phi(q_{k+1}) = 0_\lambda \end{cases} \quad . \tag{10}$$

This can be summarized as

$$g_d(v_{k+1}, q_{k+1}, v_k, q_k, \lambda_{k+1}, u_{k+1}) = 0. \tag{11}$$

Assuming the generalized coordinates q_k and velocities v_k at the time instance k to be known and given the input u_{k+1} at the time instance $k+1$, Equation (10) is solved for q_{k+1} and λ_{k+1} by substituting g_{d1} in g_{d2}, and applying a Newton-Raphson-based iterative algorithm with iteration gradient J_{NR}:

$$J_{NR} = \begin{bmatrix} K_t + \beta C_t + \gamma M_t & J^T \\ J & 0_{\lambda,\lambda} \end{bmatrix}. \tag{12}$$

Here, K_t, C_t and M_t are the tangent stiffness, damping and mass matrices obtained from the partial derivatives of the continuous g_2 equations in Equation (8) evaluated at time step $k+1$:

$$K_t = \left.\frac{\partial g_2}{\partial q}\right|_{k+1} ; \qquad C_t = \left.\frac{\partial g_2}{\partial v}\right|_{k+1} ; \qquad M_t = \left.\frac{\partial g_2}{\partial \dot{v}}\right|_{k+1}, \tag{13}$$

and β and γ are matrix coefficients function of the defined integration rule, which are given for the BDF-1 scheme by:

$$\beta = \frac{\partial v_{k+1}}{\partial q_{k+1}} = \frac{\partial \dot{v}_{k+1}}{\partial v_{k+1}} = \frac{1}{h}I_q; \qquad \gamma = \frac{\partial \dot{v}_{k+1}}{\partial q_{k+1}} = \frac{1}{h^2}I_q; \tag{14}$$

$$\frac{\partial v_{k+1}}{\partial q_k} = \frac{\partial \dot{v}_{k+1}}{\partial v_k} = -\beta; \qquad \frac{\partial \dot{v}_{k+1}}{\partial q_k} = -\gamma. \tag{15}$$

This time integration scheme will be exploited in the following section to set up a suitable solver structure for use in an ODE-base estimation framework.

3. An Explicit Linearized Approximation for Use of the Multibody Model in State-Estimation

The aim of this work is to combine MB models with a Kalman filter-based estimator to concurrently estimate the system states and unknown forces of a mechanism. These presented estimators, as will be discussed in Section 4, require the linearized time-discrete model matrices. In this section, a new approach to linearize the EOMs of a MB model starting from an ID-DAE form is presented. Subsequently, the linearized system matrices are analytically assembled.

By introducing the state vector $x \in \mathbb{R}^{n_x}$ for a time step k

$$x_k = \begin{bmatrix} v_k & q_k & \lambda_k \end{bmatrix}^T, \tag{16}$$

the ID-DAE form of the EOMs in Equation (11) can be re-written as:

$$g_d(x_{k+1}, x_k, u_{k+1}) = 0_x. \tag{17}$$

In this work we assume the function g_d of Equation (17) to be continuously differentiable in all its variables. Therefore, an explicit discrete function f_d locally exists by applying the implicit function theorem. Through a first order Taylor expansion, g_d can be approximated as

$$g_d^0 + \left.\frac{\partial g_d}{\partial x_{k+1}}\right|_0 \left(x_{k+1} - x_{k+1}^0\right) + \left.\frac{\partial g_d}{\partial x_k}\right|_0 \left(x_k - x_k^0\right) + \dots$$

$$\left.\frac{\partial g_d}{\partial u_{k+1}}\right|_0 \left(u_{k+1} - u_{k+1}^0\right) + \mathcal{O}(x_{k+1}, x_k, u_{k+1}) = 0_x; \tag{18}$$

where $(x_{k+1}^0, x_k^0, u_{k+1}^0)$ represents the linearization set point while $g_d^0 = g_d(x_{k+1}^0, x_k^0, u_{k+1}^0)$ for convenience of notation. Manipulating Equation (18) and neglecting the higher order terms, it can be made explicit as

$$x_{k+1} = x_{k+1}^0 - \left[\left.\frac{\partial g_d}{\partial x_{k+1}}\right|_0\right]^{-1} \left(g_d^0 + \left.\frac{\partial g_d}{\partial x_k}\right|_0 (x_k - x_k^0) + \left.\frac{\partial g_d}{\partial u_{k+1}}\right|_0 \left(u_{k+1} - u_{k+1}^0\right)\right). \tag{19}$$

In compact form Equation (19) becomes:

$$x_{k+1} = f_d(x_k, u_{k+1}) = x_{k+1}^0 + f_d^0(x_k, u_{k+1}) + A_{k+1}^0(x_k - x_k^0) + B_{k+1}^0(u_{k+1} - u_{k+1}^0). \tag{20}$$

$A_{k+1}^0 = \left.\frac{\partial f_d}{\partial x}\right|_{k+1}^0$ and $B_{k+1}^0 = \left.\frac{\partial f_d}{\partial u}\right|_{k+1}^0$ are the linearized system and input matrices around the linearization set point.

Starting from Equation (19) and differentiating Equation (10), the Jacobians for the backward Euler implicit time-integrator can be computed as:

$$G_{x_{k+1}} = \frac{\partial g_d}{\partial x_{k+1}} = \begin{bmatrix} \frac{\partial g_{d1}}{\partial v_{k+1}} & \frac{\partial g_{d1}}{\partial q_{k+1}} & \frac{\partial g_{d1}}{\partial \lambda_{k+1}} \\ \frac{\partial g_{d1}}{\partial v_{k+1}} & \frac{\partial g_{d2}}{\partial q_{k+1}} & \frac{\partial g_{d2}}{\partial \lambda_{k+1}} \\ \frac{\partial g_{d3}}{\partial v_{k+1}} & \frac{\partial g_{d3}}{\partial q_{k+1}} & \frac{\partial g_{d3}}{\partial \lambda_{k+1}} \end{bmatrix} = \begin{bmatrix} I_v & -\beta & 0_{v,\lambda} \\ 0_{q,v} & \gamma M_t + \beta C_t + K_t & J^T \\ 0_{\lambda,v} & J & 0_{\lambda,\lambda} \end{bmatrix}; \tag{21}$$

$$G_{x_k} = \frac{\partial g_d}{\partial x_k} = \begin{bmatrix} \frac{\partial g_{d1}}{\partial v_k} & \frac{\partial g_{d1}}{\partial q_k} & \frac{\partial g_{d1}}{\partial \lambda_k} \\ \frac{\partial g_{d2}}{\partial v_k} & \frac{\partial g_{d2}}{\partial q_k} & \frac{\partial g_{d2}}{\partial \lambda_k} \\ \frac{\partial g_{d3}}{\partial v_k} & \frac{\partial g_{d3}}{\partial q_k} & \frac{\partial g_{d3}}{\partial \lambda_k} \end{bmatrix} = \begin{bmatrix} 0_{v,v} & \beta & 0_{v,\lambda} \\ -\beta M_t & -\gamma M_t - \beta C_t & 0_{q,\lambda} \\ 0_{\lambda,v} & 0_{\lambda,q} & 0_{\lambda,\lambda} \end{bmatrix}; \tag{22}$$

$$G_{u_{k+1}} = \frac{\partial g_d}{\partial u_{k+1}} = \begin{bmatrix} \frac{\partial g_{d1}}{\partial u_{k+1}} \\ \frac{\partial g_{d2}}{\partial u_{k+1}} \\ \frac{\partial g_{d3}}{\partial u_{k+1}} \end{bmatrix} = \begin{bmatrix} 0_{v,u} \\ U_t \\ 0_{\lambda,u} \end{bmatrix}, \tag{23}$$

where U_t represents the tangent input matrix of Equation (7).

The linearized system and input matrices can be computed for any working point at the time step $k+1$ as:

$$A_{k+1} = -G_{x_{k+1}}^{-1} G_{x_k}; \tag{24}$$

$$B_{k+1} = -G_{x_{k+1}}^{-1} G_{u_{k+1}}. \tag{25}$$

These equations enable the evaluation of the linearized time-discretized explicit description of the EOMs suitable for estimation methods such as the extended Kalman filter. Their deployment of this estimation scheme is discussed in the next section.

It is important to note that the invertibility of the matrix $G_{x_{k+1}}$ is guaranteed thanks to the choice of BDF-1 differentiation scheme and that in contrast, it would not be possible. For completeness, the limitations of a forward differentiation scheme is demonstrated in the Appendix A.

4. State-Input Estimation for MB Models

The augmented discrete extended Kalman filter (ADE-KF) tailored for MB models is discussed in this section. In this section we discuss all the required components to set up a Kalman filter for assimilating the different multibody states and unknown inputs. More in details, in Section 4.1, the general form of the discrete-time system and measurement model equations are summarized. In Section 4.2, the measurement equations are augmented with the constraint equations ϕ, to deal with the constrained nature of the MB EOMs, leading to a constrained estimation problem. Moreover, in Section 4.3 the adopted approach to compute the linearized measurement matrices C and D are presented since they are required in the estimation framework and not directly available. Finally, in Section 4.4 the ADE-KF is assembled to deal with the estimation of states and unknown inputs, and the Kalman filter steps tailored for MB models are reported.

4.1. Model and Measurement Equations with Uncertainty

The system of EOMs in Equation (4) described in the previous section are assumed as deterministic. However, in practice, various noise sources will disturb the behavior of the system. For the dynamic model equations, the process noise vector $v^*_{x,k+1}$ influences the response:

$$g_d(x_{k+1}, x_k, u_{k+1}) = \tilde{v}_{x,k+1}, \tag{26}$$

where $\tilde{v}_{x,k+1}$ is associated with the noise term $v_{x,k+1}$ described by the following equation:

$$x_{k+1} = x^*_{k+1} + v_{x,k+1}. \tag{27}$$

Here, x^*_{k+1} is the deterministically varying state vector while $v_{x,k+1}$ is a zero-mean noise term with a (possibly time-varying) covariance Q_k.

The estimation framework relies on combining the prior model information with measurements $y \in \mathbb{R}^{n_y}$ on the physical process. To compare the prediction of the model with these measurements, a set of measurement equations h are necessary which are affected by measurement noise $v_{y,k+1}$:

$$y = h(\dot{v}_{k+1}, x_{k+1}, u_{k+1}) + v_{y,k+1}, \tag{28}$$

where again $v_{y,k+1}$ is assumed to be zero-mean with a (time-varying) covariance R_k. Similar to the model measurement equations y, the concept of virtual sensor (VS) $y_{vs} \in \mathbb{R}^{n_{yvs}}$ is introduced:

$$y_{vs} = h_{vs}(\dot{v}_{k+1}, x_{k+1}, u_{k+1}) + v_{vs}. \tag{29}$$

A virtual sensor represents model-based information; more specifically, it is a physical quantity that can be expressed as a function of the system variables, such as joint forces, body motion trajectories (e.g., linear and angular positions, velocities, and accelerations) and body strains/stresses. A VS equation can be treated similarly to a measurement equation but evaluated after the a posteriori Kalman filter step and hence to exploited in the estimation process itself.

4.2. The Augmented Constraint Measurement Equations

In [23] different approaches have been proposed to deal with constrained state estimation, although the authors deem it impossible to give a general guideline for constrained filtering because each individual problem is unique and dependent on how the constraints are handled.

In this work, the combination of hard and soft constraint methods is considered [23], where the constraint measurement equations are implemented in the Kalman filter scheme by augmenting the measurement equations y of Equation (28) with the constraint equations:

$$\tilde{y} = \begin{cases} h(\dot{v}_{k+1}, x_{k+1}, u_{k+1}) + v_{y,k+1} \\ \phi(q_{k+1}) + v_{\phi,k+1} \end{cases}. \tag{30}$$

Here, $v_{\phi,k+1}$ is a small zero-mean noise term added to the idealized constraint equations $\phi(q_{k+1})$ if small constraints violation is allowed (soft constraints method). In the case $v_{\phi,k+1}$ is assumed to be zero, the constraint equations are treated as perfect measurements (hard constraint method).

4.3. An Efficient Strategy for the Measurement Sensitivities Computation

To evaluate the EKF equations at each filter step, the linearized measurement sensitivity matrices C_{k+1} and D_{k+1} are required and they are defined as:

$$C_{k+1} = \left.\frac{\partial \tilde{y}}{\partial x}\right|_{k+1} = \begin{bmatrix} \frac{dy}{dx} \\ \frac{d\phi}{dx} \end{bmatrix}_{k+1} = \begin{bmatrix} \frac{dy}{dv} & \frac{dy}{dq} & \frac{dy}{d\lambda} \\ 0_{\lambda,v} & J & 0 \end{bmatrix}_{k+1} ; \tag{31}$$

$$D_{k+1} = \left.\frac{\partial \tilde{y}}{\partial u}\right|_{k+1} = \begin{bmatrix} \frac{dy}{du} \\ 0_{\lambda,u} \end{bmatrix}_{k+1} . \tag{32}$$

Generally, the measurement equations can be expressed as a function of the generalized accelerations, velocities, positions, Lagrange multipliers, and inputs, Equation (30).

However, due to the state-space description presented in Section 3, the explicit relationship of the sensor equations with respect to the generalized accelerations is lost. More in particular, when considering acceleration sensors, their sensitivity with respect to the generalized accelerations is non zero, thus it must be indirectly included. Acceleration sensors are really common in the mechanical engineering community due to their non-invasive and simple installation requirements. Within the estimation framework, they can be used either in form of measurements or in form of VSs.

Similarly, the sensitivity of the measurement equations with respect to the system inputs are not always available or nontrivial to obtain.

Therefore, an approach to explicitly obtain these dependencies is here derived by means of the acceleration level constraints.

Introducing the auxiliary variables $z \in \mathbb{R}^{3n_q+n_\lambda}$

$$z = \begin{bmatrix} \ddot{q}^T & \dot{q}^T & q^T & \lambda^T \end{bmatrix}^T = \begin{bmatrix} \dot{v}^T & v^T & q^T & \lambda^T \end{bmatrix}^T, \tag{33}$$

for a generic measurement sensor y, the sensitivities $\frac{\partial y}{\partial z}$ with respect to the auxiliary variable z can be analytically computed starting from their fundamental equations.

$$\frac{\partial y}{\partial z} = \begin{bmatrix} \frac{\partial y}{\partial \dot{v}} & \frac{\partial y}{\partial v} & \frac{\partial y}{\partial q} & \frac{\partial y}{\partial \lambda} \end{bmatrix}. \tag{34}$$

Applying the chain rule, the non-augmented linear measurement matrix $\frac{dy}{dx}$ for a generic time step can be written as

$$\frac{dy}{dx} = \frac{\partial y}{\partial z}\frac{dz}{dx}. \tag{35}$$

And the full derivative $\frac{dz}{dx}$ can be expressed as

$$\frac{dz}{dx} = \begin{bmatrix} \frac{\partial \dot{v}}{\partial v} & \frac{\partial \dot{v}}{\partial q} & \frac{\partial \dot{v}}{\partial \lambda} \\ \frac{\partial v}{\partial v} & \frac{\partial v}{\partial q} & \frac{\partial v}{\partial \lambda} \\ \frac{\partial q}{\partial v} & \frac{\partial q}{\partial q} & \frac{\partial q}{\partial \lambda} \\ \frac{\partial \lambda}{\partial v} & \frac{\partial \lambda}{\partial q} & \frac{\partial \lambda}{\partial \lambda} \end{bmatrix} = \begin{bmatrix} \frac{\partial \dot{v}}{\partial v} & \frac{\partial \dot{v}}{\partial q} & \frac{\partial \dot{v}}{\partial \lambda} \\ I_v & 0_{v,q} & 0_{v,\lambda} \\ 0_{q,v} & I_q & 0_{q,\lambda} \\ \frac{\partial \lambda}{\partial v} & \frac{\partial \lambda}{\partial q} & I_\lambda \end{bmatrix}. \tag{36}$$

To analytically compute the unknown terms of Equation (36), we introduce the acceleration level constraints $\ddot{\phi}$ defined as the second time derivative of the position level constraints. Starting from the velocity level constraints $\dot{\phi}$:

$$\dot{\phi} = \frac{\partial \phi}{\partial q} \frac{dq}{dt} = Jv,$$ (37)

the acceleration level constraint equations are obtained as follows:

$$\ddot{\phi} = \dot{J}v + J\dot{v} = \sum_{i,j=1}^{n_q} v_i \frac{\partial^2 \phi}{\partial[q_i, q_j]} v_j + J\dot{v} = 0_\lambda.$$ (38)

By means of the continuous dynamic equations g_2 of Equation (8) and the acceleration level constraint of Equation (38), a new set of equations $p \in \mathbb{R}^{n_q + n_\lambda}$ can be constructed:

$$p(z, u) = \begin{cases} p_1 = M(q)\dot{v} + f_{nl}(v, q, u) + J^T \lambda = 0_q \\ p_2 = \sum_{i,j=1}^{n_q} v_i \frac{\partial^2 \phi}{\partial[q_i, q_j]} v_j + J\dot{v} = 0_\lambda \end{cases}.$$ (39)

Equations (39) represent the governing set of equations that implicitly determines the relation between the different coordinates v, q, $\dot{v}$ and the Lagrange multipliers λ.

Therefore, the unknown terms of Equation (36) can be computed as

$$\begin{bmatrix} \frac{\partial \dot{v}}{\partial v} & \frac{\partial \dot{v}}{\partial q} & \frac{\partial \dot{v}}{\partial \lambda} \\ \frac{\partial \lambda}{\partial v} & \frac{\partial \lambda}{\partial q} & \frac{\partial \lambda}{\partial \lambda} \end{bmatrix} = -\begin{bmatrix} \frac{\partial p}{\partial \dot{v}} & \frac{\partial p}{\partial \lambda} \end{bmatrix}^{-1} \begin{bmatrix} \frac{\partial p}{\partial v} & \frac{\partial p}{\partial q} & \frac{\partial p}{\partial \lambda} \end{bmatrix} =$$

$$-\begin{bmatrix} M(q) & J^T \\ J & 0_{\lambda,\lambda} \end{bmatrix}^{-1} \begin{bmatrix} C_t & K_t & J^T \\ 2\sum_{j=1}^{n_v} \frac{\partial^2 \phi}{\partial[q, q_j]} v_j & \sum_{j=1}^{n_{\dot{v}}} \frac{\partial^2 \phi}{\partial[q, q_j]} \dot{v}_j & 0_{\lambda,\lambda} \end{bmatrix},$$ (40)

where the third order partial derivative of the constraint equations $\phi(q)$ with respect to the generalized coordinates q is zero for FNCF. Another important ingredient to fulfill the Kalman filter requirements is the measurement input matrix $\frac{dy}{du}$ of Equation (32). However, this matrix is not directly available and it is computed similarly to Equation (35) but with respect to input u:

$$\frac{dy}{du} = \frac{\partial y}{\partial z} \frac{dz}{du} + \frac{\partial y}{\partial u}.$$ (41)

Here, the full derivative $\frac{dz}{du}$ represents how the vector z varies when the input u is perturbed obtaining:

$$\frac{dz}{du} = \begin{bmatrix} \frac{\partial \dot{v}}{\partial u} & \frac{\partial v}{\partial u} & \frac{\partial q}{\partial u} & \frac{\partial \lambda}{\partial u} \end{bmatrix} = \begin{bmatrix} \frac{\partial \dot{v}}{\partial u} & 0_{v,u} & 0_{q,u} & \frac{\partial \lambda}{\partial u} \end{bmatrix}.$$ (42)

From Equation (42) can be seen that only the sensitivities of the acceleration and Lagrange multipliers are non-zero terms. A second order system is fully defined by the position coordinates, velocity coordinates, and time. An external force only influences the force balance of that system and thus the acceleration coordinates and Lagrangian multipliers while only through time-integration the velocity and position coordinates. However, these derivatives are evaluated at a single time instance. Therefore, in Equation (36), the acceleration coordinates and Lagrangian multipliers depend on the position and velocity coordinates, but not the other way around: the position and velocity coordinates do not depend on the acceleration coordinates, Lagrangian multipliers, or the external forces.

Subsequently, the remaining terms of Equation (42) can be computed as

$$\begin{bmatrix} \frac{\partial \dot{v}}{\partial u} \\ \frac{\partial \lambda}{\partial u} \end{bmatrix} = -\begin{bmatrix} \frac{\partial p}{\partial \dot{v}} & \frac{\partial p}{\partial \lambda} \end{bmatrix}^{-1} \frac{\partial p}{\partial u} = -\begin{bmatrix} M(q) & J^T \\ J & 0_{\lambda,\lambda} \end{bmatrix}^{-1} \begin{bmatrix} U_t \\ 0_{\lambda,u} \end{bmatrix}. \tag{43}$$

In Figure 1 a schematic summary of the followed steps required for the computation of the system and measurement matrices is reported.

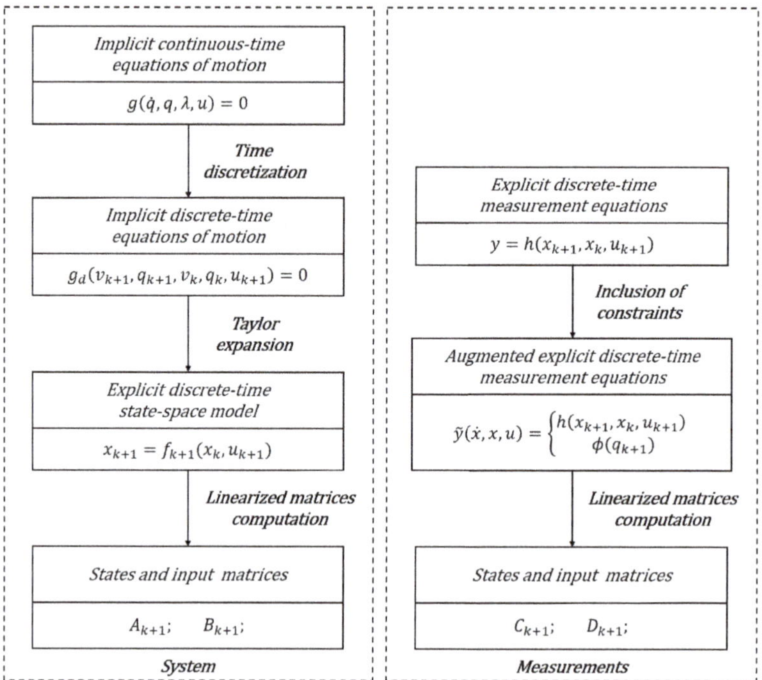

Figure 1. Block diagram representation of the system and measurement matrices computation for a generic integration step.

4.4. Augmented Discrete Extended Kalman Filter

The goal of this work is to estimate the unknown input forces, often referred to as disturbances, $d \in \mathbb{R}^{n_d}$, concurrently with the system states x through the augmented state vector $\tilde{x} \in \mathbb{R}^{n_x + n_d}$ defined as:

$$\tilde{x}_{k+1} = \begin{bmatrix} x_{k+1} \\ d_{k+1} \end{bmatrix}. \tag{44}$$

As the estimation algorithm also needs a model for the unknown inputs, a random walk model is associated with the unknown input:

$$d_{k+1} = d_k + v_{d,k}; \tag{45}$$

with $v_{d,k}$ also being zero-mean, random noise with covariance matrix $Q_{d,k}$.

In this work, the random walk model is chosen for the unknown disturbance. This approach is chosen as in general no prior information on the input is assumed to be known. However, if useful information is available, e.g., periodicity of the input, other input models can be employed, as in [24].

The augmented discrete-time state-space system of equations can be written combining Equations (26) and (45):

$$\tilde{g}_d(\tilde{x}_{k+1}, \tilde{x}_k, u_{k+1}) = \begin{cases} g_d(x_{k+1}, x_k, u_{k+1}) = \tilde{v}_{x,k} \\ d_{k+1} - d_k = v_{d,k} \end{cases}. \tag{46}$$

The linearized system matrix $\mathcal{F}$ refers to the augmented system of equations Equation (46) at the time step $k+1$ and it is assembled as

$$\mathcal{F} = \left.\frac{\partial \tilde{f}_d}{\partial \tilde{x}}\right|_{k+1} = \begin{bmatrix} A_{k+1} & B^d_{k+1} \\ 0_{x,d} & I_d \end{bmatrix}; \tag{47}$$

where $\tilde{f}_d$ is the augmented explicit ODEs associated with Equation (46), is obtained from Equation (24), while B^d_{k+1} is the linear disturbance matrix obtained similarly to B_{k+1} in Equation (25) but with respect to the unknown input d.

Subsequently, the linearized augmented measurement matrix $\mathcal{H}$ associated with the augmented state vector $\tilde{x}$ is assembled as:

$$\mathcal{H} = \left.\frac{\partial \tilde{y}}{\partial \tilde{x}}\right|_{k+1} = \begin{bmatrix} C_{k+1} & D^d_{k+1} \end{bmatrix}, \tag{48}$$

where the matrix D^d_{k+1} is computed in a similar fashion as for D_{k+1} in Equation (32) but with respect to the unknown input d.

Starting from these model, measurement and tangent matrices, an extended Kalman filter (EKF) procedure can be set up in order to estimate both the states and the unknown input forces. The application of this estimation approach is discussed in the following section.

4.5. The Adopted Extended Kalman Filter Scheme

In this work, we employ an extended Kalman filter (EKF) [25] to perform the state-estimation. With the various model and measurement equations, and their respective derivatives, defined for a MB model in the previous sections this EKF can be run efficiently. As all terms have been defined analytically it is not necessary to resort to numerical derivatives, which allows for an effective application of the EKF. Several researchers have shown the applicability of the estimators even for the strongly nonlinear case of MB systems [9,12,26].

In general, we can group the EKF scheme in two main steps: the a-priori and a posteriori steps.

- *A-priori step*: Assuming that the augmented states $\tilde{x}^+_k$ at the previous filter step and the input u_{k+1} are known, the a-priori state prediction $\tilde{x}^-_{k+1}$ and generalized accelerations $\dot{v}^-_{k+1}$ can be computed solving the ID-DAEs of Equation (17):

$$g_d(\tilde{x}^-_{k+1}, \tilde{x}^+_k, u_{k+1}) = 0_x. \tag{49}$$

Knowing the estimated state covariance matrix $\mathcal{P}^+_k$ for the previous timestep, the a-priori covariance at the current time $(k+1)$ step can be approximated from Equation (47) as

$$\mathcal{P}^-_{k+1} = \mathcal{F}\mathcal{P}^+_k\mathcal{F}^T + \tilde{Q}_k. \tag{50}$$

where $\tilde{Q}_k$ is the process noise matrix of the augmented state-space model.

The predicted measurement $\tilde{y}^-_{k+1}$ can then be evaluated from Equation (30) as:

$$\tilde{y}^-_{k+1} = \tilde{y}(\dot{v}^-_{k+1}, x^-_{k+1}, u_{k+1}). \tag{51}$$

The Kalman filter gain $\mathcal{K}$ allows achieving a desireable trade-off between the confidence in the model and the available measurements, and can be evaluated as:

$$\mathcal{K}_{k+1} = \mathcal{P}_{k+1}^{-} \mathcal{H}^{T} (\mathcal{H} \mathcal{P}_{k+1}^{-} \mathcal{H}^{T} + \tilde{R}_k)^{-1}, \tag{52}$$

where $\mathcal{H}$ is obtained from Equation (48) and $\tilde{R}_k$ is the measurement covariance matrix of the augmented measurement equations.

- *A-posteriori step*: When the real measurement y_{k+1}^{*} becomes available together with the predicted measurement $\tilde{y}_{k+1}^{-}$, the a posteriori state vector $\tilde{x}_{k+1}^{+}$ is obtained as:

$$\tilde{x}_{k+1}^{+} = \tilde{x}_{k+1}^{-} + \mathcal{K}_{k+1}(y_{k+1}^{*} - \tilde{y}_{k+1}^{-}) \tag{53}$$

The inclusion of the actual measurements also affects the posterior covariance matrix $\mathcal{P}_{k+1}^{+}$ and can be evaluated as:

$$\mathcal{P}_{k+1}^{+} = (\mathcal{I}_x - \mathcal{K}_{k+1} \mathcal{H}) \mathcal{P}_{k+1}^{-}. \tag{54}$$

In Figure 2 a block diagram scheme summarizing a generic kth step of the recursive ADE-KF is shown.

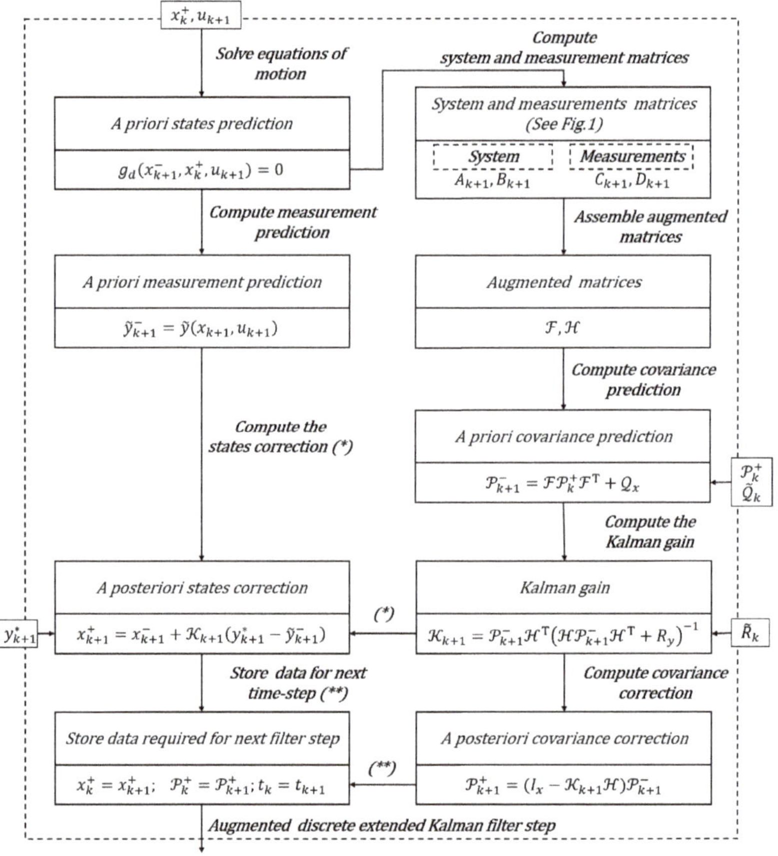

Figure 2. Block diagram representation of the recursive ADE-KF scheme for a generic filter step.

With this scheme the entire state-input estimation can be performed for an arbitrary (flexible) multibody model. In the following section we will validate it on an experimental setup.

In practical applications the performance of the estimation scheme will strongly depend on the tuning of the model covariance $\tilde{Q}$ and measurement covariance $\tilde{R}$. Unfortunately this tuning is also highly case dependent, which makes it difficult to set up general guidelines. For the particular validation case considered in this work, a detailed discussion on a possible tuning strategy is presented in the following section.

5. Validation: Joint State-Input Estimation

In this section, the presented joint state-input estimation methodology is experimentally validated on a slider-crank mechanism.

5.1. The Slider-Crank System

The slider-crank system in Figure 3 is a mechanism widely used in industry to transform a rotational motion in a linear motion as for instance in pick and place operations or in car engines.

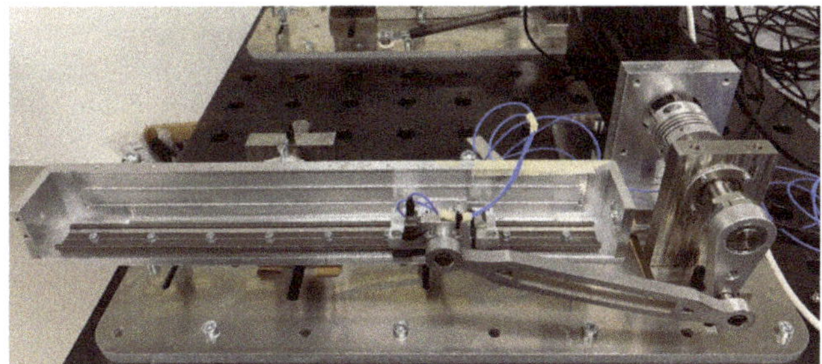

Figure 3. The slider-crank: experimental setup.

Figure 4 represents the multibody model of the experimental setup shown in Figure 3 consisting of 14 bodies: base block (BB), motor housing (MH), motor rotor (MR), motor support (MS), crank (C), crank shaft (CSh), crank support (CS), connecting rod (CR), crank-rod bearing (B_{C-CR}), slider (S), rod-slider bearing (B_{CR-S}), slider accelerometer (SA), track rail (TR) and track support (TS).

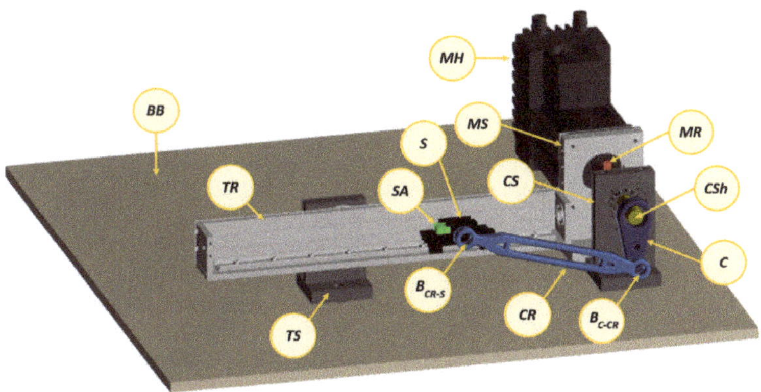

Figure 4. The slider-crank: multibody model.

The mechanical properties of the various bodies are listed in Table 1.

Table 1. Mechanical properties expressed with respect to the center of gravity of each individual body.

Body	m [kg]	J_{xx} [kg· m^2]	J_{yy} [kg· m^2]	J_{zz} [kg· m^2]
BB	3.2175×10^3	2.9225×10^2	6.274×10^2	8.714×10^2
MH	1.420×10^1	2.48647×10^{-2}	9.366×10^{-2}	9.366×10^{-2}
MR	6.670×10^{-1}	2.947×10^{-3}	2.001×10^{-3}	2.001×10^{-3}
MS	1.350	6.052×10^{-3}	3.897×10^{-3}	2.346×10^{-3}
C	1.830×10^{-1}	5.742×10^{-4}	4.930×10^{-4}	1.522×10^{-5}
CSh	4.960×10^{-1}	1.644×10^{-4}	7.068×10^{-4}	7.068×10^{-4}
CS	8.4058×10^{-1}	2.678×10^{-3}	1.954×10^{-3}	8.047×10^{-4}
B_{C-CR}	2.820×10^{-1}	-	-	-
CR	2.540×10^{-1}	1.185×10^{-2}	1.1840×10^{-2}	3.654×10^{-5}
B_{CR-S}	2.820×10^{-1}	-	-	-
S	2.562×10^{-1}	2.665×10^{-4}	1.131×10^{-4}	3.29510^{-4}
SA	2.200×10^{-2}	-	-	-
TR	1.206	2.800×10^{-1}	9.424×10^{-3}	2.800×10^{-1}
TS	4.206	1.392×10^{-2}	1.256×10^{-2}	4.963×10^{-3}

Even though the presented slider-crank mechanism is an academic example it combines several challenging phenomena such as non-linear kinematics and complex slider-track interaction phenomena. The various bodies are connected trough fixed, revolute, spherical, and universal joints to allow the desired kinematics. The slider and track bodies are connected through a contact stiffness $\mathbf{k}_c$ along the perpendicular directions to the sliding trajectory. The Pacejka friction model [27] defines the friction coefficient μ as a function of the sliding velocity Δv:

$$\mu = \mathbf{d} sin\left[\mathbf{c} tan^{-1}\left(\mathbf{b}\Delta v - \mathbf{e}\left(\mathbf{b}\Delta v - tan^{-1}(\mathbf{b}\Delta v)\right)\right)\right], \tag{55}$$

where the friction force f_f is evaluated as

$$f_f = \mu f_n = \mu(\mathbf{k}_c\delta + \mathbf{c}_c\dot{\delta}), \tag{56}$$

where f_n is the resultant normal force acting on the slider and, δ is the local compliance between the slider and track rail interfaces projected onto the normal direction to the sliding direction. The adopted (identified) Pacejka friction model parameters are listed in Table 2.

Table 2. Pacejka friction model parameters.

$\mathbf{k}_c$ [N/m]	$\mathbf{c}_c$ [Ns/m]	$\mathbf{b}$ [s/m]	$\mathbf{c}[-]$	$\mathbf{d}[-]$	$\mathbf{e}[-]$
9.7854×10^6	1.196	5.036×10^2	1.5708	2.653×10^{-2}	-9.8534

The system motion is driven by a brushless servomotor "MAC3000"' with integrated controller MAC00-B4 from JVL (www.jvl.dk, accessed on 1 March 2021). It includes a motor encoder sensor allowing the measurement of the rotation angle and velocity of the motor shaft, indicated by θ and $\dot{\theta}$ respectively. As can be seen in Figures 3 and 4, the slider is equipped with a mono-axial MEMS accelerometer (3711D1FB200G) from PCB (www.pcb.com) to measure its translational acceleration along the sliding direction indicated by $\ddot{Y}$. The entire system is controlled via the motor using a closed-loop PID controller targeting a desired rotational motor velocity while adapting the determined motor torque T.

5.2. Results

After the setup of the MB model of the slider-crank mechanism, it is embedded into the ADE-KF estimation framework presented in Section 4.

The input torque delivered by the motor is assumed unknown, and jointly estimated with the model states as introduced in Section 4.4. It is assumed that all model uncertainties is dominated by the augmented state, representing the unknown input, while the model is considered perfect.

The a posteriori Kalman filter step is computed using the augmented measurements discussed in Section 4.2. These combine the angular motor velocity $\dot{\theta}$ together with the model constraint equations ϕ. The angular motor velocity is directly available on many mechatronic drives since they are equipped with relatively accurate encoders for control or monitoring purposes.

For the validation of our estimation framework, summarized in Figure 5, we compare the estimates (virtual sensors) of the motor position θ, motor velocity $\dot{\theta}$, slider acceleration $\ddot{Y}$ and motor torque T with measurements directly obtained from the experimental setup. Besides the motor encoder, an accelerometer on the slider is employed and motor torques can be directly extracted from the drive unit.

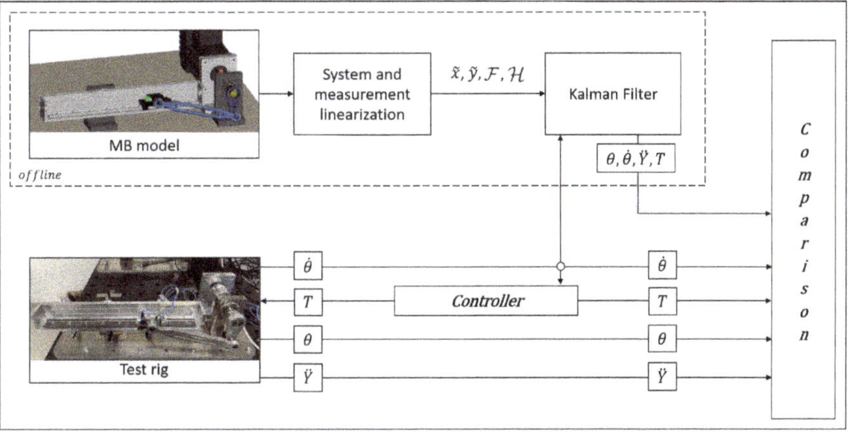

Figure 5. Diagram of the coupled state-input estimation scheme and signal comparisons. θ and $\dot{\theta}$ are the motor angle and angular motor velocity respectively; $\ddot{Y}$ is the translational slider acceleration; T is the motor torque.

The performed experiments span 9.4 s and are executed for three levels of desired angular motor velocity, which are provided to the motor controller as desired set points: 40, 50 and 60 rad/sec. Note that, due to the non-ideal behavior of the system and the limitations of the PID controller, the desired set point results in practice in a varying angular velocity.

The measured and the estimated motor angle θ, rotational velocity $\dot{\theta}$, and the slider translational acceleration $\ddot{Y}$ are compared in Figure 6.

Three subsets of this full timespan are shown in Figure 7 to better appreciate the transient effects during the start-up and the two velocity transitions.

It is clear from Figures 6 and 7 that the ADE-KF with the MB model tracks these various (derivative) states very well, underlining the well represented system kinematics.

In Figure 8 the estimated motor torque is compared to the measured motor torque.

This comparison shows a good prediction over the full time series (on the left) and on the angular velocity transitions (zoom-in figures on the right) thanks to the proposed methodology.

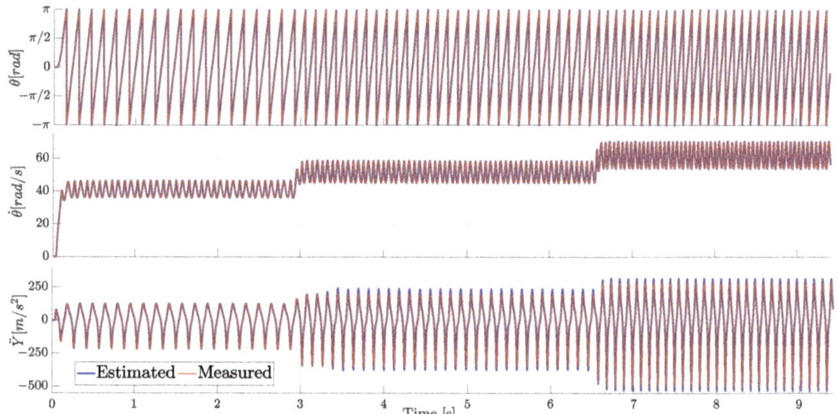

Figure 6. Comparison of the measured and estimated motor angle θ (**top**), motor angular velocity $\dot\theta$ (**middle**) and translational slider acceleration $\ddot Y$ (**bottom**) for the full time series.

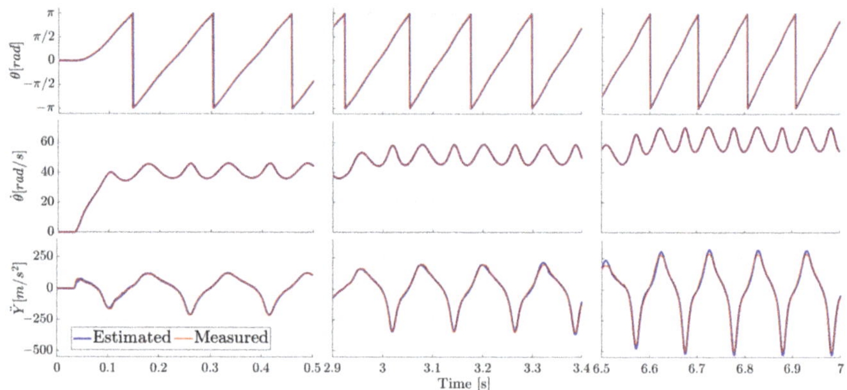

Figure 7. Comparison of the measured and estimated motor angle θ (**top row**), motor angular velocity $\dot\theta$ (**middle row**) and translational slider acceleration $\ddot Y$ (**bottom row**). Zoom-in per column on the velocity transitions.

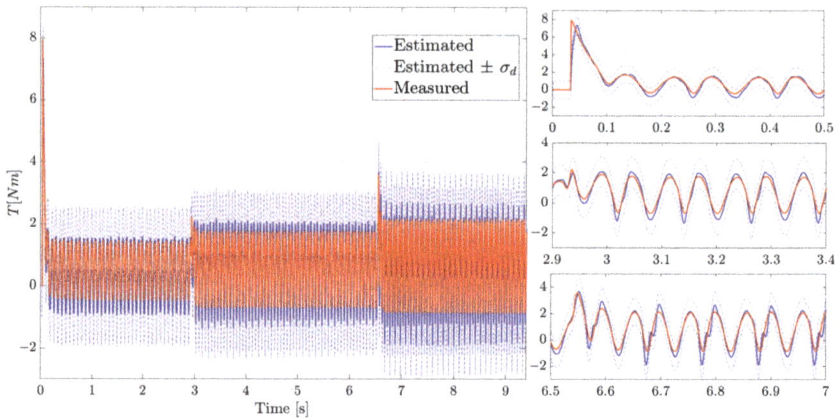

Figure 8. Comparison of the measured and estimated motor torque; on the left, full time series; on the right, the zoom-in on the velocity transitions are shown.

In Table 3 the root mean square error of the estimated virtual sensors and input torque are reported, underlying the relatively high accuracy of the estimated quantities. It is defined as $Error_{RMS} = \sqrt{\frac{\sum_k [\chi_m(k) - \chi_e(k)]^2}{n_k}}$, where χ_m and χ_e are the measured and estimated variables respectively, while n_k is the total number of data samples.

Table 3. Accuracy of the estimated quantities in terms of root mean square error.

	θ [rad]	$\dot{\theta}$ [rad/s]	$\ddot{Y}$ [m/s^2]	T [Nm]
$Error_{RMS}$	0.005	0.051	13.651	0.334

The choice of performing the experiments for a relatively long timespan was made to demonstrate the filter stability in terms of both mean value and covariance prediction. For these kind of applications, where the uncertainty is dominated by difficult to model load effects (friction, etc.), the choice of focussing the model covariance on the input load allows effectively propagating the uncertainties. The dashed blue curves in Figure 8 represent the estimated expected variation of the augmented average state estimate within the 70% of confidence. It is expressed in terms of the standard deviations $\sigma_d = \sqrt{\mathcal{P}_d^+}$ computed for each kth filter step where $\mathcal{P}_d^+$ is the diagonal term of the a posteriori estimated covariance associated with the augmented state (disturbance). It is important to notice that the experimental motor torque (in red) remains bounded by the estimated input uncertainty therefore being an accurate estimation of the real covariance.

These results for the estimated torques show a significantly larger error than those obtained for the (derivative) states. For multibody problems in general, the state-estimates can be expected to be dominated by the kinematics of the system, which are generally well known. For the load estimates however, the dynamics of the system will play a crucial role. Besides key dynamic quantities such as the system inertia which can be modelled very accurately, other effects such as friction forces also influence this outcome. In this work we employed the friction model from Equation (56), but any error on this model is also propagated to the torque estimates. Due to the complex nature of the interface conditions for the slider and the rail (and the other bearings in the system), some errors are to be expected here. Key for future research will therefore be to examine how these complex load phenomena can be accounted for as effectively as possible. Moreover, by only employing one physical measurement ($\dot{\theta}$) in the ADE-KF scheme, it is guarantied that the accuracy of the estimated input torque would be less accurate for a poorly identified model. This can be observed in [15] where the same validation case was considered. Here, Angeli et al. proposed a deep learning methodology to reduce the initial MB equations from redundant to minimal coordinates where the resulting equations are then employed in an augmented extended Kalman filter scheme. Alternatively to what is presented in the current contribution, the supposed unknown motor torque is estimated employing the slider acceleration ($\ddot{Y}$) and no slider-track friction model was considered which has led to slightly less accurate input and virtual sensors estimation as compared to the currently proposed approach and model.

An important aspect in estimation problems is the choice of the Kalman filter parameters such as the process and measurement noise covarince matrices, $\tilde{Q}_k$ and $\tilde{R}_k$. It is recurrent while dealing with KF-based estimators that the accuracy of the estimated quantities is highly influenced by the selection of those parameters. However, general rules are not available since the filter parameters and influence strongly depend on the application case. Therefore, it is common to resort to a tuning step as the process of investigating and selecting these parameters. In the context of this work, the adopted strategy is described in the following section.

5.3. Kalman Filter Tuning

To attain the best accuracy from the presented estimation scheme, the model and measurement covariances need to be judiciously selected. In this work we have started with the selection of the measurement covariance matrix $\tilde{R}_k$ associated with the augmented measurement equations $\tilde{y}$. In lack of other information, We assume this measurement covariance constant over time. The covariance results from the combination of two main measurement contributions: the (motor) angular velocity $\dot{\theta}$ and the augmented constraint measurements ϕ, which reads as

$$\tilde{R}_k = \begin{bmatrix} R_{\dot{\theta}} & 0_\lambda^T \\ 0_\lambda & R_\phi \end{bmatrix}, \tag{57}$$

where $R_{\dot{\theta}}$ is generally tuned based on reference noise measurements while for this application, since no noise reference was available, the author has chosen a value which is representative of the encoder measurement noise: $R_{\dot{\theta}} = 10^{-2}$ (rad^2/s^2). Moreover, the authors have experienced that the influence of the measurement noise parameter $R_{\dot{\theta}}$ is relative to the value of the process noise covariance $\tilde{Q}_k$.

R_ϕ instead is linked to the mathematical and physical meaning of the constraint equations. In MB applications, we can distinguish two types of constraint equations: the ones that come from the inherent coordinates formulations, i.e., ϕ_c (e.g., Euler parameters should be unit vector and rotation matrix should be orthogonal), and the ones that come from joints definition, i.e., ϕ_j, as for instance the spherical and/or revolute joints. In this work, for the latter a small noise term is allowed (i.e., all diagonal terms of R_{ϕ_j} are set to 10^{-9} while the off-diagonal terms are set to zero) representing the joint imperfections typical of real systems, whereas for the former, they are treated as perfect measurements (i.e., $R_{\phi_c} = 0$), otherwise the kinematics and the mathematical principles that are used to describe the MB system are no longer valid.

Similarly to the augmented measurement covariance $\tilde{R}_k$, the augmented process noise matrix $\tilde{Q}_k$ is assumed constant for all filter steps and it can be written as combination of the system and augmented states contributions as

$$\tilde{Q}_k = \begin{bmatrix} Q_x & 0 \\ 0 & Q_d \end{bmatrix}. \tag{58}$$

As we assume the model to be practically perfect compared to the high uncertainty on the unknown inputs, the process noise matrix Q_x associated with the system states is set to zero.

The remaining parameter Q_d associated with the unknown input torque (disturbance) is obtained in a brute-force optimization fashion. Since in this case there is only a single value to be chosen, an exhaustive search is therefore not computationally prohibitive. In this regard a grid of Q_d values have been selected, going from 10^{-5} to 10^5 sampled exponentially in eleven increments, leading to a corresponding eleven performed estimations. The choice of the Q_d is based on the L-curve method proposed by [28] where its nearly optimal value (Not optimal in absolute sense due to discrete evaluation points Q_d and the user defined $R_{\dot{\theta}}$) is such that the best trade-off is achieved between the *Error norm* and *Smoothing error*. These norms are defined as

$$Error\ norm = \sum_{k=1}^{n_k} ||y_k^* - \tilde{y}_k^-||_2 \tag{59}$$

and,

$$Smoothing\ norm = \frac{\sum_{k=1}^{n_k} ||d(k)||_2}{\sum_{k=1}^{n_k} ||T(k)||_2} \tag{60}$$

where $d(k)$ and $T(k)$ are the estimated (disturbance) and measured torque respectively, while n_k is the total number of sampling point k. On the left side of Figure 9 the L-curve for the evaluated Q_d grid points is shown where the blue marked point is the chosen value corresponding to $Q_d = 10^0$. This value have been used for the results shown in Figures 6–8. As expected, it is observed that the error norm decreases with an increasing smoothing error which corresponds to an increasing Q_d till a saturation is reached. This occurs when a further increase of process noise value does not show significant improvements to the estimation results ($Q_d > 10^0$).

On the right side of Figure 9 the estimated torque with the chosen Q_d (marked in blue) is compared to a sub-optimal value (marked in orange) on the velocity transitions zones while in Figure 10 the full time series are compared.

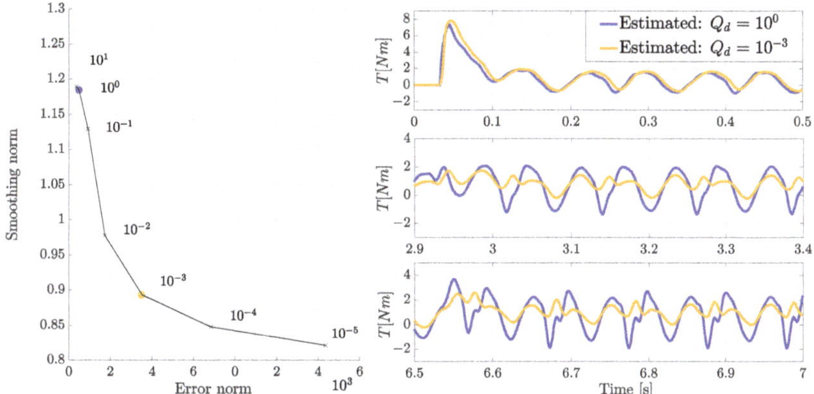

Figure 9. The L-curve plot for different process variance Q_d (**left figure**) and zoom-in comparison on the velocity transition of the measured and estimated motor torque using two different values of process variance Q_d (**right figure**).

In Figures 9 and 10, it can be seen that despite a relatively similar tracking of the input (more delayed), at low rotational speed using a smaller covariance than the "optimal" one, the estimation accuracy degrades over time with a worse tracking of estimated torque.

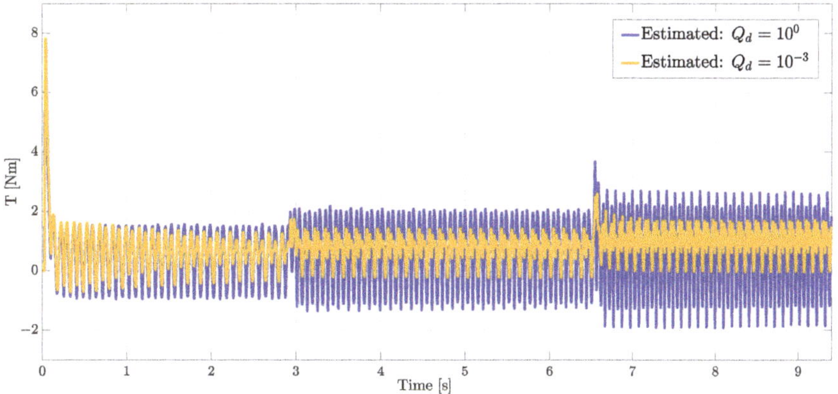

Figure 10. Comparison of the estimated motor torque using two different values of process covariance Q_d of the augmented state.

Although this study does not address the full scope of noise covariance tuning, the author deemed it sufficient to explain the methodological developments considered in this contribution. Further research on covariance tuning might be necessary to obtain a more holistic approach.

6. Discussion and Conclusions

This work presents a new estimation methodology tailored for MB models to enable the definition of virtual sensors for various system states and inputs.

Through the choice of a general MB modeling approach various key physical effects can be accurately accounted for, ranging from nonlinear kinematics to complex dynamic effects. The developed framework allows using these multibody models in an estimation framework without particular additional modeling assumptions or reformulations. More specifically, no constraint elimination methods are required to employ the defined MB model into the estimation framework, reducing the preparation time and the user effort to setup the estimation problem while ensuring the correct physical and mathematical interpretation of the system under investigation. As the proposed methodology has no particular assumptions with respect to the multibody model formulation it can be easily extended to any of the commonly available (flexible) MB approaches, e.g., FNCF, FFR-CMS, or GCMS. However, to fully benefit from the proposed approach the equations of motion and tangent stiffness matrices of the system should be analytically available to efficiently assemble the linearized system and measurement matrices.

In the present work, we exploited the FNCF MB approach, as this methodology inherently enables an easy and efficient evaluation of the different tangent model matrices required for the estimation framework.

Finally, the developed methodology has been experimentally validated on a slider-crank mechanism. Very high accuracy is obtained for the estimated states with respect to the available measurements. Good accuracy is also obtained for the estimated input torque, but due to the larger dynamic model errors in e.g., friction effects the resulting errors are higher than those obtained for the states. The validation has been performed over a relatively large time-span which also demonstrates the capability of the presented framework to obtain long-term stable estimates with a bounded uncertainty, in the form of a bounded covariance.

The presented methodology has some drawbacks since a large number of states (including the Lagrange multipliers) and measurements (including the constraints equations) are employed. These lead to a computationally less efficient approach as compared to other state of the art techniques (e.g., using minimal coordinates [15]). A possible solution to mitigate this issue might come from a wise selection of the number of bodies and constraints equations to construct the high-fidelity model. For instance, in this contribution, the choice of redundant number of bodies and constraints was made in purpose to stress the potential of the proposed methodology while dealing with redundant set of DAEs equations. More precisely, not all bodies and hence constraints were required to achieve the estimated motor torque with the same level of accuracy (e.g., the motor housing and the different supports). Nevertheless, if computational efficiency is not a limiting factor, i.e., if an online estimation is not required, this approach has the potential to enable a very generic deployment of (flexible) multibody-based state-input estimation.

Future work will focus on how these methodologies can be employed to obtain more accurate descriptions for key dynamic effects such as the friction present in these multibody systems.

Author Contributions: Conceptualization and implementation, R.A. and M.V.; Methodology, R.A.; in-house MB research code, M.V.; validation, R.A.; resources, W.D.; instrumentation and data acquisition, R.A.; writing, review and editing, R.A., M.V. and F.N.; supervision J.C., F.N. and W.D. All authors have read and agreed to the published version of the manuscript.

Funding: This work has been carried out within the framework of the Flanders Make ICON project: Virtual Sensing on Flexible systems using distributed parameter models (VSFlex). The research was partially supported by Flanders Make, the strategic research center for the manufacturing industry. Internal Funds KU Leuven are gratefully acknowledged for their support. VLAIO (Flanders Innovation & Entrepreneurship Agency) is also acknowledged for its support.

Institutional Review Board Statement: Not applicable.

Informed Consent Statement: Not applicable.

Data Availability Statement: The data presented within this study are resulting from activities within the acknowledged projects and are available therein.

Conflicts of Interest: The authors declare no conflict of interest.

Abbreviations

The following abbreviations are used in this manuscript:

TPA	Transfer Path Analysis
KF	Kalman Filter
ADE-KF	Augmented Discrete Extended Kalman Filter
MB	MultiBody
FE	Finite Element
EOM	Equation Of Motion
(I-), (E-) DAE	(Implicit), (Explicit) Differential Algebraic Equation
(I-), (E-) ODE	(Implicit), (Explicit) Ordinary Differential Equation
MBRC	MultiBody Research Code
FNCF	Flexible Natural Coordinates Formulation
BDF	Backward Differentiation Formula
VS	Virtual Sensor
MEMS	Micro Electro-Mechanical Systems
PID	Proportional Integrative Derivative
$\mathbb{Z}$	integer numbers set
$\mathbb{R}$	real numbers set
$a \in \mathbb{R}$	scalar
$a \in \mathbb{R}^{n_a}$	column vector
$\begin{bmatrix} a_1^T & a_2^T \end{bmatrix}^T \in \mathbb{R}^{n_{a_1}+n_{a_2}}$	vertical vector concatenation
$A \in \mathbb{R}^{n_1 \times n_2}$	matrix
$I_a \in \mathbb{Z}^{n_a \times n_a}$	identity matrix
$0_{a_1} \in \mathbb{Z}^{n_{a_1}}$	zero vector
$0_{a_1,a_2} \in \mathbb{Z}^{n_{a_1} \times n_{a_2}}$	zero matrix
$\square^T$	transpose operator
$\square^{-1}$	inverse matrix operator
$\square^-$	a priori prediction
$\square^+$	a posteriori prediction
$\square_k = \square(t = t_k)$	kth time step
$\square_n$	natural coordinates
$\dot{\square} = \frac{d\square}{dt}, \ddot{\square} = \frac{d^2\square}{dt^2}$	time derivatives
$\frac{da_1}{da_2} \in \mathbb{R}^{n_{a_1} \times n_{a_2}}$	total derivative
$\frac{\partial a_1}{\partial a_2} \in \mathbb{R}^{n_{a_1} \times n_{a_2}}$	partial derivative
$\frac{\partial^2 a_1}{\partial a_2 \partial a_3} \in \mathbb{R}^{n_{a_1} \times n_{a_2} \times n_{a_3}}$	second partial derivative
$\|\square\|_2$	2-norm operator

Appendix A. Influence of the Forward Differentiation Scheme to the Linearization of the EOMs

In Section 3, the single step backward differentiation formula (BDF-1) was introduced to derive the fully explicit discrete-time linearized form of the EOMs (Equation (19)). In this appendix, the influence of another differentiation scheme, namely the single step forward differentiation scheme, is reported, demonstrating the limitations introduced by such differentiation formula to the proposed linearization approach.

The single step forward differentiation scheme can be written as

$$\begin{cases} \dot{v}_k = \frac{1}{h}(v_{k+1} - v_k) \\ v_k = \frac{1}{h}(q_{k+1} - q_k) \end{cases}, \tag{A1}$$

with h representing the constant time step size as shown in Figure A1.

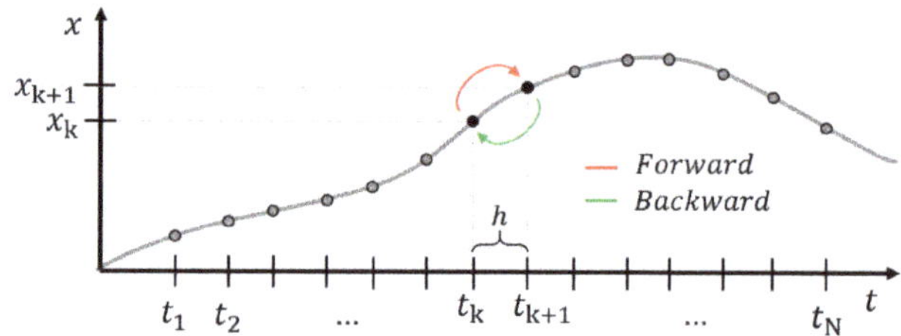

Figure A1. Graphical interpretation of the forward and backward linearization for a generic state-time evolution x.

By substituting Equation (A1) into Equation (8) the discrete-time EOMs g'_d are obtained:

$$\begin{cases} g'_{d1} = v_k - \frac{1}{h}(q_{k+1} - q_k) = 0_v \\ g'_{d2} = M(q_k)\frac{v_{k+1}-v_k}{h} + f_{nl}(q_k) + J^T(q_k)\lambda_k = 0_q \\ g'_{d3} = \phi(q_k) = 0_\lambda \end{cases}. \tag{A2}$$

Similarly to Equation (21) the Jacobian for the forward Euler time-integrator can be computed as:

$$G'_{x_{k+1}} = \frac{\partial g'_d}{\partial x_{k+1}} = \begin{bmatrix} \frac{\partial g'_{d1}}{\partial v_{k+1}} & \frac{\partial g'_{d1}}{\partial q_{k+1}} & \frac{\partial g'_{d1}}{\partial \lambda_{k+1}} \\ \frac{\partial g'_{d1}}{\partial v_{k+1}} & \frac{\partial g'_{d2}}{\partial q_{k+1}} & \frac{\partial g'_{d2}}{\partial \lambda_{k+1}} \\ \frac{\partial g'_{d3}}{\partial v_{k+1}} & \frac{\partial g'_{d3}}{\partial q_{k+1}} & \frac{\partial g'_{d3}}{\partial \lambda_{k+1}} \end{bmatrix} = \begin{bmatrix} 0_{v,v} & -\beta' & 0_{v,\lambda} \\ \beta' M'_t & -\gamma' M'_t + \beta' C'_t & 0_{q,\lambda} \\ 0_{\lambda,v} & 0_{\lambda,q} & 0_{\lambda,\lambda} \end{bmatrix}; \tag{A3}$$

Here, C'_t and M'_t are the tangent damping and mass matrices obtained from the partial derivatives of the continuous g_2 equations in Equation (8) evaluated at time step k:

$$C'_t = \left.\frac{\partial g_2}{\partial v}\right|_k; \qquad\qquad M'_t = \left.\frac{\partial g_2}{\partial \dot{v}}\right|_k, \tag{A4}$$

and β' and γ' are matrix coefficients function of the defined integration rule, which are given for the forward Euler scheme by:

$$\beta' = \frac{\partial v_k}{\partial q_{k+1}} = \frac{\partial \dot{v}_k}{\partial v_{k+1}} = \frac{1}{h}I_q; \qquad\qquad \gamma' = \frac{\partial \dot{v}_k}{\partial q_{k+1}} = \frac{1}{h^2}I_q; \tag{A5}$$

$$\frac{\partial v_k}{\partial q_k} = \frac{\partial \dot{v}_k}{\partial v_k} = -\beta'; \qquad\qquad \frac{\partial \dot{v}_k}{\partial q_k} = -\gamma'. \tag{A6}$$

As it can be seen from Equation (A3), the matrix $G'_{x_{k+1}}$ is singular and therefore not invertible to compute the linearized system matrix A of Equation (24) required for the KF-based estimation framework. Based on the above demonstration, the forward Euler differentiation scheme is not applicable to the proposed linearization approach.

This can be better understood looking at the fundamental assumptions behind the choice of the differentiation scheme. If a forward differentiation scheme is chosen, it practically means that Equation (19) will be derived starting from x_k looking forward in time to x_{k+1} as graphically depicted in Figure A1, but due to the implicit nature of the problem there are not enough information in x_k to invert the problem with respect to x_{k+1}.

References

1. Forrier, B.; Naets, F.; Desmet, W. Broadband Load Torque Estimation in Mechatronic Powertrains Using Nonlinear Kalman Filtering. *IEEE Trans. Ind. Electron.* **2018**, *65*, 2378–2387. [CrossRef]
2. Van der Seijs, M.V.; De Klerk, D.; Rixen, D.J. General framework for transfer path analysis: History, theory and classification of techniques. *Mech. Syst. Signal Process.* **2016**, *68–69*, 217–244. [CrossRef]
3. Naets, F.; Croes, J.; Desmet, W. An online coupled state/input/parameter estimation approach for structural dynamics. *Comput. Methods Appl. Mech. Eng.* **2015**, *283*, 1167–1188. [CrossRef]
4. Naets, F.; Cuadrado, J.; Desmet, W. Stable force identification in structural dynamics using Kalman filtering and dummy-measurements. *Mech. Syst. Signal Process.* **2015**, *50–51*, 235–248. [CrossRef]
5. Shabana, A.A. *Dynamics of Multibody Systems*; Cambridge University Press: Cambridge, UK, 2013. [CrossRef]
6. Géradin, M.; Cardona, A. *Flexible Multibody Dynamics: A Finite Element Approach*; John Wiley & Sons: New York, NY, USA, 2001; Volume 4.
7. Bauchau, O.A. *Flexible Multibody Dynamics*; Springer Science and Business Media: Berlin, Germany, 2011; Volume 176. [CrossRef]
8. Hiller, M.; Kecskeméthy, A. Dynamics of Multibody Systems with Minimal Coordinates. In *Computer-Aided Analysis of Rigid and Flexible Mechanical Systems*; Seabra Pereira, M.F.O., Ambrósio, J.A.C., Eds.; Springer: Dordrecht, The Netherlands, 1994; pp. 61–100. [CrossRef]
9. Cuadrado, J.; Dopico Dopico, D.; Perez, J.; Pastorino, R. Automotive observers based on multibody models and the extended Kalman filter. *Multibody Syst. Dyn.* **2012**, *27*, 3–19. [CrossRef]
10. Pastorino, R.; Richiedei, D.; Cuadrado, J.; Trevisani, A. State estimation using multibody models and non-linear Kalman filters. *Int. J. Non-Linear Mech.* **2013**, *53*, 83–90. [CrossRef]
11. Sanjurjo, E.; Dopico, D.; Luaces, A.; Naya, M.Á. State and force observers based on multibody models and the indirect Kalman filter. *Mech. Syst. Signal Process.* **2018**, *106*, 210–228. [CrossRef]
12. Sanjurjo, E.; Naya, M.; Blanco, J.L.; Moreno, J.L.; Gimenez, A. Accuracy and efficiency comparison of various nonlinear Kalman filters applied to multibody models. *Nonlinear Dyn.* **2017**, *88*. [CrossRef]
13. Palomba, I.; Richiedei, D.; Trevisani, A. Kinematic state estimation for rigid-link multibody systems by means of nonlinear constraint equations. *Multibody Syst. Dyn.* **2017**, *40*, 1–22. [CrossRef]
14. Angeli, A.; Desmet, W.; Naets, F. Deep learning for model order reduction of multibody systems to minimal coordinates. *Comput. Methods Appl. Mech. Eng.* **2021**, *373*, 113517. [CrossRef]
15. Angeli, A.; Desmet, W.; Naets, F. Deep learning of multibody minimal coordinates for state and input estimation with Kalman filtering. *Multibody Syst. Dyn.* **2021**, 1–19. [CrossRef]
16. Risaliti, E.; Tamarozzi, T.; Vermaut, M.; Cornelis, B.; Desmet, W. Multibody model based estimation of multiple loads and strain field on a vehicle suspension system. *Mech. Syst. Signal Process.* **2019**, *123*, 1–25. [CrossRef]
17. Vermaut, M.; Tamarozzi, T.; Naets, F.; Desmet, W. Development of a flexible multibody simulation package for in-house benchmarking. In Proceedings of the ECCOMAS Thematic Conference on Multibody Dynamics, Barcelona, Catalonia, Spain, 29 June–2 July 2015; pp. 1560–1571.
18. De Jalon, J.G.; Bayo, E. *Kinematic and Dynamic Simulation of Multibody Systems: The Real-Time Challenge*; Springer: New York, NY, USA, 2012. [CrossRef]
19. Vermaut, M.; Naets, F.; Desmet, W. A flexible natural coordinates formulation (FNCF) for the efficient simulation of small-deformation multibody systems. *Int. J. Numer. Methods Eng.* **2018**, *115*, 1353–1370. [CrossRef]
20. Pechstein, A.; Reischl, D.; Gerstmayr, J. A Generalized Component Mode Synthesis Approach for Flexible Multibody Systems With a Constant Mass Matrix. *J. Comput. Nonlinear Dyn.* **2012**, *8*, 011019. [CrossRef]
21. Brenan, K.E.; Campbell, S.L.; Petzold, L.R. *Numerical Solution of Initial-Value Problems in Differential-Algebraic Equations*; Siam: Philadelphia, PA, USA, 1996; Volume 14. [CrossRef]
22. Blockmans, B. Model Reduction of Contact Problems in Flexible Multibody Dynamics. Ph.D. Thesis, KU Leuven, Leuven, Belgium, 2018.
23. Simon, D. Kalman filtering with state constraints: A survey of linear and nonlinear algorithms. *IET Control Theory Appl.* **2010**, *4*, 1303–1318. [CrossRef]
24. Kirchner, M.; Croes, J.; Cosco, F.; Desmet, W. Exploiting input sparsity for joint state/input moving horizon estimation. *Mech. Syst. Signal Process.* **2018**, *101*, 237–253. [CrossRef]
25. Simon, D. *Optimal State Estimation: Kalman, H Infinity, and Nonlinear Approaches*; John Wiley & Sons: Hoboken, NJ, USA, 2006.
26. Tamarozzi, T.; Risaliti, E.; Rottiers, W.; Desmet, W. Noise, ill-conditioning and sensor placement analysis for force estimation through virtual sensing. In Proceedings of the International Conference on Noise and Vibration Engineering (ISMA2016), Leuven, Belgium, 19–21 September 2016; KU Leuven: Leuven, Belgium, 2016; pp. 1741–1756.

27. Pacejka, H. The Wheel Shimmy Phenomenom: A Theoretical and Experimental Investigation with Particular Reference to the Nonlinear Problem (Analysis of Shimmy in Pneumatic Tires due to Lateral Flexibility for Stationary and Nonstationary Conditions). Ph.D. Thesis, Delft University of Technology, Delft, The Netherlands, 1966.
28. Hansen, P.C.; O'Leary, D.P. The use of the L-curve in the regularization of discrete ill-posed problems. *SIAM J. Sci. Comput.* **1993**, *14*, 1487–1503. [CrossRef]

Article

Virtual Sensoring of Motion Using Pontryagin's Treatment of Hamiltonian Systems

Timothy Sands

Sibley School of Mechanical and Aerospace Engineering, Cornell University, Ithaca, NY 14850, USA; tas297@cornell.edu

Abstract: To aid the development of future unmanned naval vessels, this manuscript investigates algorithm options for combining physical (noisy) sensors and computational models to provide additional information about system states, inputs, and parameters emphasizing deterministic options rather than stochastic ones. The computational model is formulated using Pontryagin's treatment of Hamiltonian systems resulting in optimal and near-optimal results dependent upon the algorithm option chosen. Feedback is proposed to re-initialize the initial values of a reformulated two-point boundary value problem rather than using state feedback to form errors that are corrected by tuned estimators. Four algorithm options are proposed with two optional branches, and all of these are compared to three manifestations of classical estimation methods including linear-quadratic optimal. Over ten-thousand simulations were run to evaluate each proposed method's vulnerability to variations in plant parameters amidst typically noisy state and rate sensors. The proposed methods achieved 69–72% improved state estimation, 29–33% improved rate improvement, while simultaneously achieving mathematically minimal costs of utilization in guidance, navigation, and control decision criteria. The next stage of research is indicated throughout the manuscript: investigation of the proposed methods' efficacy amidst unknown wave disturbances.

Keywords: virtual sensoring; physical sensors; smart/intelligent sensors; sensor technology and applications; sensing principles; signal processing in sensor systems

check for
updates

Citation: Sands, T. Virtual Sensoring of Motion Using Pontryagin's Treatment of Hamiltonian Systems. *Sensors* **2021**, *21*, 4603. https://doi.org/10.3390/s21134603

Academic Editors: Javier Cuadrado and Miguel Ángel Naya Villaverde

Received: 5 June 2021
Accepted: 27 June 2021
Published: 5 July 2021

1. Introduction

Inertial measurement units provide continuous and accurate estimates of motion states in between sensor measurements. Future unmanned naval vessels depicted in Figure 1a require very accurate motion measurement units including active sensor systems and inertial algorithms when active sensor data is unavailable. State observers are duals of state controllers used for establishing decision criteria to declare accurate positions and rates and several instantiations are studied here when fused with noisy sensors, where theoretical analysis of the variance resulting from noise power is presented and validated in over ten-thousand Monte Carlo simulations.

The combination of physical sensors and computational models to provide additional information about system states, inputs, and/or parameters, is known as virtual sensoring. Virtual sensoring is becoming more and more popular in many sectors, such as the automotive, aeronautics, aerospatial, railway, machinery, robotics, and human biomechanics sectors. Challenges include the selection of the fusion algorithm and its parameters, the coupling or independence between the fusion algorithm and the multibody formulation, magnitudes to be estimated, the stability and accuracy of the adopted solution, optimization of the computational cost, real-time issues, and implementation on embedded hardware [1].

The proposed methods stem from Pontryagin's treatment of Hamiltonian systems, rather than utilization of classical or modern optimal estimation and control concepts applied to future naval vessels as depicted in (Figure 1) [2–4].

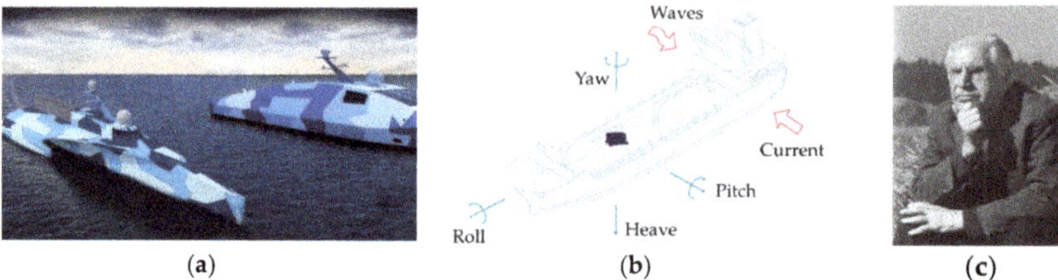

Figure 1. Representative motion measurement units for future ships depicted in (**a**) with measurement bases depicted in (**b**) are proposed to be augmented by virtual sensing by minimization of Hamiltonian systems by the principles of Pontryagin depicted in (**c**). Future unmanned U.S. Navy vessels [3] Medium Unmanned Surface Vessel (MUSV) concept renderings in (**a**) from shipbuilder Austal USA. Photo Credit: Austal USA. Boat motion monitoring [4] uses measurement bases depicted in (**b**) whose graphic is from cited reference modified by author. Photo (**c**) of Lev Pontryagin from the archive of the Steklov Mathematical Institute [2] used with permission (30 June 2021).

Typical motion reference units conveniently have accuracies on the order of 0.05 (in meters and degree for translation and rotation, respectively, as depicted in Figure 1b for representative naval vessels as depicted in Figure 1a). These figures of merit are aspirational for the virtual sensor that must provide accurate estimates whether active measurements are available to augment the algorithm. Lacking active measurements, the algorithm is merely an inertial navigation unit, while with active measurements, the algorithm becomes an augmented virtual sensor. This manuscript investigates virtual sensing by evaluating several options for algorithms, resulting estimated magnitudes, accuracy of each solution, optimization of resulting costs of motion, and sensitivity to variations like noise and parameter uncertainty of the translational and rotational motion models investigated (both simplified and high-fidelity). Algorithms are compared using various decision criteria to compare approaches for consideration of usage as motion reference units potentially aided by global navigation systems.

Noting the small size of motion measurement units, simple algorithms are preferred to minimize computational burdens that can increase unit size. Motion estimation and control algorithms to be augmented by sensor measurements are based on well-known mathematical models of translation and rotation from physics, both presented in equations. In 1834, the Royal Society of London published two celebrated papers by William R. Hamilton on Dynamics in the Philosophical Transactions. Ref. [5] The notions were slowly adopted, and not presented relative to other thoughts of the age for nearly seventy years [6], but quickly afterwards, the now-accepted axioms of translational and rotational motion were self-evidently accepted by the turn of the twentieth century [7–10] as ubiquitous concepts. Half a century later [11,12], standard university textbooks elaborate on the notions to the broad scientific community. Unfortunately, the notions arose in an environment already replete with notions of motion estimation and control based on classical proportional, rate, and integral feedback, so the fuller utilization of the first principals languished until exploitation by Russian mathematician Pontryagin [13]. Pontryagin proposed to utilize the first principles as the basis for treating motion estimation and control as the classical mathematical feedback notions were solidifying in the scientific community. Decades later, the first-principal utilization proposed by Pontryagin are currently rising in prominence as an improvement to classical methods [14]. After establishing performance benchmarks [15] for motion estimation and control of unmanned underwater vehicles, the burgeoning field of deterministic artificial intelligence [16,17] articulates the assertion of the first-principles as "self-awareness statements" with adaption [18,19] or optimal learning [20] used to achieved motion estimation and control commands. The key difference between the usage of first principals presented here follows. Classical methods impose the form of the estimation and control (typically negative feedback with gains) and they have very recently been applied to railway vehicles [21], biomechanical applications [22], and remotely operated undersea vehicles [23], electrical vehicles [24], and even residential heating energy consumption [25] and multiple access channel usage by wireless sensor networks [26]. Deterministic artificial intelligence uses first principals and optimization for all quantities but asserts a desired trajectory. Meanwhile the proposed methods in this manuscript leave the trajectory "free" and calculate an optimal state and rate trajectory for fusion with sensor data and calculates optimal decision criteria for estimation and controls in the same formulation.

This manuscript seeks to use the same notion, assertion of the first principals (via Pontryagin's formulation of Hamiltonian systems) in the context of inertial motion estimation fused with sensor measurements (that are presumed to be noisy). Noise in sensors is a serious issue elaborated by Oliveiera et al. [27] for background noise of acoustic sensors and by Zhang et al. [28] for accuracy of pulse ranging measurement in underwater multi-path environments. Barker et al. [29] evaluated impacts on doppler radar measurements beneath moving ice. Thomas et al. [30] proposes a unified guidance and control framework for Autonomous Underwater Vehicles (AUVs) based on the task priority control approach, incorporating various behaviors such as path following, terrain following, obstacle avoidance, as well as homing and docking to stationary and moving stations. Zhao et al. [31] very recently pursued the presently ubiquitous pursuit of optimality via stochastic artificial intelligence using particle swarm optimization genetic algorithm, while Anderlini et al. [32] used real-time reinforcement learning. Sensing the ocean environment parallels the current emphasis in motion sensing, e.g., Davidson et al.'s [33] parametric resonance technique for wave sensing and Sirigu et al.'s [34] wave optimization via the stochastic genetic algorithm. Motion control similarly mimics the efforts of motion sensing and ocean environment sensing, e.g., Veremey's [35] marine vessel tracking control, Volkova et al.'s [36] trajectory prediction using neural networks, and the new guidance algorithm for surface ship path following proposed by Zhang et al. [37]. Virtual sensory will be utilized in this manuscript where noisy state and rate sensors are combined to provide smooth, non-noisy, accurate estimates of state, rate, and acceleration, while no acceleration sensors were utilized. A quadratic cost was formulated for acceleration, since accelerations are directly tied to forces and torques and therefore fuels.

> " ... *condition of the physical world can either be "directly" observed (by a physical sensor) or indirectly derived by fusing data from one or more physical sensors, i.e., applying virtual sensors".* [38]

Thus, the broad context of the field is deeply immersed in a provenance of classical feedback driving a current emphasis on optimization by stochastic methods. Meanwhile this study will iterate options utilizing analytic optimization including evaluation of the impacts of variations and random noise in establishing the efficacy of each proposed approach. Analytical predictions are made of the impacts of applied noise power, and Monte Carlo analysis agrees with the analytical predictions. Developments presented in this manuscript follow the comparative prescription presented in [39], comparing many (eleven) optional approaches permitting the reader to discern their own preferred approach to fusion of sensor data with inertial motion estimation:

1. Validation of simple yet optimal inertial motion algorithms for both translation and rotation derived from Pontryagin's treatment of Hamiltonian systems when fused with sensor data that is assumed to be noisy.
2. Validation of high-fidelity optimal (nonlinear, coupled) internal motion algorithms for rotation with translation asserted by logical extension derived from Pontryagin's treatment of Hamiltonian systems when fused with sensor data that is assumed to be noisy;
3. Validation of three approaches for sensor data fused with the proposed motion estimation algorithm (not using classical feedback in a typical control topology): *pinv*, *backslash*, and *LU inverses* derived from Pontryagin's treatment of Hamiltonian systems when fused with sensor data that is assumed to be noisy;
4. Comparison of each proposed fused implementation algorithm to three varieties of classical feedback motion architectures including linear-quadratic optimal tracking regulators, classical proportional plus velocity feedback tuned for performance spec-

ification and manually tuned proportional plus integral plus derivative feedback topologies, where these classical methods are utilized as benchmarks for performance comparisons when fused with sensor data that is assumed to be noisy.

5. Comparisons are made based on motion state and velocity errors, algorithm parameter estimation errors, and quadratic cost functions, which map to fuel used to create translational and rotational motion.

6. Vulnerability to variation is evaluated using ten-thousand Monte Carlo simulations varying state and rate sensor noise power and algorithm plant model variations, where noise power is tailored to the simulation discretization, permitting analytic prediction of the impacts of variations to be compared to the simulations provided.

7. Sinusoidal wave action is programmed in the same simulation code to permit future research, and inclusion of such is indicated throughout the manuscript.

Appendix A, Table A1 contains a consolidated list of variables and acronyms in the manuscript.

2. Materials and Methods

Inertial navigation algorithms use physics-based mathematics to make predictions of motion states (position, rate, acceleration, and sometimes jerk). The approach taken here is to utilize the mathematical relationships from physics in a feedforward sense to produce optimal, nonlinear estimates of states that when compared to noisy sensor measurements yield corrected real-time optimal, smooth, and accurate estimates of state, rate, and acceleration. Sensors are modeled as ideal with added Gaussian noise and the smooth estimates will be seen to exhibit none of the noise. The optimization of the estimates will be derived using Pontryagin's optimization.

Motion control algorithms to be augmented by sensor measurements are based on well-known mathematical models of translation and rotation from physics, both presented in Equation (1), where both high-fidelity motion models are often simplified to identical double-integrator models where nonlinear coupling cross-products of motion are simplified, linearized, or omitted by assumption. The topologies are provided in Figure 2. Centrifugal acceleration is represented in Equation (1) by $-m\omega \times (\omega \times r')$. Coriolis acceleration is represented in Equation (1) by $-2m\omega \times v'$. Euler acceleration is represented in Equation (1) by $m\dot{\omega} \times r'$. In this section, double-integrator models are optimized by Pontryagin's treatment of Hamiltonian systems, where the complete (not simplified, linearized, or omitted) nonlinear cross-products of motion are accounted for using feedback decoupling. Efficacy of feedback decoupling of the full equations of motion is validated by disengaging this feature in a single simulation run to reveal the deleterious effects of the coupled motion when not counteracted by the decoupling approach.

$$\tau = I\dot{\omega} + \underbrace{\omega \times I\omega}_{\substack{\text{rotation due} \\ \text{to rotating} \\ \text{reference}}} \quad \leftrightarrow \quad F = ma' + \underbrace{m\dot{\omega} \times r' - 2m\omega \times v' - m\omega \times (\omega \times r')}_{\text{translation due to rotating reference}} \tag{1}$$

where

F, τ external force and torque, respectively
m, I mass and mass moment of inertia, respectively
$\omega, \dot{\omega}$ angular velocity and acceleration, respectively
r', v', a' position and velocity, and acceleration relative to rotating reference
$\tau = I\dot{\omega}$ and $F = m\,a'$ are double integrator plants
$\omega \times I\omega$ cross-product rotational motion due to rotating reference frame
$m\dot{\omega} \times r'$ cross-product translation motion due to rotating reference frame
$-2m\omega \times v'$ cross-product translation motion due to rotating reference frame
$-m\omega \times (\omega \times r')$ cross-product translation motion due to rotating reference frame.

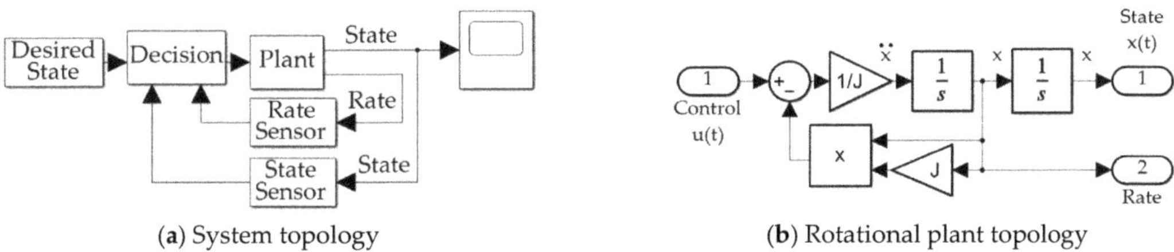

(**a**) System topology (**b**) Rotational plant topology

Figure 2. SIMULINK simulation program topologies used to generate the results in Section 3: (**a**) Overall system topology used to simultaneously produce state and rate estimates integrated with noisy sensors and additionally optimal control calculations; (**b**) Euler's moment from Equation (1) elaborated in [5–12] describing rotational motion (notice the nonlinear coupled motion).

2.1. Problem Scaling and Balancing

Consider problems whose solution must simultaneously perform mathematical operations on very large numbers and very small numbers. Such problems are referred to as poorly conditioned. Scaling and balancing problems are one potential mitigation where equations may be transformed to operate with similarly ordered numbers by scaling the variables to nominally reside between zero and unity. Scaling problems by common, well-known values permits single developments to be broadly applied to a wide range of state spaces not initially intended. Consider problems simultaneously involving very large and very small values of time ($\bar{t}$), mass ($\bar{m}$)/mass moments of inertia ($\bar{I}$), and/or length ($\bar{r}$). Normalizing by a known value permit variable transformation such that newly defined variables are of similar order, e.g., $t \equiv \frac{\bar{t}}{t_f}$, $I \equiv \frac{\bar{I}}{I_{system}} = J \equiv \frac{\bar{J}}{J_{system}}$, $m \equiv \frac{\bar{m}}{m_{system}}$, $r \equiv \frac{\bar{r}}{r'}$ where r indicates generic displacement units like $x, y,$ or $angle$. Such scaling permits problem solution with a transformed variable mass and inertia of unity value, initial time of zero and final time of unity, and state and rate variables that range from zero to unity making the developments here broadly applicable to any system of particular parameterization.

2.2. Scaled Problem Formulation

The problem is formulated in terms of standard form described in Equations (2)–(8), where $x(\cdot), v(\cdot)$ are the decision variables. The endpoint cost $E\left(x\left(t_f\right)\right)$ is also referred to as the Mayer cost. The running cost $F(x(t), u(t))$ is also referred to as the Lagrange cost (usually with the integral). The standard cost function $J[x(\cdot),\ u(\cdot)]$ is also referred to as the Bolza cost as the sum of the Mayer cost and Lagrange cost. Endpoint constraints $e\left(x\left(t_f\right)\right)$ are equations that are selected to be zero when the endpoint is unity.

$$x^T = [x, v], \quad u = [u] \tag{2}$$

$$\text{Minimize} \quad J[x(\cdot),\ v(\cdot),\ u(\cdot)] = E\left(x\left(t_f\right)\right) + \int_0^{t_f} F(x(t), u(t))dt = \frac{1}{2}\int_0^{t_f} u^2 dt \tag{3}$$

$$\text{Subject to} \quad \dot{x}_1 = f_1(x(t), u(t)) = v \tag{4}$$

$$\dot{x}_2 = f_2(x(t), u(t)) = \dot{v} = u \tag{5}$$

$$(x_0, v_0) = (0, 0) \tag{6}$$

$$\left(x_{f-1}, v_f, t_{f-1}\right) = (0, 0, 0) \tag{7}$$

$$e\left(x\left(t_f\right)\right) = 0 \tag{8}$$

where

$J[x(\cdot),\ u(\cdot)]$ cost function
$x^T = [x, v]$ state vector of motion state x and rate v with initial condition

$$x^T = [x, v] \, (x_0, v_0) \text{ and final conditions } \left(x_{f-1}, v_f, t_{f-1}\right) = (0, 0, 0)$$

$u = [u]$ decision vector

H Hamiltonian operator corresponding to system total energy

λ^T adjoint operators, also called co-states (corresponding to each state)

v^T endpoint costates

$e\left(x\left(t_f\right)\right)$ endpoint constraints.

2.3. Hamiltonian System: Minimization

The Hamiltonian in Equation (8) is a function of the state, co-state, and decision criteria (or control) and allows linkage of the running costs $F(x, u)$ with a linear measure of the behavior of the system dynamics $f(x, u)$. Equation (9) articulates the Hamiltonian of the problem formulation described in Equations (2)–(5). Minimizing the Hamiltonian with respect to the decision criteria vector per Equation (10) leads to conditions that must be true if the cost function is minimized while simultaneously satisfying the constraining dynamics. Equation (11) reveals the optimal decision u will be known if the rate adjoint can be discerned.

$$H = F(x, u) + \lambda^T f(x, u) \tag{9}$$

$$H = \frac{1}{2}u^2 + \lambda_x v + \lambda_v u \tag{10}$$

$$\frac{\partial H}{\partial u} = 0 \rightarrow u + \lambda_v = 0 \tag{11}$$

2.4. Hamiltonian System: Adjoint Gradient Equations

The change of the Hamiltonian with respect to the adjoint λ maps to the time-evolution of the corresponding state in accordance with Equations (12) and (13).

$$\dot{\lambda}_x = -\frac{\partial H}{\partial x} = 0 \rightarrow \lambda_x(t) = a \tag{12}$$

$$\dot{\lambda}_v = -\frac{\partial H}{\partial v} = \lambda_x \rightarrow \dot{\lambda}_v = \lambda_x(t) = a \rightarrow \lambda_v(t) = -at - b \tag{13}$$

The rate adjoint was discovered to reveal the optimal decision criteria, and the adjoint equations reveal the rate adjoint is time-parameterized with two unknown constants still to be sought. Together, Equations (11)–(13) form a system of differential equations to be solved with boundary conditions (often referred to as a two-point boundary value problem in mathematics).

2.5. Terminal Transversality of the Enpoint Lagrangian

The endpoint Lagrangian $\overline{E}$ in Equation (14) adjoins the endpoint function endpoint cost $E\left(x\left(t_f\right)\right)$ and the endpoint constraints functions $e\left(x\left(t_f\right)\right)$ in Equation (8) and provides a linear measure for endpoint conditions in Equation (7). The endpoint Lagrangian $\overline{E}$ exists at the terminal (final) time alone. The transversality condition in Equation (15) specifies the adjoint at the final time is perpendicular to the cost at the end point. In this problem, the endpoint cost $E\left(x\left(t_f\right)\right) = 0$. These Equations (16) and (17) are often useful when seeking a sufficient number of equations to solve the system.

$$\overline{E} = E + v^T e = v^T e = v_x \left(x_f - 1\right) + v_v \left(v_f - 0\right) = v_x \left(x_f - 1\right) + v_v v_f \tag{14}$$

$$\frac{\partial \overline{E}}{\partial x_f} = \lambda_x \left(t_f\right) \tag{15}$$

$$\frac{\partial \overline{E}}{\partial x_f} = \lambda_x\left(t_f\right) = v_x \tag{16}$$

$$\frac{\partial \overline{E}}{\partial v_f} = \lambda_v\left(t_f\right) = v_v \tag{17}$$

2.6. New Two-Point Boundary Value Problem

For the two-state system, four equations are required with four known conditions to evaluate the equations. In this instance, two Equations (3)–(10) have been formulated for state dynamics, two more Equations (18) and (19) have been formulated for the adjoints, and two more Equations (20) and (21) have been formulated for the adjoint endpoint conditions. Four known conditions, Equations (22)–(25) have also been formulated. Combining Equations (11) and (13) produce Equation (26).

$$\dot{x} = v \tag{18}$$

$$\dot{v} = u \tag{19}$$

$$\dot{\lambda}_x = 0 \tag{20}$$

$$\dot{\lambda}_v = -\lambda_x \tag{21}$$

$$x(0) = 0 \tag{22}$$

$$v(0) = 0 \tag{23}$$

$$x(1) = 1 \tag{24}$$

$$v(1) = 0 \tag{25}$$

Evaluating Equation (27) with Equation (23) produces the value $c = 0$. Evaluating Equation (28) with Equation (22) produces the value $d = 0$. Evaluating Equation (27) with Equation (25) produces Equation (29), while evaluating Equation (28) with Equation (24) produces Equation (30).

$$\dot{v} = -\lambda_v(t) = at + b \tag{26}$$

$$v = \int \dot{v}dt = \frac{1}{2}at^2 + bt + c \tag{27}$$

$$x = \int vdt = \frac{1}{6}at^3 + \frac{1}{2}bt^2 + ct + d \tag{28}$$

$$v(1) = \frac{1}{2}a + b = 0 \tag{29}$$

$$x(1) = \frac{1}{6}a + \frac{1}{2}b = 1 \tag{30}$$

Solving the system of two Equations (29) and (30) produces $a = -12$ and $b = 6$. Substituting Equation (26) into Equation (11) with a and b produces Equation (31), and substitution of a and b into Equations (27) and (28), respectively, produce Equations (32) and (33) the solution of the trajectory optimization problem.

$$u^*(t) = -12t + 6 \tag{31}$$

$$v^*(t) = -3t^2 + 6t \tag{32}$$

$$x^*(t) = -2t^3 + 3t^2 \tag{33}$$

Equations (31)–(33) constitute the optimal solution for quiescent initial conditions and the state final conditions (zero velocity and unity scaled position). To implement a *form of feedback* (not classical feedback), consider leaving the initial conditions non-specific in variable-form as described next.

2.7. Real-Time Feedback Update of Boundary Value Problem Optimum Solutions

Classical methods utilize feedback of asserted form $u = -Kx$ for state variable x, where the decision criteria (for control or state estimation/observer) and gains K are solved to achieve some stated performance criteria. Such methods are used in Section 3 and their results are established as benchmarks for comparison. So-called modern methods utilize optimization problem formulation to eliminate classical gain tuning substituting optimal gain selection but retaining the asserted form of the decision criteria. Such methods are often referred to as "linear-quadratic optimal" estimators or controllers. These estimators are also presented as benchmarks for comparison, where the optimization problem equally weights state errors and estimation accuracy.

Alternative use of feedback is proposed here (whose simulation is depicted in Figure 3b). Rather than classical feedback topologies asserting $u = -Kx$ utilization of state feedback in formulating the estimator or control's decision criteria, this section proposes re-labeling the current state feedback as the new initial conditions of the two-point boundary-value problems used to solve for optimal state estimates or control decision criteria in Equations (22) and (23). The solution of (26)–(28) using the initial values of (22) and (23) manifest in values of the integration constants: $a = -12$ and $b = 6$. As done in real-time optimal control, the values of the integration constants are left "free" in variable form, and their values are newly established for each discrete instance of state feedback (re-labeled as new initial conditions). This notion is proposed in Proposition 1, whose proof expresses the form of the online calculated integration constants that solve the new optimization problem. The two constants $\hat{a}$ and $\hat{b}$ are utilized in the same decision Equation (31) where the estimates replace the formerly solved values of the boundary value problem resulting in Equation (40).

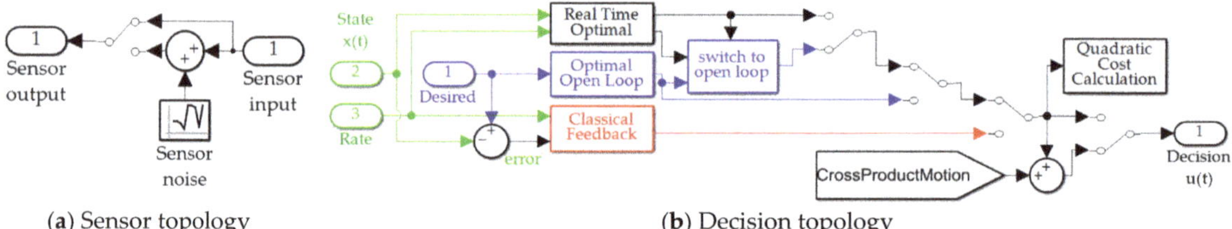

(**a**) Sensor topology   (**b**) Decision topology

Figure 3. Simulink systems for noisy sensors and decision criteria (guidance or control) subsystem: (**a**) noisy sensor subsystem; (**b**) decision topology (guidance or control).

Proposition 1. *Feedback may be utilized not in closed form to solve the constrained optimization problem in real time.*

$$x = \frac{1}{6}at^3 + \frac{1}{2}bt^2 \tag{34}$$

$$v = \frac{1}{2}at^2 + bt \tag{35}$$

$$x_f = \frac{1}{6}at_f^3 + \frac{1}{2}bt_f^2 \tag{36}$$

$$v_f = \frac{1}{2}at_f^2 + bt_f \tag{37}$$

Proof of Proposition 1. Implementing Equations (34)–(37) in matrix form as revealed in Equation (38) permits solution for the unknown constants as a function of time as displayed in Equation (39), and subsequent use of the unknown constants form the new optimal solution from the current position and velocity per Equation (40).

$$\begin{bmatrix} \frac{t_0^3}{6} & \frac{t_0^2}{2} & t_0 & 1 \\ \frac{t_0^2}{2} & t_0 & 1 & 0 \\ \frac{1}{6} & \frac{1}{2} & 1 & 1 \\ \frac{1}{2} & 1 & 1 & 0 \end{bmatrix} \begin{Bmatrix} a \\ b \\ c \\ d \end{Bmatrix} = \begin{Bmatrix} x_0 \\ v_0 \\ 1 \\ 0 \end{Bmatrix} \tag{38}$$
$$\underbrace{\phantom{\begin{bmatrix} \frac{t_0^3}{6} \end{bmatrix}}}_{T} \quad \underbrace{\phantom{\begin{Bmatrix} a \end{Bmatrix}}}_{p} \quad \underbrace{\phantom{\begin{Bmatrix} x_0 \end{Bmatrix}}}_{q}$$

$$\begin{Bmatrix} \hat{a} \\ \hat{b} \\ \hat{c} \\ \hat{d} \end{Bmatrix} = \begin{bmatrix} \frac{t_0^3}{6} & \frac{t_0^2}{2} & t_0 & 1 \\ \frac{t_0^2}{2} & t_0 & 1 & 0 \\ \frac{1}{6} & \frac{1}{2} & 1 & 1 \\ \frac{1}{2} & 1 & 1 & 0 \end{bmatrix}^{-1} \begin{Bmatrix} x_0 \\ v_0 \\ 1 \\ 0 \end{Bmatrix} \tag{39}$$

$$u^* \equiv \hat{a}t + \hat{b} \tag{40}$$

In Section 3, estimation of $\hat{a}$ and $\hat{b}$ becomes singular due to the inversion in Equation (39) as approaching the terminal endpoint, where switching to Equations (31)–(33) is implemented as depicted in Figure 4d to avoid the deleterious effects of singularity when applying Proposition 1. The cases with switching at singular conditions are suffixes with "*with switching*" in the respective label.

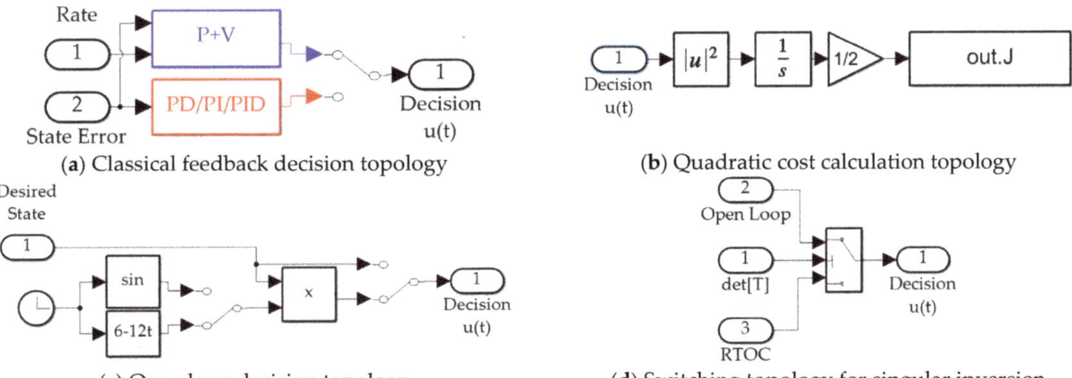

Figure 4. SIMULINK subsystems: (**a**) implementation of classical feedback methods; (**b**) calculation of quadratic cost; (**c**) open loop decision topologies (sinusoidal and optimal); (**d**) switching function to disengage use of real-time parameter updates when the matrix in Equation (39) is rank deficient. Notice in subfigure (**c**) sinusoidal wave input is coded using the identical time-index of the rest of the simulation. The next stages of future research will utilize this identical simulation to investigate efficacy of the proposed virtual sensing amidst unknown wave actions.

2.8. Feedback Decoupling of Nonlinear, Coupled Motion Due to Cross Products

The real-time feedback update of boundary value problem optimum solutions is often used in the field of real-time optimal control, but a key unaddressed complication remains the nonlinear, coupling cross-products of motion due to rotating reference frames. Here, a feedback decoupling scheme is introduced, allowing the full nonlinear problem to be addressed by the identical scaled problem solution presented, and such is done without simplification, linearization, or reduction by assumption. In proposition 2, feedback decoupling is proposed to augment the optimal solution already derived. The resulting modified decision criteria in Equation (42) is utilized in simulations presented in Section 3 of this manuscript, but a single case omitting Proposition 2 is presented to highlight the efficacy of the approach.

Proposition 2. *The real-time optimal guidance estimation and/or control solution may be extended from the double-integrator to the nonlinear, coupled kinetics by feedback decoupling as implemented in Equation (41).*

$$\tau = I\dot{\omega} + \underbrace{\omega \times I\omega}_{\substack{\text{rotation due} \\ \text{to rotating} \\ \text{reference}}} \tag{41}$$

Proof of Proposition 2. For nonlinear dynamics of translation or rotation as defined in Equation (1), where the double-integrator is augmented by cross-coupled motion due to rotating reference frames, the same augmentation may be added to the decision criteria in Equation (40) using feedback of the current motion states in accordance with Equation (42). The claim is numerically validated with simulations of "cross-product decoupling" that are nearly indistinguishable from open loop optimal solution, and a single case "without cross-product decoupling" is provided for comparison.

$$u^* \equiv \hat{a}t + \hat{a} + \omega \times I\omega \tag{42}$$

2.9. Analytical Prediction of Impacts of Variations

Assuming Euler discretization (used in the validating simulations) for output y, index i and integration solver timestep h Equation (43) would seem to indicate a linear noise output relationship. Equation (44) indicates the relationship for quiescent initial conditions indicating the results of a style draw. In a Monte Carlo sense (to be simulated) of a very large number n, Equation (45) indicates expectations from theory Equation (46) in simulation for scaled noise entry to the simulation to correctly reflect the noise power of the noisy sensors in the discretized computer simulation. Equation (46) was used to properly enter the sensor noise in the simulation (Figures 2a and 3a).

$$\dot{y}(t) = \frac{y_{i+1} - y_i}{h} = n_i \to y_{i+1} = y_i + hn_i \tag{43}$$

$$y_1 = \underbrace{y_0}_{0} + hn_0 \tag{44}$$

$$\frac{1}{N}\sum_{i=1}^{N} y_{1i}^2 = \sigma_y^2 \to \frac{1}{N}\sum_{i=1}^{N}(hn_{o,i})^2 = h^2\frac{1}{N}\sum_{i=1}^{N} n_{o,i}^2 \to \sigma_y^2 = h^2\sigma_n^2 \tag{45}$$

$$\text{let } \sigma_{sim}^2 = \frac{\sigma_n^2}{h} \to \sigma_y^2 = h^2\sigma_{sim}^2 = h^2\frac{\sigma_n^2}{h} = h\sigma_n^2 \to \sigma_{sim}^2 = \frac{\sigma_n^2}{h} \tag{46}$$

Assuming this implementation of noise power for a given Euler (ode1) discretization in SIMULINK, $1 - \sigma$ error ellipse may be calculated as Equation (47) for the system in canonical form in accordance with [40] and was implemented in Figure 3a and depicted on "scatter plots" in Section 3's presentation of results of over ten-thousand Monte Carlo simulations.

$$\sigma_{n_{state}} = \sqrt{\frac{\omega_n^2 + 4\zeta^2}{4\zeta\omega_n}} \quad \sigma_{n_{rate}} = \sqrt{\frac{\omega_n^3 + 4\zeta^2\omega_n}{4\zeta}} \tag{47}$$

2.10. Numerical Simulation in MATLAB/SIMULINK

Validating simulations were performed in MATLAB/SIMULINK Release R2021a with Euler integration solver (ode1) and a fixed time step of 0.01 s, whose results are presented in Section 3, while this subsection displays the SIMULINK models permitting the reader to duplicate the results presented here. Sensor noise was added per Section 2.8. The classical feedback subsystem is displayed in Figure 4a. The optimal open loop subsystem implements Equation (31), and is elaborated in Figure 4b,c. The real time optimal subsystem implements Equations (42) and (31) augmented by feedback decoupling as in Equation (42). The "switch to open loop" subsystem switches when the matrix inverted in Equation (39) is singular indicated by a zero valued determinant and is elaborated in Figure 4d. The quadratic cost calculation computes Equation (3) and is elaborated in Figure 4b, while the cross-product motion feedback implements the cross product of Equation (42). The P + V subsystem and PD/PI/PID subsystems depicted in Figure 4a implement classical methods not re-derived here, but whose computer code is presented in Appendix B, Algorithms A1 and A2.

Figure 5 displays the SIMULINK subsystems used to implement the three instantiations of real-time optimization (labeled RTOC from provenance in optimal control) where the switching displayed in Figure 3b permits identical simulation experiments to be performed with all conditions fixed, varying only the proposed implementation. The subsystems execute Equation (39) with three variations of matrix inversion: (1) MATLAB's backslash "\", (2) Moore-Penrose pseudoinverse (*pinv*), (3) *LU-inverse*.

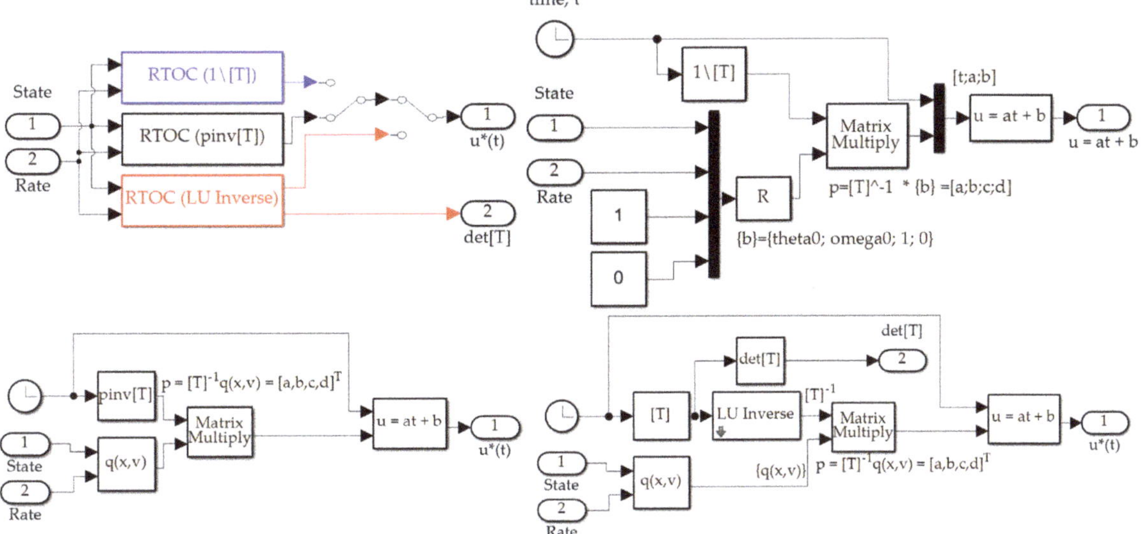

Figure 5. SIMULINK subsystems; (**a**) real-time optimal decision switching topology; (**b**) real-time optimal calculation of Equation (39) using *backslash\inverse*; (**c**) real-time optimal calculation of Equation (39) using *pinv inverse*; (**d**) real-time optimal calculation of Equation (39) using *LU inverse*.

Section 2.10 presented SIMULINK subsystems used to implement the equations derived in the section. Table 1 displays the software configuration used to simulate the equations leading to the results presented immediately afterwards in Section 3.

Table 1. Software configuration for simulations reported in Section 3.

Software Version	Integration Solver	Step-Size
MATLAB R2021a	Euler (ode1)	0.01 secs

3. Results

Section 2 derived several options for estimating state, rate, and control simultaneously as outputs of Pontryagin's treatment of the problem formulated as Hamiltonian systems. Section 2.9 described implementation of sensor noise narratively, while Figure 3 illustrated the topological elaboration using SIMULINK including state and rate sensor with added Gaussian noise whose noise power was set in accordance with Section 2.9. SIMULINK subsystems were presented to aid repeatability (with callback codes in the Appendix B). Those subsystems were used to run more than ten-thousand simulations: a nominal simulation run for each technique with the remainder utilized to evaluate vulnerability to variations as described in Section 2.9. In Section 3.1, benchmarks of performance are established using classical methods for state and rate errors and optimum cost calculated in Section 2. Sections 3.2–3.4 describe real-time optimal utilization of feedback to establish online estimates of the solution of the modified boundary value problem described in Section 2. Each section respectively evaluates the three methods compared: *backslash\inverse*, *pinv inverse*, and *LU inverse*.

General lessons from the results include:

1. Classical feedback estimation methods are very effective at achieving very low estimation errors, but at higher costs utilizing the estimates in the decision criteria (guidance or control).
2. *Backslash\inverse* is relatively inferior to all other inverse methods
3. Singular switching generally improves state and rate estimation and costs;
4. *LU inverse* and *pinv inverse* methods perform alike with disparate strengths and weaknesses relative to each other.
5. Choosing the *pinv inverse* method as the chosen recommendation, Monte Carlo analysis reveals the residual sensitive to parameter variation is indistinguishable from the inherent sensitivity of the optimal solution when using the singular switching technique. Meanwhile, substantial vulnerability to parameter variation is revealed when singular switching is not used.
6. Lastly, omitting the complicating cross-products in the problem results in an order of magnitude high estimation errors and several orders of magnitude higher parameter estimation error. Therefore, cross-product motion decoupling is strongly recommended for all instantiations of state and rate estimation.

3.1. Benchmark Classical Methods

Classical methods as presented in [41,42] with nonlinear decoupling loops as proposed in Equation (42) depicted in Figure 3b were implemented in SIMULINK according to Figure 4a. Computer code implementing these classical methods is presented in Appendix B, Algorithm 1. Estimation was executed by feedback of proportional plus integral plus derivative (PID), proportional plus derivative (PD), and also proportional plus velocity, and the results displayed in Figure 6, establishing the benchmark for state and rate tracking. Table 2 displays quantitative data corresponding to Figure 6's qualitative displays. Notice the optimal estimation of state, rate, and decision criteria is also included in Figure 6 and Table 2, since the optimal cost benchmark is established by Pontryagin's treatment in Equation (31).

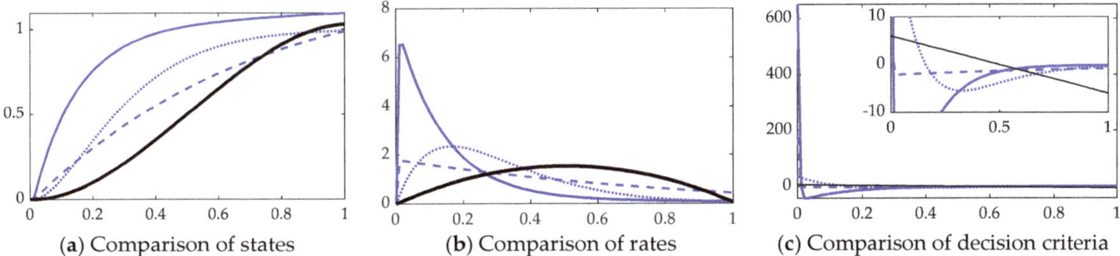

| (a) Comparison of states | (b) Comparison of rates | (c) Comparison of decision criteria |

Figure 6. Comparison of classical methods with scaled time on the abscissae and respective ordinates titled in the subplot captions: thinner solid blue line is manually tuned PID, dashed line is linear quadratic optimal PD, dotted line is proportional plus velocity tuned to performance specification, thicker solid black line is open loop optimal per Pontryagin (Equation (31) provided for anticipated comparison). (**a**) States, (**b**) rates, and (**c**) decision criteria). Notice the solid black line representing the optimal open loop solution in subfigure (**c**) is initially positive to create movement in the desired direction, and the control switches exactly at the halfway point to a negative input, slowing progress towards the final desired end states (position and rate states). The ranges of the zoomed view in the inset are indicated by the respective scales.

Table 2. Comparison of classical decision methods.

Decision Method	Final State Error	Final Rate Error	Decision Criteria/Control Effort
Classical feedback: proportional + integral + derivative (manually tuned)	0.0949	0.0850	2192
Classical feedback: proportional + derivative (linear quadratic optimal)	−0.0120	0.4341	152.7
Classical feedback: proportional + velocity (tuned to performance specs)	−0.0082	0.0555	30.82
Open loop optimal	0.0296	0.06	6

3.2. Real-Time Optimal Methods with Backslash

This section displays the results of real-time optimal estimation using the *backslash\inverse* depicted in Figure 5b with and without singular switching displayed in Figure 4d to inverse the $[T]$ in Equation (39). The results are compared to open loop optimal results per Equation (31) displayed in Figure 4c. Figure 7 reveals real-time optimal state estimation performs relatively poorly using the MATLAB *backslash\inverse*, but performance is restored to near-optimal performance when augmented with singular switching. State and rate errors are restored to essentially optimal values, while cost is restored to very near the optimal case as evidenced by the quantitative results displayed in Table 3.

Table 3. Comparison of real-time optimal decision methods using *backslash* matrix inversion.

Decision Method	Final State Error	Final Rate Error	Decision Criteria/Control Effort
Open loop optimal	0.0296	0.06	6
Real-time optimal using *backslash*	0.2849	2.0762	2.863
Real-time optimal using *backslash* with singular switching	0.0296	0.06	6.0012

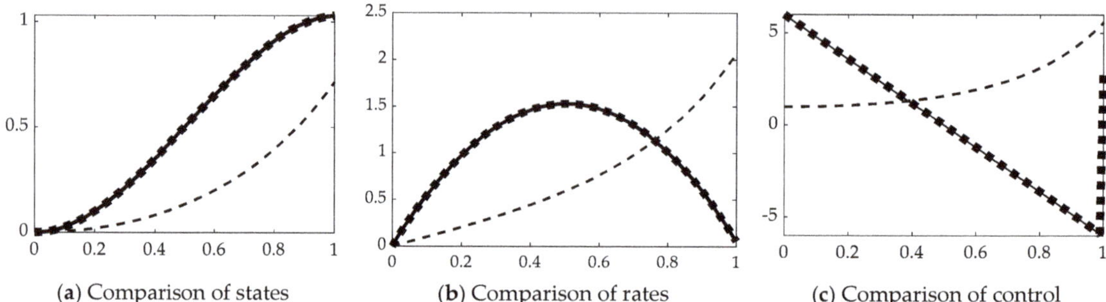

(**a**) Comparison of states

(**b**) Comparison of rates

(**c**) Comparison of control

Figure 7. Comparison of real-time optimal methods with scaled time on the abscissae and respective ordinates titled in the subplot captions: solid line is open loop optimal (benchmark), dashed line is real-time optimal using *backslash*, thick dotted line is real-time optimal using *backslash* with singular switching. (**a**) States, (**b**) rates. Notice the solid black line representing the optimal open loop solution in subfigure (**c**) is initially positive to create movement in the desired direction, and the control switches exactly at the halfway point to a negative input, slowing progress towards the final desired end states (position and rate states). The ranges of the zoomed view in the inset are indicated by the respective scales.

Figure 8 and Table 4 reveal the estimation performance of the constants of integration solving the modified two-point boundary value problem (BVP) using state and rate feedback to reset the initial conditions of the BVP. Oddly, despite relatively superior performance estimating the state and rates when using singular switching augmentation, parameter estimation is far inferior.

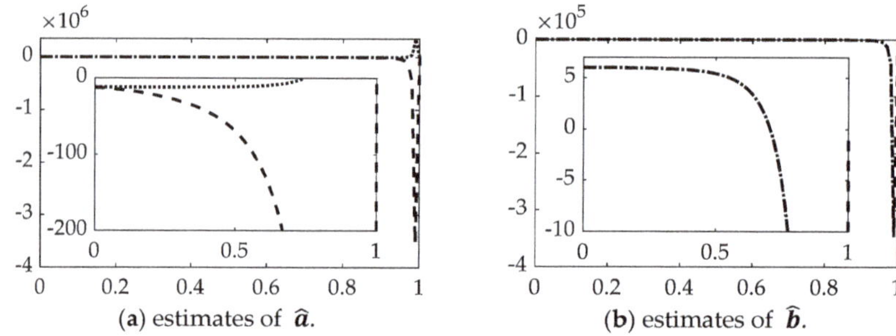

(**a**) estimates of $\hat{a}$.

(**b**) estimates of $\hat{b}$.

Figure 8. Comparison of real-time optimal methods with scaled time on the abscissae and respective ordinates titled in the subplot captions: dashed line is real-time optimal using *backslash*, dotted line is real-time optimal using *backslash* with singular switching. (**a**) Estimates of $\hat{a}$, (**b**) estimates of $\hat{b}$. The ranges of the zoomed view in the inset are indicated by the respective scales.

Table 4. Comparison of real-time optimal decision methods using *backslash* matrix inversion.

Decision Method	Mean $\hat{a}$ Error	Mean $\hat{b}$ Error
Open loop optimal	−12	6
Real-time optimal using *backslash*	21.2	−26.6
Real-time optimal using *backslash* with singular switching	4142	−4121

Section 3.1 presented the results of classical and optimal methods as benchmarks for performance. Meanwhile, Section 3.2 presented the results of implementing real-time optimal estimation with *backslash\inverse* with and without singular switching compared to the optimal benchmark. Next, Section 3.3 presents results using *pinv inverse* with and without singular switching.

3.3. Real-Time Optimal Methods with Pinv

Inversion of the $[T]$ matrix in Equation (39) was also accomplished by the Moore–Penrose pseudoinverse equation: $[T]^{-1} \cong [T]^{\dagger} \equiv \left([T]^{T}[T]\right)^{-1}[T]^{T}$. All other facets of the problem are left identical, while only the method of matrix inversion is modified resulting in state and rates estimates and comparison of control in Figure 9 with corresponding quantitative results in Table 5. Parameter estimation accuracy is displayed in Figure 10 and Table 6.

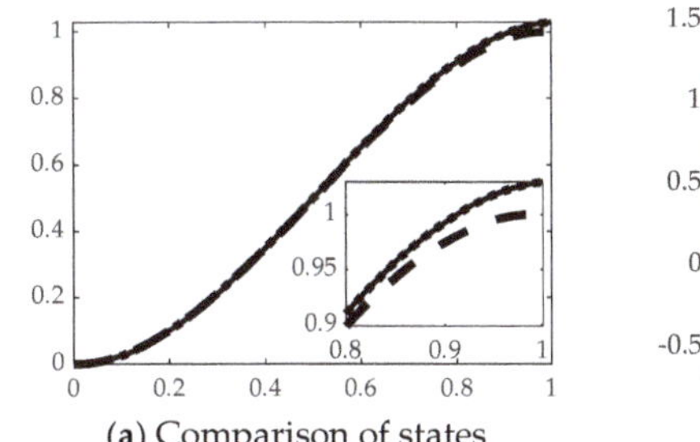

(a) Comparison of states

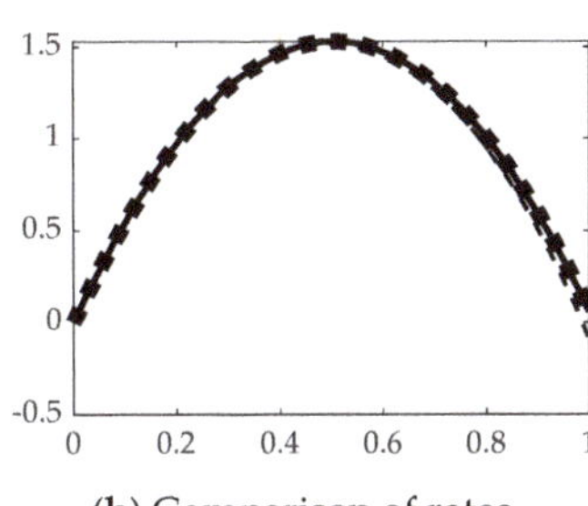

(b) Comparison of rates

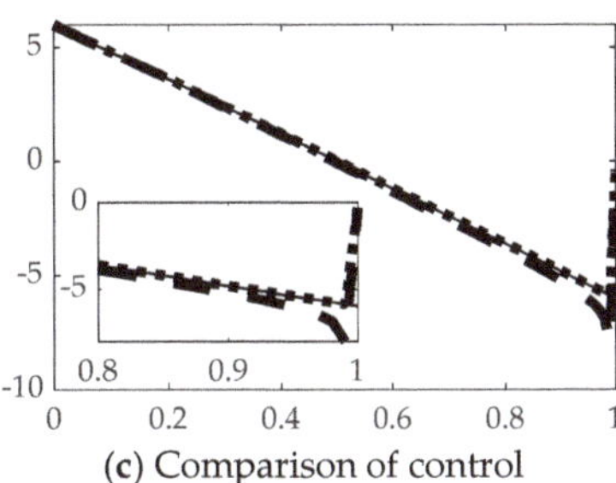

(c) Comparison of control

Figure 9. Comparison of real-time optimal methods with scaled time on the abscissae and respective ordinates titled in the subplot captions: solid line is open loop optimal (benchmark), dashed line is real-time optimal using backslash, thick dotted line is real-time optimal using *pinv* with singular switching. (a) States, (b) rates, and (c) decision criteria or control. The ranges of the zoomed view in the inset are indicated by the respective scales.

Table 5. Comparison of real-time optimal decision methods using *pinv* matrix inversion.

Decision Method	Final State Error	Final Rate Error	Decision Criteria/Control Effort
Open loop optimal	0.0296	0.060	6
Real-time optimal using *pinv*	0	−0.1088	6.6914
Real-time optimal using *pinv* with singular switching	0.0296	0.0600	6.0012
−*without* cross-product decoupling	0.3381	−0.5936	6.0012

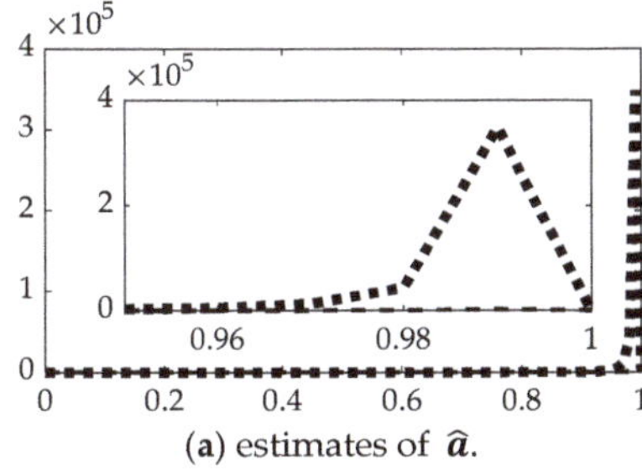

(a) estimates of $\hat{a}$.

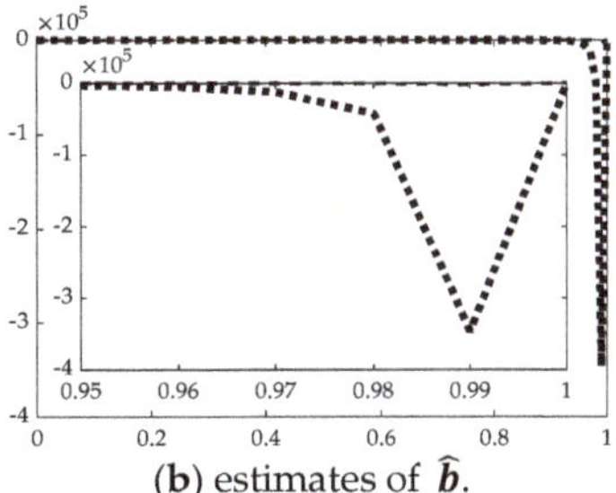

(b) estimates of $\hat{b}$.

Figure 10. Comparison of real-time optimal methods with scaled time on the abscissae and respective ordinates titled in the subplot captions: dashed line is real-time optimal using backslash, dotted line is real-time optimal using *pinv* with singular switching. (a) Estimates of $\hat{a}$, (b) estimates of $\hat{b}$. The ranges of the zoomed view in the inset are indicated by the respective scales.

Table 6. Comparison of real-time optimal decision methods using *p-inv* matrix inversion.

Decision Method	Mean $\hat{a}$ Error	Mean $\hat{b}$ Error
Open loop optimal	−12	6
Real-time optimal using *p-inv*	21	−26
Real-time optimal using *p-inv* with singular switching	4101	−4080

3.4. Real-Time Optimal Methods with Lu-Inverse

Inversion of the [T] matrix in Equation (39) was also accomplished by the *LU-inverse*, which first creates a pivoted version T_p and then inverts the product of a lower triangular matrix, [L] and an upper triangular matrix [U] based on [T]: $[T]^{-1} \cong [T_p]^{-1} \equiv ([L][U])^{-1}$. All other facets of the problem are left identical, while only the method of matrix inversion is modified resulting in state and rates estimates and comparison of control in Figure 11 with corresponding quantitative results in Table 7. Parameter estimation accuracy is displayed in Figure 12 and Table 8.

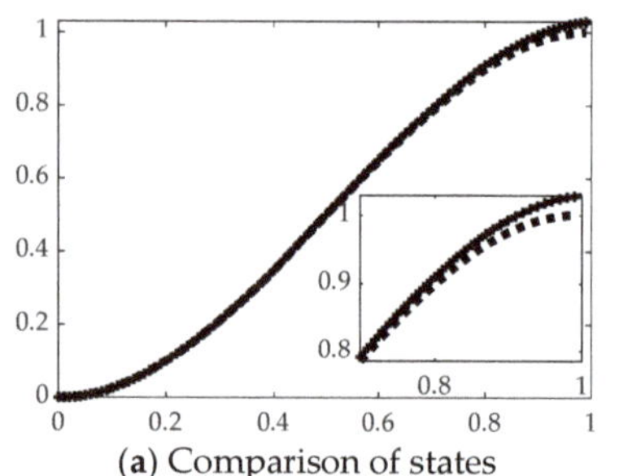

(a) Comparison of states

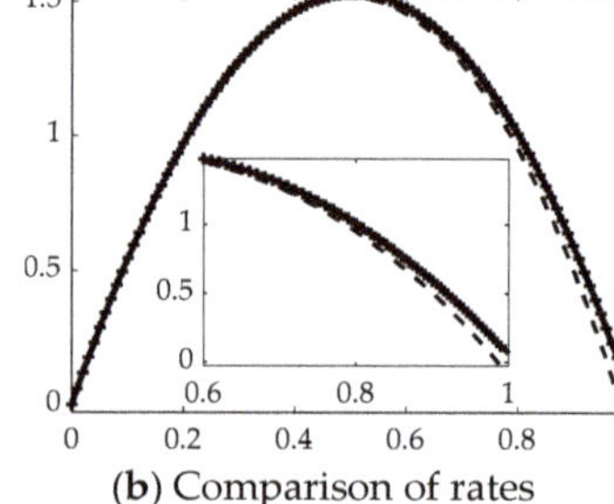

(b) Comparison of rates

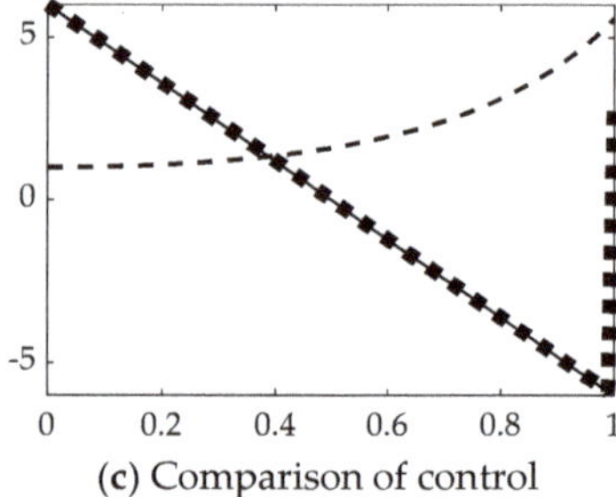

(c) Comparison of control

Figure 11. Comparison of real-time optimal methods with scaled time on the abscissae and respective ordinates titled in the subplot captions: solid line is open loop optimal (benchmark), dashed line is real-time optimal using backslash, thick dotted line is real-time optimal using *LU-inverse* with singular switching. (**a**) States, (**b**) rates, and (**c**) decision criteria or control. The ranges of the zoomed view in the inset are indicated by the respective scales.

Table 7. Comparison of real-time optimal decision methods using *LU-inverse* matrix inversion.

Decision Method	Final State Error	Final Rate Error	Decision Criteria/Control Effort
Open loop optimal	0.0296	0.060	6
Real-time optimal using *LU-inverse*	0.0030	−0.087	6.371
Real-time optimal using *LU-inverse* with singular switching	0.0284	0.1188	5.8283

Table 8. Comparison of real-time optimal decision methods using *LU-inverse* matrix inversion.

Decision Method	Mean $\hat{a}$ Error	Mean $\hat{b}$ Error
Open loop optimal	−12	6
Real-time optimal using *LU-inverse*	21	21
Real-time optimal using *LU-inverse* with singular switching	4142	4142

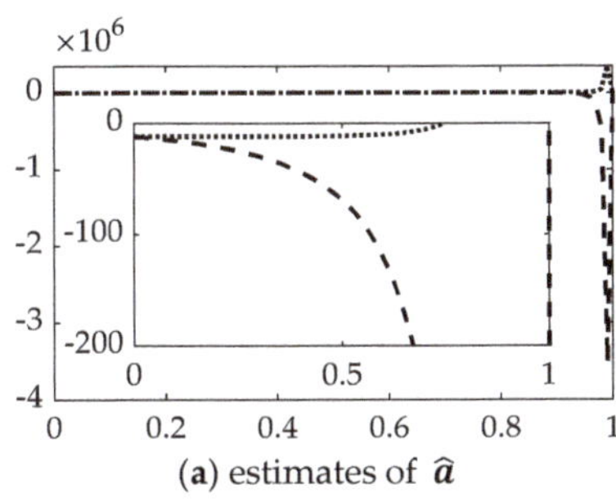
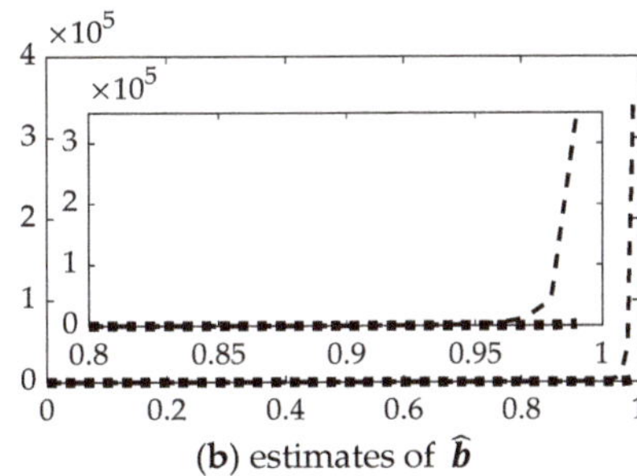

(a) estimates of $\widehat{a}$ (b) estimates of $\widehat{b}$

Figure 12. Comparison of real-time optimal methods with scaled time on the abscissae and respective ordinates titled in the subplot captions: dashed line is real-time optimal using backslash, dotted line is real-time optimal using *LU-inverse* with singular switching. (**a**) Estimates of $\hat{a}$, (**b**) estimates of $\hat{b}$. The ranges of the zoomed view in the inset are indicated by the respective scales.

3.5. Monte Carlo Analysis Using Pinv (with Singular Switching) and Open Loop Optimal with Cross-Product Deoupling

Over ten-thousand simulation runs were performed with 10% uniformly random variations in plant parameters (mass and mass moment of inertia). Noise was added to state and rate sensors with zero mean and standard deviation 0.01, and the results are displayed in the "scatter" plots in Figure 13 with corresponding quantitative results displayed in Table 9. Feedback implemented by resetting the initial condition of the reformulated boundary value problem (when implemented with singular switching) yielded optimal results when augmented with cross-product decoupling.

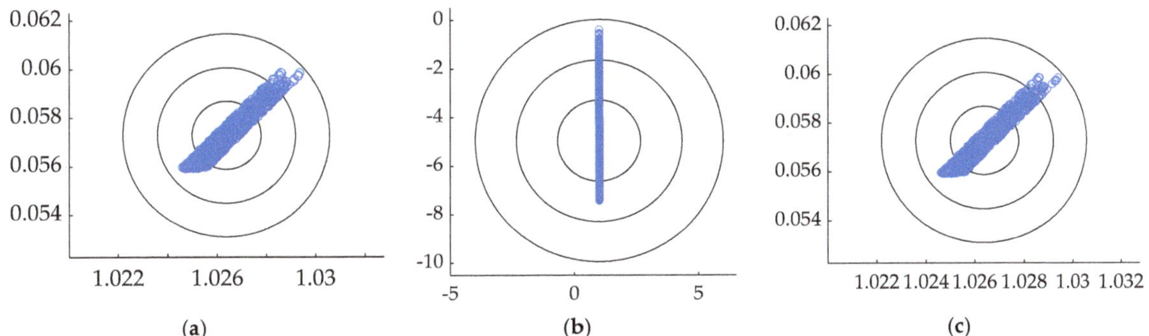

(a)   (b)   (c)

Figure 13. Comparison of the impacts of system variations on real-time optimal using (**a**) open loop optimal, (**b**) *pinv* without switching, (**c**) *pinv* with switching. Scaled state on the abscissae and scaled rate on the ordinates.

Table 9. Impact of variations with real-time optimal decision methods using *pinv* inversion.

Decision Method	Mean Final State Error	Mean Final RateError
Open loop optimal	0.0264	0.0573
Real-time optimal using *pinv*	0.0041	−4.960
Real-time optimal using *pinv* with singular switching	0.0264	0.0573

3.6. Comparison of Results

Section 3.1 presented the benchmark results produced by classical methods and open loop optimal mathematical solutions. Section 3.2 presented results utilizing MATLAB's *backslash\inversion*, while Section 3.3 included results using Moore–Penrose pseudoinverse, *pinv*. Section 3.4 presented results using *LU-inverse*. Section 3.5 revealed robustness to variations in plan parameters with both state and rate sensor noise. This section consolidates the results into a single table of raw data depicted in Table 10. This data will be used to produce percent performance improvement as figures of merit in the Discussion (Section 4).

Table 10. Comparison of real-time optimal decision methods.

Decision Method	Mean $\hat{a}$ Error	Mean $\hat{b}$ Error	Mean Final Position Error	Mean Final Rate Error	Decision Criteria Cost
Classical feedback: PID (manually tuned)	–	–	0.0949	0.0850	**2192**
Classical feedback: PD (linear quadratic optimal)	–	–	−0.0120	0.4341	152.7
Classical feedback: P+V (tuned to performance specs)	–	–	−0.0082	0.0555	30.82
Open loop optimal	0	0	0.0264	**0.0573**	6.000
Real-time optimal using *Backslash* (\) *inverse*	9.2	−20.6	0.2849	2.0762	**2.863**
Real-time optimal using *Backslash* (\) with switching	4130	−4115	0.0296	0.0600	6.0012
Real-time optimal using *LU-inverse*	9	15	0.0041	**−4.960**	6.371
Real-time optimal using *LU-inverse* with switching	4142	4136	0.0264	0.0573	5.8283
Real-time optimal using *pinv-inverse*	21	−20	**0.0000**	−0.1088	6.6914
Real-time optimal using *pinv-inverse* with switching	4130	−4074	0.0296	0.0600	6.0012
− *without* cross-product decoupling	−47,766	47,449	**0.3381**	−0.59357	6.0012

4. Discussion

State and rate estimation algorithms fused with noisy sensor measurements using several of the proposed methodologies achieve state-of-the art accuracies with optimality that is analytic and deterministic rather than stochastic, and therefore use very simple equations with necessarily low computational burdens. Simple relationships with small numbers of multiplications and additions are required to be comparable to the simplicity of classical methods, but optimum results are produced that exceed the modern notion of linear-quadratic optimal estimation. Implementation of non-standard feedback achieves robustness with the additional computational cost of a matrix inverse, and therefore three optional inversion methods were compared. General lessons taken with manually tuned PID as a benchmark for state and rate estimation errors, while optimal loop optimal cost is the benchmark for the cost of utilization of state estimations for guidance and control:

1. Classical feedback estimation methods (tuned per computer code is presented in Appendix B, Algorithm A1) are very effective at achieving very low estimation errors, but at higher costs utilizing the estimates in the decision criteria (guidance or control).

    a. Linear-quadratic optimal estimation achieved 87% better state estimates, but over 400% poorer rate estimates compared to classical PID with costs over 2000% open-loop optimal costs.

    b. Classical position plus velocity estimation achieved 90% improved state estimation with over 30% better rate estimation, but cost of implementation remains high (over 400% higher than the optimal benchmark).

2. Open loop optimal estimation established the mathematical benchmark for cost, and achieved 72% improved state estimation and 33% rate estimation errors.

3. *Backslash\inverse* is relatively inferior to all other inverse methods, producing 200% poorer state estimation and over 2000% poorer rate estimates with 52% reduced costs compared to the optimal benchmark;

4. Singular switching generally improves state and rate estimation and costs;

    a. Singular switching with *backslash\inverse* produced 69% improvement in state estimation and 29% improvement in rate estimation with roughly optimal costs.

    b. Singular switching with *LU-inverse* produced 72% improvement in state estimation and 33% improvement in rate estimation with roughly optimal costs (3% better than optimal . . . a numerical curiosity).

    c. Singular switching with *pinv inverse* produced 69% improvement in state estimation and 29% improvement in rate estimation with roughly optimal costs (approximately identical improvement percentages to *LU-inverse* with singular switching).

5. *LU inverse* and *pinv inverse* methods perform alike with disparate strengths and weaknesses relative to each other.
6. Choosing the *pinv inverse* method as the chosen recommendation, Monte Carlo analysis reveals the residual sensitive to parameter variation is indistinguishable from the inherent sensitivity of the optimal solution when using the singular switching technique. Meanwhile, substantial vulnerability to parameter variation is revealed when singular switching is not used.
7. Lastly, omitting the complicating cross-products in the problem results in an order of magnitude higher estimation errors and several orders of magnitude higher parameter estimation error. Therefore, cross-product motion decoupling is strongly recommended for all instantiations of state and rate estimation.

4.1. Notes on Percentages of Performance Improvements

The choice of benchmarks for establishing percentage performance improvements leads to seemingly exaggerated numbers. The current selection of benchmarks emphasizes the strengths of the respective methods: classical feedback estimation methods are designable to achieve high accuracy but suffer from high effort by the decision criteria associated with their use. Optimal methods as instantiated here emphasize minimization of decision effort, so the benchmark for control effort is selected as optimal open loop rather than classical feedback (e.g., manually tuned PID). Percent degradation over thirty thousand percent results when compared to the optimal value (of six) as a benchmark. If the calculation had instead used the manually tuned classical PID as a benchmark, the optimal effort would exhibit an improvement over ninety-nine percent.

The final line in Table 11 illustrates the extreme penalty of not using feedback decoupling of the vector cross-products in Equation (1) representing translation due to the rotating reference. The penalty embodies the deleterious effects of neglecting treatment of the nonlinear, coupled, full six-degree-of-freedom system of equations.

Table 11. Percent improvement comparison of real-time optimal decision methods percent performance improvement.

Decision Method	Mean Final Position Error Percent Improvement	Mean Final Rate Error Percent Improvement	Decision Criteria/Control EffortPercent Compared to Optimal
Classical PID (manually tuned)	–	–	+36,433%
Classical PD (linear quadratic optimal)	87%	−411%	+2445%
Classical P+V (tuned to performance specs)	91%	35%	+414%
Open loop optimal	**72%**	**33%**	**–**
Real-time optimal using *Backslash* (\) *inverse*	−200%	−2343%	−52%
Real-time optimal using *Backslash* (\) with switching	**69%**	**29%**	**0%**
Real-time optimal using *LU-inverse*	96%	−5735%	+6%
Real-time optimal using *LU-inverse* with switching	**72%**	**33%**	**−3%**
Real-time optimal using *pinv-inverse*	100%	−28%	+12%
Real-time optimal using *pinv-inverse* with switching	**69%**	**29%**	**0%**
−*without* cross-product decoupling	−256%	−598%	0%

4.2. Notes on Executability

Table 12 displays simulation run time for the eleven methods compared in the manuscript. Run-time was established by establishing the run start-time with the *tic* command, while simulation stop-time was the very first command (*toc*) following each simulation run. Simply neglecting the cross-product terms results in the fastest calculation. All the methods evaluated are of the same order of magnitude of computational burden. Simulations are depicted graphically in Figures 2–5 and alphanumerically in Appendix B.

Table 12. Simulation run-time comparison.

Decision Method	Simulation Run-Time	Percent Improvement
Classical PID (manually tuned)	1.4342	- -
Classical PD (linear quadratic optimal)	1.4325	-0.1
Classical P + V (tuned to performance specs)	1.3372	-6.8
Open loop optimal	1.3511	-5.8
Real-time optimal using *Backslash* (\) *inverse*	1.3908	-3.0
Real-time optimal using *Backslash* (\) with switching	1.3954	-2.7
Real-time optimal using *LU-inverse*	1.3814	-3.7
Real-time optimal using *LU-inverse* with switching	1.3829	-3.6
Real-time optimal using *pinv-inverse*	1.3407	-6.5
Real-time optimal using *pinv-inverse* with switching	1.3361	-6.8
$-$*without* cross-product decoupling	1.3273	-7.5

4.3. Future Research

Notice in Figure 3c sinusoidal wave input is coded using the identical time-index of the rest of the simulation. The next stages of future research will utilize this identical simulation to investigate efficacy of the proposed virtual sensing amidst unknown wave actions. Secondly, hardware validation of key facets of this research is a logical next step.

5. Conclusions

Using variations of mathematical optimization to provide state, rate, and decision/control provides virtual sensing information useful as sensor replacements. In this instance, arbitrary position and rate sensors were modeled as ideal sensors, plus Gaussian random noise and algorithms were presented and compared that provide very smooth (not noisy) signals for position, rate, and acceleration (manifest in the decision/control). There was no acceleration sensor, so the notion of sensor replacement is manifest for acceleration, while the position and rate information was provided by the selected algorithm acting as a vital sensor. Real-time optimal (nonlinear) state estimation using the Moore–Penrose pseudoinverse (implemented in MATLAB using the *pinv* command) was revealed to be the most advised approach with very highly accurate estimates and essentially mathematically optimally low costs of utilization. The real-time optimal inverse calculation becomes poorly conditioned as the end-state is approached due to rank deficiency in the matrix inversion, so switching to the open loop optimal in the very end was implemented when the determinant of the matrix became nearly zero.

Funding: This research received no external funding. The APC was funded by the author.

Institutional Review Board Statement: Not applicable.

Informed Consent Statement: Not applicable.

Data Availability Statement: While the simulation codes used to produce these results are presented in the manuscript and the appendix, data supporting reported results can be obtained by contacting the corresponding author.

Conflicts of Interest: The author declares no conflict of interest.

Appendix A

This appendix contains a table of variable and acronym definitions used in the manuscript.

Table A1. Variables and acronyms.

Variable/Acronym	Definition
F, τ	external force and torque, respectively
m, I	mass and mass moment of inertia, respectively
$\omega, \dot{\omega}$	angular velocity and acceleration, respectively
$r', v', a\prime$	position and velocity, and acceleration relative to rotating reference
$\tau = I\dot{\omega}$ & $F = ma\prime$	double integrator plants
$\omega \times I\omega$	cross-product translation motion due to rotating reference frame
$m\dot{\omega} \times r'$	cross-product translation motion due to rotating reference frame
$-2m\omega \times v'$	cross-product translation motion due to rotating reference frame
$-m\omega \times (\omega \times r')$	cross-product translation motion due to rotating reference frame
$E\left(x\left(t_f\right)\right)$	endpoint cost also known as "Mayer cost"
$F(x(t), u(t))$	running cost also known as "Lagrange cost"
$J[x(), u()]$	cost function
$e\left(x\left(t_f\right)\right)$	endpoint constraints
$x^T = [x, v]$	state vector of motion state x and rate v with initial condition (x_0, v_0) and final conditions $\left(x_{f-1}, v_f, t_{f-1}\right) = (0, 0, 0)$
$u = [u]$	control vector
H	Hamiltonian operator corresponding to system total energy
λ^T	adjoint operators, also called co-states (corresponding to each state)
v^T	endpoint costates
a, b, c, d	constants of integration
$\hat{a}, \hat{b}$	estimates of constants of integration
t	time variable
x_f, v_f	states evaluated at the final endpoint time t_f
x_0, v_0	states evaluated at the initial time t_0
$\cdot$	Dot notation for time-derivative $\frac{d}{dt}$
$\frac{1}{s}$	notation for integration
I	mass moment of inertia, normalized to unity by scaling in Section 2.1
y_i	output at discrete time index i
i	discretization time-interval (timestep)
σ_y^2	theoretical variance of output
σ_{sim}^2	variance predicted in simulation of proposed methods
ζ, ω_n	critical damping ratio and natural frequency
P+V	proportional plus velocity
PD,PI,PID	proportional plus derivative, integral respectively
det	determinant
RTOC	real-time optimal calculation
$\backslash$	backslash matrix inversion (a MathWorks technique)
$pinv$	Moore–Penrose pseudoinverse matrix inversion
LU	matrix inversion by factoring and inverting a row-pivoted variant

Appendix B

The appendix contains computer codes inserted into the InifFcn callback and StopFcn call back, respectively. Together with figures, implementation of Equations (1), (31)–(46), this is the only computer code needed to achieve the results presented in this manuscript.

Algorithm A1. InitFcn callbacks for SIMULINK model depicted in Figures

Includes classical feedback tuning methods

```
clear all; close all; clc;

% Initialize variables and matrices for assembly of results
timestep = 0.01; theta0 = 0; omega0 = 0; thetaD = 1; omegaD = 0; J = 1;
MeanAngleNoise = 0; STDAngleNoise = 0.01; VarianceAngleNoise = (STDAngleNoise)^2/sqrt(timestep);
MeanRateNoise = 0; STDRateNoise = 0.01; VarianceRateNoise=(STDRateNoise)^2/sqrt(timestep);

% PID: Rule of thumb
Kp = 1; Kd = 6.5; Ki = 5;
A = [0 1; 0 0]; B = [0;1]; C = [10]; D = 0; Q = ones(2,2); R = 1*ones(1);

% PD: Ziegler-Nichols tuning for manually tuning PD
%[NUM,DEN] = ss2tf(A,B,C,D); G = tf(NUM,DEN); C = tf([Kd Kp],[1]); sisotool(G,C)

% PD: Linear Quadratic Regulator
%K = lqr(A,B,Q,R); Kp = K(1); Kd = K(2); Ki = 0; % J = 76 (ish)
Kp = 1; Kd = 1.7321; Ki = 0;

% P + V: Tuning for performance specification
tr = 0.3; ts = 2; wn = 1.8/tr; zeta = 4.6/(ts*2.4);
Kp = wn^2;
Kv = 2*zeta*wn;
```

Algorithm A2. StopFcn callbacks for SIMULINK model depicted in Figures

---

```
% Printing scripts

theta = squeeze(out.theta); omega = squeeze(out.omega);
t = squeeze(out.tout); J = squeeze(out.J); u = squeeze(out.u);

qxv = squeeze(out.qxv); a = squeeze(qxv(1,:)); b = squeeze(qxv(2,:));
thetaText = ['\theta(t) J=', num2str(J(end)), ': ', '\theta(t_f)= ', num2str(theta(end))];
omegaText = ['\omega(t) J=', num2str(J(end)), ': ', '\omega(t_f)= ', num2str(omega(end))];
ControlText = ['\mu_u_(_t_)=', num2str(mean(u)) ];

figure(1); plot(t,theta, ':', 'LineWidth', 2); hold on; plot(t,omega, '-', 'LineWidth', 2); hold off;
legend(thetaText, omegaText, 'FontSize',16, 'FontName','Palatino Linotype');
set(gca, 'FontSize',16, 'FontName','Palatino Linotype');grid;

figure(2); plot(t,a, ':', 'LineWidth', 2); hold on; plot(t,b, ':', 'LineWidth', 2); hold off;
legendtexta=['\mu_a=', num2str(mean(a))]; legendtextb=['\mu_b=', num2str(mean(b))]
legend(legendtexta,legendtextb,'FontSize',16, 'FontName','Palatino Linotype');
set(gca, 'FontSize',16, 'FontName','Palatino Linotype');grid; axis([0,1,-12.1,6.1])

figure(3); plot(t,u,'LineWidth', 2);
legend(ControlText, 'FontSize',16, 'FontName','Palatino Linotype');
set(gca, 'FontSize',16, 'FontName','Palatino Linotype');grid;
```

References

1. Special Issue "Combining Sensors and Multibody Models for Applications in Vehicles, Machines, Robots and Humans". Available online: https://www.mdpi.com/journal/sensors/special_issues/multibody_sensors (accessed on 29 May 2021).
2. International Conference "Optimal Control and Differential Games" Dedicated to the 110th Anniversary of L. S. Pontryagin, Russia, 12–14 December 2018; Steklov Mathematical Institute of RAS: Moscow. Photo Originally Published in the Newspaper "Soviet Russia", April 16, 1989. Used with mild Permission (29 June 2021) from the Head of the Information and Publishing Sector, Department of Computer Networks and Information Technology, Steklov Mathematical Institute of Russian Academy of Sciences. Available online: http://www.mathnet.ru/eng/conf1287 (accessed on 24 June 2021).
3. Dunn, B. The Future for Unmanned Surface Vessels in the US Navy. Georgetown Security Studies Review. 28 October 2020. Available online: https://georgetownsecuritystudiesreview.org/2020/10/28/the-future-for-unmanned-surface-vessels-in-the-us-navy/ (accessed on 24 June 2021).
4. Boat Motion Monitoring. Available online: https://www.sbg-systems.com/applications/ship-motion-monitoring/ (accessed on 29 May 2021).
5. Hamilton, W.R. *On a General Method in Dynamics*; Royal Society: London, UK, 1834; pp. 247–308.
6. Merz, J. *A History of European Thought in the Nineteenth Century*; Blackwood: London, UK, 1903; p. 5.
7. Whittaker, E. *A Treatise on the Analytical Dynamics of Particles and Rigid Bodies*; Cambridge University Press: New York, NY, USA, 1904; p. 1937.
8. Church, I.P. *Mechanics of Engineering*; Wiley: New York, NY, USA, 1908.
9. Wright, T. *Elements of Mechanics Including Kinematics, Kinetics, and Statics, with Applications*; Nostrand: New York, NY, USA, 1909.
10. Gray, A. *A Treatise on Gyrostatics and Rotational Motion*; MacMillan: London, UK, 1918; ISBN 978-1-4212-5592-7.
11. Rose, M. *Elementary Theory of Angular Momentum*; John Wiley & Sons: New York, NY, USA, 1957; ISBN 978-0-486-68480-2.
12. Greenwood, D. *Principles of Dynamics*; Prentice-Hall: Englewood Cliffs, NJ, USA, 1965; ISBN 9780137089741.
13. Pontryagin, L.S.; Boltyanski, V.G.; Gamkrelidze, R.S.; Mishchenko, E.F. *The Mathematical Theory of Optimal Processes*; CRC Press: Boca Raton, FL, USA, 1962.
14. Sands, T. Optimization Provenance of Whiplash Compensation for Flexible Space Robotics. *Aerospace* **2019**, *6*, 93. [CrossRef]
15. Sands, T.; Bollino, K.; Kaminer, I.; Healey, A. Autonomous Minimum Safe Distance Maintenance from Submersed Obstacles in Ocean Currents. *J. Mar. Sci. Eng.* **2018**, *6*, 98. [CrossRef]
16. Sands, T. Development of deterministic artificial intelligence for unmanned underwater vehicles (UUV). *J. Mar. Sci. Eng.* **2020**, *8*, 578. [CrossRef]
17. Sands, T. Control of DC Motors to Guide Unmanned Underwater Vehicles. *Appl. Sci.* **2021**, *11*, 2144. [CrossRef]
18. Sands, T.; Kim, J.J.; Agrawal, B.N. Spacecraft fine tracking pointing using adaptive control. In Proceedings of the 58th International Astronautical Congress, Hyderabad, India, 24–28 September 2007; International Astronautical Federation: Paris, France, 2007.
19. Sands, T.; Kim, J.J.; Agrawal, B.N. Improved Hamiltonian adaptive control of spacecraft. In Proceedings of the IEEE Aerospace, Big Sky, MT, USA, 7–14 March 2009; IEEE Publishing: Piscataway, NJ, USA, 2009; pp. 1–10.
20. Smeresky, B.; Rizzo, A.; Sands, T. Optimal Learning and Self-Awareness Versus PDI. *Algorithms* **2020**, *13*, 23. [CrossRef]
21. Escalona, J.L.; Urda, P.; Muñoz, S. A Track Geometry Measuring System Based on Multibody Kinematics, Inertial Sensors and Computer Vision. *Sensors* **2021**, *21*, 683. [CrossRef] [PubMed]
22. Cuadrado, J.; Michaud, F.; Lugrís, U.; Pérez Soto, M. Using Accelerometer Data to Tune the Parameters of an Extended Kalman Filter for Optical Motion Capture: Preliminary Application to Gait Analysis. *Sensors* **2021**, *21*, 427. [CrossRef]
23. Trslić, P.; Omerdic, E.; Dooly, G.; Toal, D. Neuro-Fuzzy Dynamic Position Prediction for Autonomous Work-Class ROV Docking. *Sensors* **2020**, *20*, 693. [CrossRef] [PubMed]
24. Gruosso, G.; Storti Gajani, G.; Ruiz, F.; Valladolid, J.D.; Patino, D. A Virtual Sensor for Electric Vehicles' State of Charge Estimation. *Electronics* **2020**, *9*, 278. [CrossRef]
25. Yoon, S.; Choi, Y.; Koo, J.; Hong, Y.; Kim, R.; Kim, J. Virtual Sensors for Estimating District Heating Energy Consumption under Sensor Absences in a Residential Building. *Energies* **2020**, *13*, 6013. [CrossRef]
26. Chau, A.; Dawson, J.; Mitchell, P.; Loh, T.H. Virtual Sensing Directional Hub MAC (VSDH-MAC) Protocol with Power Control. *Electronics* **2020**, *9*, 1219. [CrossRef]
27. Oliveira, A.J.; Ferreira, B.M.; Cruz, N.A. A Performance Analysis of Feature Extraction Algorithms for Acoustic Image-Based Underwater Navigation. *J. Mar. Sci. Eng.* **2021**, *9*, 361. [CrossRef]
28. Zhang, Z.; Wang, H.; Yao, H. Pulse Ranging Method Based on Active Virtual Time Reversal in Underwater Multi-Path Channel. *J. Mar. Sci. Eng.* **2020**, *8*, 883. [CrossRef]
29. Barker, L.D.L.; Whitcomb, L.L. Performance Analysis of Ice-Relative Upward-Looking Doppler Navigation of Underwater Vehicles Beneath Moving Sea Ice. *J. Mar. Sci. Eng.* **2021**, *9*, 174. [CrossRef]
30. Thomas, C.; Simetti, E.; Casalino, G. A Unifying Task Priority Approach for Autonomous Underwater Vehicles Integrating Homing and Docking Maneuvers. *J. Mar. Sci. Eng.* **2021**, *9*, 162. [CrossRef]
31. Zhao, W.; Wang, Y.; Zhang, Z.; Wang, H. Multicriteria Ship Route Planning Method Based on Improved Particle Swarm Optimization–Genetic Algorithm. *J. Mar. Sci. Eng.* **2021**, *9*, 357. [CrossRef]
32. Anderlini, E.; Husain, S.; Parker, G.G.; Abusara, M.; Thomas, G. Towards Real-Time Reinforcement Learning Control of a Wave Energy Converter. *J. Mar. Sci. Eng.* **2020**, *8*, 845. [CrossRef]

33. Davidson, J.; Kalmár-Nagy, T. A Real-Time Detection System for the Onset of Parametric Resonance in Wave Energy Converters. *J. Mar. Sci. Eng.* **2020**, *8*, 819. [CrossRef]
34. Sirigu, S.A.; Foglietta, L.; Giorgi, G.; Bonfanti, M.; Cervelli, G.; Bracco, G.; Mattiazzo, G. Techno-Economic Optimisation for a Wave Energy Converter via Genetic Algorithm. *J. Mar. Sci. Eng.* **2020**, *8*, 482. [CrossRef]
35. Veremey, E.I. Optimal Damping Concept Implementation for Marine Vessels' Tracking Control. *J. Mar. Sci. Eng.* **2021**, *9*, 45. [CrossRef]
36. Volkova, T.A.; Balykina, Y.E.; Bespalov, A. Predicting Ship Trajectory Based on Neural Networks Using AIS Data. *J. Mar. Sci. Eng.* **2021**, *9*, 254. [CrossRef]
37. Zhang, Z.; Zhao, Y.; Zhao, G.; Wang, H.; Zhao, Y. Path-Following Control Method for Surface Ships Based on a New Guidance Algorithm. *J. Mar. Sci. Eng.* **2021**, *9*, 166. [CrossRef]
38. Martin, D.; Kühl, N.; Satzger, G. Virtual Sensors. *Bus Inf. Syst. Eng.* **2021**, *63*, 315–323. [CrossRef]
39. Sands, T. Comparison and Interpretation Methods for Predictive Control of Mechanics. *Algorithms* **2019**, *12*, 232. [CrossRef]
40. Brown, R.; Hwang, P. *Introduction to Random Signals and Applied Kalman Filtering*, 3rd ed.; Wiley: New York, NY, USA, 1997.
41. Ogata, K. *Modern Control Systems*, 5th ed.; Prentice Hall: New York, NY, USA, 2010; ISBN 978-0-13-615673-4.
42. Lurie, B.; Enright, P. *Classical Feedback Control with Nonlinear Multi-Loop Systems*, 3rd ed.; CRC Press: Boca Raton, FL, USA, 2019; ISBN 978-1-1385-4114-6.

Article

Haptic Devices Based on Real-Time Dynamic Models of Multibody Systems

Nicolas Docquier *, Sébastien Timmermans and Paul Fisette

Mechatronic, Electrical Energy, and Dynamic Systems (MEED), Institute of Mechanics,
Materials and Civil Engineering (iMMC), Université catholique de Louvain, 1348 Louvain-la-Neuve, Belgium;
sebastien.timmermans@uclouvain.be (S.T.); paul.fisette@uclouvain.be (P.F.)
* Correspondence: nicolas.docquier@uclouvain.be

Abstract: Multibody modeling of mechanical systems can be applied to various applications. Human-in-the-loop interfaces represent a growing research field, for which increasingly more devices include a dynamic multibody model to emulate the system physics in real-time. In this scope, reliable and highly dynamic sensors, to both validate those models and to measure in real-time the physical system behavior, have become crucial. In this paper, a multibody modeling approach in relative coordinates is proposed, based on symbolic equations of the physical system. The model is running in a ROS environment, which interacts with sensors and actuators. Two real-time applications with haptic feedback are presented: a piano key and a car simulator. In the present work, several sensors are used to characterize and validate the multibody model, but also to measure the system kinematics and dynamics within the human-in-the-loop process, and to ultimately validate the haptic device behavior. Experimental results for both developed devices confirm the interest of an embedded multibody model to enhance the haptic feedback performances. Besides, model parameters variations during the experiments illustrate the infinite possibilities that such model-based configurable haptic devices can offer.

Keywords: multibody dynamics; symbolic generation; real-time computation; human-in-the-loop; haptic devices

Citation: Docquier, N.; Timmermans, S.; Fisette, P. Haptic Devices Based on Real-Time Dynamic Models of Multibody Systems. *Sensors* **2021**, *21*, 4794. https://doi.org/10.3390/s21144794

Academic Editors: Miguel Ángel Naya Villaverde and Javier Cuadrado

Received: 7 June 2021
Accepted: 12 July 2021
Published: 14 July 2021

Publisher's Note: MDPI stays neutral with regard to jurisdictional claims in published maps and institutional affiliations.

1. Introduction

Multibody formalisms, which appeared in the seventies, were intended to produce the equations of motion of so-called multibody systems (MBS) of any size and of any kind, with the aim of studying their motion. For several decades, multibody models were mainly used to study mechanical systems with a purely predictive (development of a new device) or corrective (improvement of an existing device) target. The "real-time computation" character was completely absent for obvious limitations related to the performance of computers at the time.

The massive and amazing arrival of increasingly more powerful processors and of faster and larger memories (RAM) encouraged research teams to produce very compact multibody formalisms (e.g., based on so-called recursive techniques or Order-N methods), to exploit programming techniques (such as symbolic generation) and to take maximum advantage of state-of-the-art computer architectures (vector or parallel processors).

All this contributed to exploring the possibilities of using multibody models in the context of real-time computing. Already at the beginning of the nineties for instance, the inverse dynamic computation of robot actuator torques was carried out to enrich their internal controllers with a feedforward component (e.g., predicted-torque control). Computational algorithms in haptic rendering may consider various approaches [1].

Among demanding applications (in terms of real-time computation point of view), one finds a series of multibody systems (vehicle simulator, remotely-actuated surgical robot, etc.) whose haptic feedback (i.e., the force or torque to be sent back to the human)

123

should reproduce, as accurately as possible, the dynamics of the system with which it is in interaction. Among the ingredients required to develop such devices one finds, beside the real-time multibody model, a series of sensors (at the human–machine and machine–machine interfaces), powerful computer processors, and an operating meta-system. The latter must play the role of a real conductor responsible for synchronizing and interfacing the involved hardware and software components.

This work presents the developments realized in this sense by the UCLouvain multibody team, who decided to make the most to couple the symbolic multibody models (issuing from their symbolic ROBOTRAN (www.robotran.eu, accessed on 13 July 2021) software) within the ROS (www.ros.org, accessed on 13 July 2021) architecture (Robot Operating System). The latter is used to interface the models with a set of well-chosen sensors pursuing two objectives: (i) first to validate the multibody models with respect to experimental physics and (ii) second to handle specific haptic devices whose underlying physical system is dynamic in nature.

Two very distinct applications are presented to illustrate the approach. The first one—of research type—concerns the development of a digital piano keyboard with haptic feedback, based on the real-time multibody model of the grand piano action. The second application—rather educational—concerns a car simulator whose 3D multibody model allows to feedback pertinent sensory information, namely, the steering torque and a 3D visualization with realistic visual clues. The model can be tuned in real-time by instantaneously modifying the suspension settings but also the contact characteristics of the tire/ground forces.

In all cases, the haptic specifications and requirements of the envisaged applications have guided our choices for (i) the electromechanical design itself, (ii) the level of refinement and computational efficiency of the multibody model, (iii) the type and conditioning of the sensors, and (iv) the way of interfacing the whole through a suitable meta-architecture (ROS). The results confirm the advantages and assets of the symbolic multibody approach and the interest of embedding such models in haptic devices to best capture highly dynamic effects (a technique that was not conceivable with the technologies of the past in a context of haptic feedback).

The paper is organized as follows. After a state-of-the-art on multibody formalisms and haptic devices (Section 2), the multibody approach and its symbolic implementation are treated in Section 3. Section 4 addresses the different aspects of the haptic problem, still focusing on multibody applications. Section 5 illustrates the approach proposed for the two above-mentioned applications and, more succinctly and illustratively, for other systems developed in our laboratory. A concluding section closes this work, pointing out the interesting perspectives of this research. Having a immense potential, haptics technology is still in its beginning stage [2].

2. State-of-the-Art

2.1. Multibody Formalisms

Multibody dynamics, a scientific discipline that emerged in the early seventies (see, e.g., in [3,4]), is concerned with the kinematic and dynamic study of "polyarticulated" mechanical systems (referred to as MultiBody Systems, or MBS in short) such as transmission mechanisms, humanoid robots, road and railway vehicles, human body, cranes, etc. (Figure 1).

Beside experimental investigations, the analysis of the motion of these MBS requires their mathematical modeling, based on the fundamental laws of mechanics, of which the Newton equations (for body translations) and Euler equations (for body rotations) are common starting points, in the same way as the Lagrange equations or the virtual work and virtual power principles.

To obtain the equations describing the motion of MBS, several choices of variables (referred to as generalized coordinates q) are possible. Some schools have favored absolute or nodal coordinates (denoting the configuration of each body of the MBS with respect to an inertial frame) [4–6], others preferred using relative coordinates (representing the configuration of each body with respect to another body of the MBS) [7,8] or natural coordinates (representing the absolute configuration of specific material points on each body of the MBS) [9].

Figure 1. Multibody applications range from small piano action, through road and railway vehicles, to huge port cranes (UCLouvain).

Concerning the bodies themselves, the formalisms are distinguished according to whether they consider them rigid or flexible, in which case deformation equations are superimposed to the equations of the overall motion of the MBS (see. e.g., in [6,8]).

Regardless of the formalism used, the complexity of real applications is such that the automatic generation of their multibody models on a computer has quickly become essential. Indeed, it allows obtaining reliable and sufficiently generic models to deal with the numerous MBS families (see above). A distinction is made between the numerical approach (e.g., Adams-MSC or Samcef-Mecano software) and the so-called symbolic approach (e.g., Neweul-M2, Maplesoft, and ROBOTRAN software) which still presently cohabit among the available multibody computer programs.

The first approach requires reconstructing the model through a series of numerical subroutines at each step of the computation (e.g., integration time step). This reconstruction inevitably drains a series of useless computations linked to the numerous zeros of multibody models (inherent to the data or to the tree-like morphology of the physical system). Symbolic multibody programs avoid this algorithmic reconstruction: the whole dynamic model fits in a single file. Above all, symbolic engines can eliminate unnecessary operations like multiplication by zero, condensation of trigonometric formulas or even deletion of complete unnecessary equations provided by generic multibody formalisms [10]. Their superiority in terms of FLOPS is established and leads to much lower computation times, favorable to the real-time performances required here. In sum, symbolic multibody-dedicated programs allow to produce compact equations leading to very efficient and portable models, written in the desired language (e.g., C, Python, Matlab).

Multibody dynamics then opened up to other physics disciplines (e.g., hydraulics, pneumatics, mechatronics, and granular media) and to specific analysis methods (control, optimization, and HIL real-time applications), the challenge being essentially to ensure a numerically robust and computationally efficient coupling between the mathematical models of the disciplines involved (see, e.g., in [11]). The scientific literature of the last twenty years is full of developments in this field of multibody-multiphysics coupling, most often guided by a specific type of application (see, e.g., in [12]).

For systems requiring real-time (or even faster than real-time) computation of their multibody model, as in the case of haptic devices whose internal kinematics or dynamics cannot be disregarded, specific developments have been carried out by the scientific community. They allow, among others, to achieve feedforward computed-torque control of walking robots, to increase the precision of surgical tools or to improve the haptic feedback of driving simulators.

In this context, one of the current challenges concerns the inclusion in haptic feedback devices of dynamic models that are (i) sufficiently representative of the system (number of bodies, joints, nature of internal forces, etc.), (ii) accurate (correct identification of parameters and force laws), and (iii) efficient (i.e., faster than real-time). All this cannot be achieved without a strong interaction between the virtual model, the real device and the external world, through the use of both virtual and real sensors, in order to claim high-level haptic performances.

2.2. Hardware Implementation

On one hand, validation of a multibody model thanks to specific sensors is widespread and ranges from many applications with various sensors, for example, a railway track geometry measurement based on IMUs, cameras, and encoders [13]; model estimation of a vehicle suspension system with strain gauges [14]; gait analysis through IMUs compared to optical motion capture [15]; validation of a axial piston pump model via force and pressure sensors [16]; and acceleration and force piezoelectric sensors to test a wind turbine flexible multibody model [17].

On the other hand, a growing number of researches has been carried out to use sensors in real-time to measure the physical system behavior. This information is crucial to properly feed the multibody model.

For example, in 2016, Torres-Moreno et. al focused on an online dynamic estimation of the four-bar mechanism [18]. The kinematics of the system is retrieved through IMUs measurements and optical encoders. They proposed an extended Kalman filter approach that allows to deal with constrained multibody system. The estimators are run in real-time and experiments show good agreements with simulation results.

Running a multibody model in real-time requires sufficient computational power on the considered hardware. While it is possible to perform the online computation on a classic computer with *Intel* processor and *Windows* OS [19], it may not be the most robust solution. Besides, if one aims to perform an embedded simulation, the hardware is often limited. In this regard, some researches have achieved to run multibody models on other kinds of platforms, such as FPGA [20] or ARM-based systems [21]. In particular, the above-mentioned symbolic approach has been proved to be a good candidate to achieve real-time computation without having to resort to simplifications within the model [22].

Concerning the software implementation, new tools have been provided by the scientific community to combine integrated devices and simulators, mainly in the robotic field. The so-called *middleware* platforms are able to manage the communications between the different interfaces and facilitate the code modularity and reuse. Rivera presented a recent summary of the middleware platforms in [23], while using a Gazebo/ROS-based environments for his application. The ROS environment is now widely used for real-time robotics, for example, in [24] to represent a complex robot real-time dynamic simulator.

In the recent past, our multibody software ROBOTRAN has been coupled with the YARP interface for humanoid applications [25], combining the real robot with the simulated one. Force–torque sensors and IMUs have been used in reality and successfully matched with their simulation equivalent.

Despite all these researches, the real-time multibody models are still rarely embedded in haptic devices, the goal being to improve the feedback for "rigid–rigid interaction" type applications, which only represent one of the multiple components of the Haptic rendering discipline [26]. Haptic devices often consider simplified dynamics to synthesize the haptic feedback, for example in the REPLALINK prototype [27] or in a piano action model [28] (see also Section 5.1). Advanced haptic devices embedding a full multibody model begin to appear in the literature [29,30], exhibiting promising prospects thanks to the conjoint evolution of modeling, computational and technological tools.

3. Multibody Formalism

Among the modeling options presented in Section 2.1, we have opted for a relative coordinate formalism (Section 3.1) and a symbolic implementation of multibody equations (Section 3.2). This combination offers the advantage of producing models that can be easily interfaced, either with other disciplines (in a multiphysics perspective) or within a middelware platform (to couple devices and software in a real-time process).

3.1. Modeling MBS Using Relative Coordinates

When choosing relative coordinates, a MBS can be built on the basis of its topology characterized by the successive bodies and joints, as shown in Figure 2, on the left for a tree-like MBS (such as a serial robot) and on the right for a MBS containing kinematic loops (such as a multi-link car suspension). The relative angular (*rad*) or linear (*m*) displacements within the joints are called the generalized coordinates, q, of the MBS (also called *joint coordinates*). These are independent for a tree-like MBS but not for a closed-loop one, each loop of bodies leading to geometrical constraints, denoted as $h(q) = 0$, between the joint coordinates present in the loop. The distinction between tree-like and closed-loop MBS is essential because it is at the root of the relative coordinate formalisms as synthetically explained below.

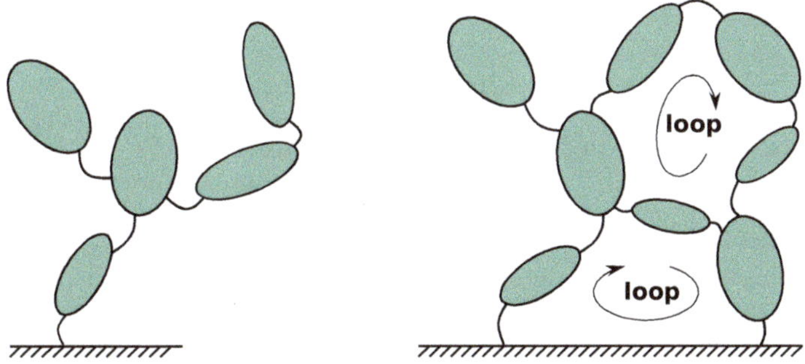

Figure 2. Tree-like (**left**) and closed-loop (**right**) MBS topologies.

Tree-Like MBS Modeling

Basically (see in [8] for more details), in the case of a tree-like MBS containing n joint coordinates, these confer n motion degrees of freedom (DOF) to the system, leading to a set of n equations of motion.

Various techniques can be envisaged to obtain them (d'Alembert principle, Lagrange equations, Newton–Euler equations recursively computed on the topology [8]). In any case, one ends up with a form equivalent to the following one ("." stands for $\frac{d}{dt}$ and ".." for $\frac{d^2}{dt^2}$):

$$M(q)\,\ddot{q} + c(q, \dot{q}, frc, trq) = Q(q, \dot{q}), \tag{1}$$

in which $\dot{q}$ and $\ddot{q}$, respectively, stand for the generalized relative velocities and accelerations; $M(q)$ represents the system mass matrix; $c(q, \dot{q})$ gather the centripetal, Coriolis, gyroscopic terms as well as the external forces $frc(q, \dot{q})$ and torques $trq(q, \dot{q})$; and $Q(q, \dot{q})$ denotes the efforts within the joints (actuators or passive elements).

Equation (1) can be straightforwardly solved using linear algebra to obtain the relative accelerations $\ddot{q}$: this refers to *Direct Dynamics*. This computation has a $O(n^3)$ computational complexity due to the factorization of the mass matrix. Among other things, direct dynamics allows, via the time integration of the accelerations, to determine the time history of the positions $q(t)$ and velocities $\dot{q}(t)$ and thus to predict the motion of the MBS subjected to internal or external forces, starting from initial conditions $q(0), \dot{q}(0)$.

As an illustration, direct dynamics is widely used for road and railway vehicles to predict, analyze, or optimize their dynamic performances. Direct dynamics is also used in various driving simulators to make haptic feedback more accurate and realistic (see Section 5.2).

Inverse Dynamics is another way of using Equation (1) for applications where the MBS trajectory is known ($q(t)$, $\dot{q}(t)$, and $\ddot{q}(t)$) and for which the unknowns are now the joint forces and torques $Q(t)$ and not the accelerations anymore. Equation (1) can then be written as follows:

$$Q(q,\dot{q}) = \phi(q,\dot{q},\ddot{q},frc,trq) \tag{2}$$

with

$$\phi = M(q)\,\ddot{q} + c(q,\dot{q},frc,trq)$$

As the explicit computation of the mass matrix M is not necessary anymore to compute Q via (2), ϕ can be generated without explicitly computing it, which leads to an efficient $O(n)$ computational complexity of the formalism.

For instance, inverse dynamics is useful to compute the actuator torques of robots for evaluating a feedforward component in their motion controller.

Closed-Loop MBS Modeling

A multibody system can be subject to different types of constraints between its generalized coordinates, q, which can be holonomic (i.e., algebraic, like assembly constraints), non-holonomic (expressed at the level of velocities, like for a slip-free rolling), and/or rheonomic (when they depend explicitly on time t). In the present context of haptic devices modeling, we will limit ourselves to the holonomic case, focusing on kinematic loop constraints; this does not exclude the case of other algebraic constraints that would be treated in the same way by the following formalism.

When dealing with closed-loop MBS (as depicted on the right of Figure 2), m loop constraints between the generalized coordinates q, denoted $h(q) = 0$, must be solved in parallel to the equations of motion (1) or (2), in which an additional term is required to take the constraint forces and torques into account, according to the well-established Lagrange multipliers technique. For the direct dynamics, the final form reads

$$\begin{aligned} M(q)\,\ddot{q} + c(q,\dot{q},frc,trq) &= Q(q,\dot{q}) + J^t(q)\,\lambda \\ h(q) = 0\;;\; \dot{h}(q,\dot{q}) = J(q)\,\dot{q} &= 0\;;\; \ddot{h}(q,\dot{q},\ddot{q}) = 0 \end{aligned} \tag{3}$$

And in a similar way, for the inverse dynamics the equations becomes

$$\begin{aligned} Q(q,\dot{q}) &= \phi(q,\dot{q},\ddot{q},frc,trq) - J^t(q)\,\lambda \\ h(q) &= 0\;;\; \dot{h}(q,\dot{q}) = J(q)\,\dot{q} = 0\;;\; \ddot{h}(q,\dot{q},\ddot{q}) = 0 \end{aligned} \tag{4}$$

in which $J(q) \triangleq \frac{\partial h}{\partial q^t}$ denotes the constraint Jacobian matrix and λ are the Lagrange multipliers associated with the constraints $h(q)$. In both cases, the number of DOF equals $n - m$, that is the number of generalized coordinates q minus the number of independent constraints h (The case of redundant constraints is treated within Robotran, but only for "classical" multibody simulations, and not for the proposed haptic coupling. As there were no redundant constraints for the applications treated in this paper, this topic has not been discussed.).

Let us illustrate closed-loop morphologies in the field of robotics. An anthropomorphic serial robot with 6 actuated joints (Figure 3 (left)) possesses 6 DOFs. On the other hand, a parallel structure such as the Stewart platform (Figure 3 (right)) has 18 joints embedded into multiple loops of bodies, but its upper-plate only possesses 6 DOFs and its motion is ensured by 6 actuators forming the legs of the platform.

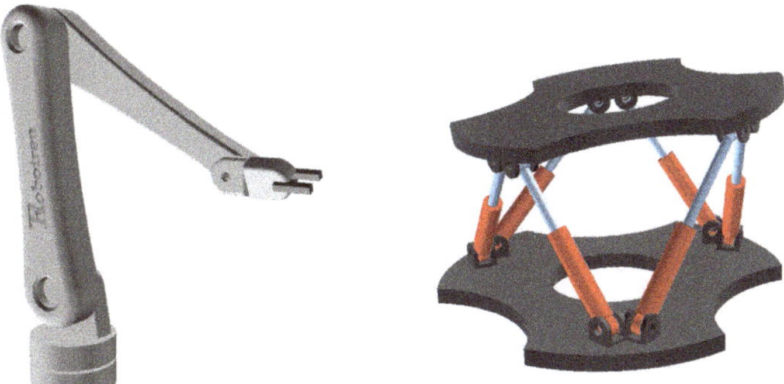

Figure 3. Serial (**left**) versus parallel (**right**) manipulators.

Equations (3) or (4) form a DAE system, as they are made of differential and algebraic equations, whose direct solving is not trivial. A practical and elegant way to solve them is to transform them into ordinary differential equations (ODE), using the so-called *Coordinate Partitioning Method* briefly presented below (see in [4,8] for more details).

The starting point of the method refers to the number of degrees of freedom $(n - m)$ of the system, to which $(n - m)$ generalized coordinates can match. They are called *independent* coordinates, denoted u, while the others, denoted v, are *dependent*. Therefore, we can partition the column vector q as follows:

$$q = \begin{pmatrix} u \\ v \end{pmatrix} \tag{5}$$

By applying this partitioning to all the matrices and vectors appearing in Equation (3) or (4), and by expressing the dependent coordinates via the solving of the constraints at position, velocity and acceleration levels,

$$\begin{aligned} h(q) = 0 &\Rightarrow v = f(u), \\ \dot{h}(q, \dot{q}) = 0 &\Rightarrow \dot{v} = g(q, \dot{u}), \\ \ddot{h}(q, \dot{q}, \ddot{q}) = 0 &\Rightarrow \ddot{v} = h(q, \dot{q}, \ddot{u}), \end{aligned} \tag{6}$$

it is possible, after a few matrix manipulations, to end up with a reduced set of equations of motion in terms of the independent coordinates u:

$$M_r(u), \ddot{u} + c_r(u, \dot{u}, frc, trq) = Q_r(u, \dot{u}) \tag{7}$$

$$Q_r(u, \dot{u}) = \phi_r(u, \dot{u}, \ddot{u}), \tag{8}$$

for the direct and inverse dynamics, respectively.

The coordinate partitioning technique amounts to projecting the equations of motion in the MBS motion manifold, with the advantage of getting rid rigorously of the algebraic constraint equations. Therefore, one ends up with a ODE system of minimum size (n-m) which exactly corresponds to the number of DOF of the MBS (ex. $n - m = 6$ for the Stewart platform of Figure 3).

It is worth mentioning that the resolution of the constraints (6) can be very accurate using an iterative (for h) or an algebraic (for $\dot{h}$ and $\ddot{h}$) algorithm, at each evaluation of the dynamic equations.

Driven Motion

When dealing with MBS in which the time evolution of some of the generalized coordinates u is fully imposed (such as for instance in robotics when motion control is achieved), the corresponding joints are qualified as *driven*. The time history of the associated coordinates is thus a prescribed function of time t. To take them into account in the above formalism, we can apply a second partitioning, but on u in the present case, that is,

$$u = \begin{pmatrix} u_u \\ u_d \end{pmatrix},\tag{9}$$

in which u_u represents the new subset of independent coordinates and u_d stands for the driven coordinates viewed, from the MBS point of view, as known functions of time at position, velocity, and acceleration levels:

$$u_d = f(t) \; ; \; \dot{u}_d = \frac{du_d}{dt} \; ; \; \ddot{u}_d = \frac{d^2 u_d}{dt^2}\tag{10}$$

Note that driven variables can also be used to deal with rheonomic constraints. Indeed, they allow to describe an explicit time dependence of any function in which they appear, as in an algebraic constraint for example.

Introducing the partitioning (9) into the equations of motion (7) gives

$$\begin{pmatrix} M_r^{uu} & M_r^{ud} \\ M_r^{du} & M_r^{dd} \end{pmatrix} \begin{pmatrix} \ddot{u}_u \\ \ddot{u}_d \end{pmatrix} + \begin{pmatrix} c_r^u \\ c_r^d \end{pmatrix} = \begin{pmatrix} Q_r^u \\ Q_r^d \end{pmatrix}\tag{11}$$

The first line of (11) can be solved with respect to $\ddot{u}_u$ and represents the final form of a *direct dynamics* problem (to be time integrated with respect to the u_u variables for instance):

$$M_r^{uu} \ddot{u}_u + c_r^u = Q_r^u - M_r^{ud} \ddot{u}_d\tag{12}$$

Let us note that this equation can also be formulated in an inverse dynamics form, in a similar way to Equation (8).

The second line of (11) can be put aside if we are only interested in analysing the motion of the MBS Equations (12). However, those equations can be very useful to compute efforts (Q_r^d) associated with the driven coordinates (ex. a motor torque, a force in an mechanical assembly or a reaction force in a bearing):

$$Q_r^d = M_r^{du} \ddot{u}_u + M_r^{dd} \ddot{u}_d + c_r^d\tag{13}$$

In a certain manner, this equation represents an inverse dynamics computation carried out in parallel with—or as a postprocess of—the direct dynamics (12).

Implementation in a Haptic Context

The multibody computation workflow varies slightly whether direct dynamics or inverse dynamics are considered (Figure 4). For direct dynamics simulations, a numerical integrator computes the time evolution of the independent variables. Trajectory sensors (i.e., position, velocity, or acceleration sensors) define the motion of *driven joints* (Step A), such as the angular position of a steering wheel for a driver simulator or the displacement of a piano key (see Section 5). When the trajectories of all independent joints are measured, the inverse dynamics formulation is used. Filtering, differentiating or integrating the sensor signal may be necessary.

The algebraic constraints (6) of closed-loop MBS are used to compute dependent positions and velocities (step B). Once the configuration of the system is fully defined, external and internal forces are computed, for instance, the tyre/ground interaction for a car or the contact interaction between the parts of a piano action (step C). At this stage, inputs coming from force/torque sensors can be considered in the dynamics. The dynamic Equations (3) or (4) are computed and, for closed-loop systems, the coordinate partitioning technique is applied (step D) to get the reduced set of Equation (7). In the case of direct dynamics, a linear system of equations must be solved (step E) to compute the independent acceleration that will be handled by the time integrator. For the inverse dynamics, joint forces/torques are derived directly from the dynamics (step F) using (8).

Forces or torques associated with driven joints are computed (step F) using (13), which enables computing a feedback that takes the dynamics into account, like the wheel steering torque on the driver or the reaction of the piano key on the musician finger. In addition, any kinematic information of the system, such as the position or velocity of any point, can also be computed (step G) and be used for the system feedback using so-called *multibody sensors*.

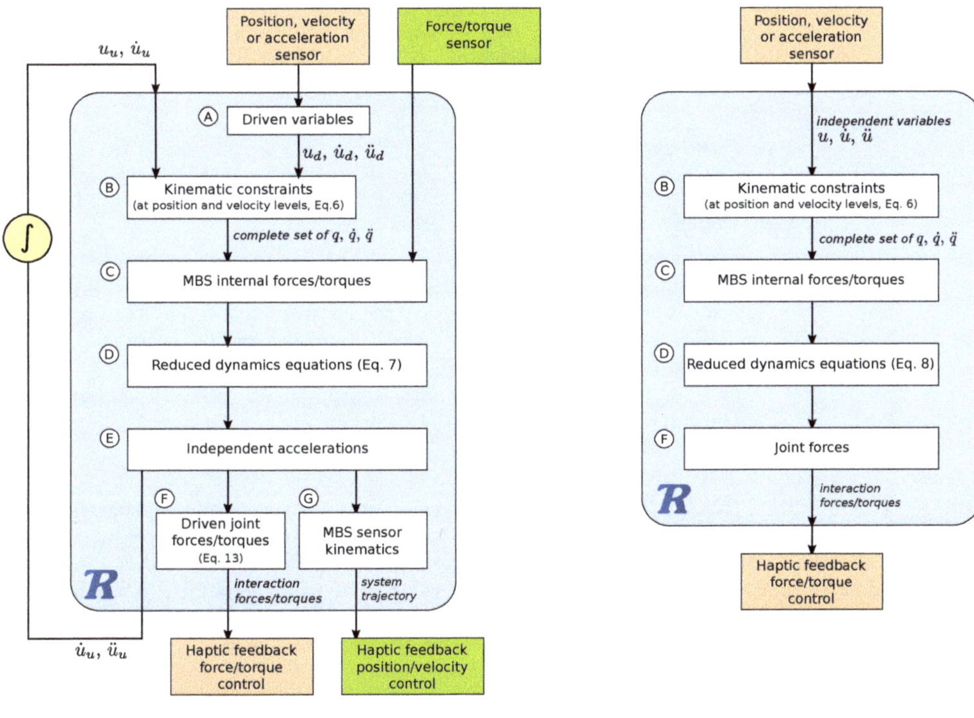

Figure 4. Flow diagram of direct and inverse dynamics in view of coupling them with haptic devices.

3.2. Symbolic Multibody Model

The aim of the symbolic generation of multibody models is to take advantage of both the genericity inherent to a numerical generation of multibody equations and of the simplification potentialities of their manual writing. So-called *symbolic* multibody programs manipulate arithmetical operators ($+, -, *, /$) and alphanumeric strings representing the data (m, c, K, etc., see Figure 5) to generate analytical equations in C, Python, or MATLAB, making the most of arithmetical and trigonometric simplifications. For a generic 3D MBS, the symbolic generation is performed only once (in less than one second with ROBOTRAN for MBS containing a hundred DOF) prior to any numerical analysis.

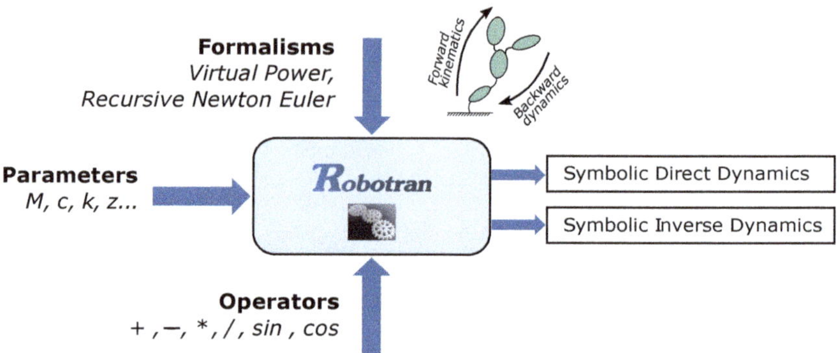

Figure 5. Symbolic Generation Process.

The computational superiority of the symbolic vs. the numerical approach is particularly attractive since it gives the opportunity to

- significantly speed up the computation of MBS models; this represents a very valuable asset for applications requiring real-time computing, such as those targeted here, and
- easily couple multibody models—being encapsulated in symbolic files—to other disciplines (such as optimization, control, electromechanical dimensioning) at software but also at hardware levels.

These two advantages are widely exploited in the applications underlying this work as detailed in the next sections.

ROBOTRAN Symbolic Generation

This section summarizes the key elements of the symbolic generation underlying the multibody software ROBOTRAN, the details of which can be found in reference [10], dedicated to this precise topic. There do obviously exist commercial general-purpose symbolic packages: why do we not use these to generate multibody equations?

The first reason relates to the amount of computer memory required to generate large symbolic models (up to 400 DOFs in our case). The combined use of dynamic memory allocation (via the use of C-pointers) and of the linked lists technique, allows us to finely control the ROBOTRAN memory requirement during the generation process.

The second reason concerns the possible simplifications of mathematical alphanumeric expressions. Although the simplification capabilities of commercial packages are extremely powerful, the symbolic condensation of multibody equations relies on a set of rules and tricks which are specific to MBS, namely, the simplification of complex trigonometric formulae and kinematic expressions, the pure and lasting elimination of unnecessary recursive equations and the parallelization of the latter, among others.

Arithmetic and Trigonometric Simplifications

To simplify multibody expressions containing unnecessary null data or similar terms which cancel each other, any encountered expression is recursively reorganized by ROBO-TRAN on the basis of a so-called *multibody priority rule* (e.g., a force symbol has a lesser priority than a mass symbol but has a higher priority than a generalized coordinate) as illustrated in the following priority "$<$" relationship:

$$m^i < d^{ij} < F_{ext}^i < q^i < sin, cos < b * c < (a - d) \tag{14}$$

in which we have intentionally classified expressions with respect to their physical meaning (ex.: mass, length, force, generalized coordinates, sine and cosine, etc.) in addition to their arithmetical nature.

For purely illustrative purposes, here are some excerpts of the symbolic generation of the inverse dynamics of a mechanism (articulation 2). *qk*, *qdk*, and *qddk*, respectively, represent the generalized coordinate, velocity and acceleration associated with joint k, *mi*, and *bij* are, respectively, masses and barycentric vectors. Dij and Ljk are geometric vectors. Pre-computations:

$$S1 = sin(q1); S2 = sin(q2); C1 = cos(q1); S3 = sin(q3);$$

$$q1p2 = q1 + q2; q2p3 = q2 + q3; q1p2p3 = q1p2 + q3;$$

$$S1p2 = sin(q1p2); S2p3 = sin(q2p3); S1p2p3 = sin(q1p2p3);$$

$$mb2 = m2 + m3; b11 = m1 * L11; b21 = m2 * L12 + m3 * D13; b31 = m3 * L13; b32 = m3 * L23;$$

Inverse dynamics (excerpt):

$$Q(2) = b11 * qdd1 * C1 \ldots - b31 * qd1 * qd1 * S1p2p3 \ldots - 2 * b21 * qd1 * qd2 * S1p2 \ldots - mb2 * g.$$

When dealing with a given symbolic expression (whatever its length and complexity), the ROBOTRAN symbolic engine *recursively* refers to the above priority rule (14) to ensure that the final form of any expression will be *purged* of zero parameters and of consecutive identical terms—or sub-expressions—with opposite signs.

As regards trigonometric expressions, which can be very numerous in MBS due to the 3D rotations, an internal process is activated to drastically simplify them, as illustrated by the following example (in which C and S stand for the *sine* and *cosine* functions, respectively, and *Sij* denotes the *sine* of the sum of the generalized coordinates q^i and q^j).

$$C2 * C4 * C56 * C56 * S8 + C2 * C4 * S56 * S56 * S8 + C2 * S4 * S56 * C8$$

$$+S2 * C4 * S56 * C8 - S2 * S4 * C56 * C56 * S8 - S2 * S4 * S56 * S56 * S8$$

is automatically simplified on-line by ROBOTRAN into:

$$C24 * S8 + S24 * S56 * C8$$

leading to far more efficient trigonometric computation.

Generation of Recursive Schemes

The formalism governing the generation of the direct or inverse dynamics (Equations (3) and (4)), uses a recursivity principle to compute all the variables along the tree-like structure of the MBS. For instance, the kinematic relation which expresses the absolute angular velocity ω^3 of a given body 3 with respect to that of its parent body 2, ω^2 is written in a vector form as

$$\omega^3 = \omega^2 + \Omega^{23} \tag{15}$$

where Ω^{23} represents the *relative* angular velocity vector of body 2 with respect to body 3.

Among the three components of this vector (and of all vectors recursively computed in the same manner as vector ω^3 in (15)), one surprisingly notices that some of them are squarely useless for the final form of the equations of motion (3) or (4). This can represent up to 30% of the intermediate equations such as some components of the above vector Equation (15): the ROBOTRAN symbolic engine can detect all of them before printing, via the use of a double-linked list containing the full set of recursive equations of type (15), and by tracing their individual utility for the final expected result, such as the mass matrix $M(q)$ or the generalized accelerations $\ddot{q}$.

Constrained MBS: Fully Symbolic Models

Let us rewrite the semi-explicit form (12) in an explicit way in terms of the independent accelerations $\ddot{u}_u$:

$$\ddot{u}_u = f(u_u, \dot{u}_u, u_d, \dot{u}_d, \ddot{u}_d, frc, trq) \tag{16}$$

Thanks to the current ROBOTRAN symbolic engine capabilities, particularly in terms of memory management, it is possible to generate the independent accelerations $\ddot{u}_u$ according to Equation (16) in a *fully symbolic* manner (Figure 6) and in the form of a unique set of recursive equations of type (15), which successfully compute (in C or Python):

- the constraints and their resolution at position, velocity, and acceleration levels;
- the external forces and torques (interfaced with possible external user constitutive equations);
- the dynamics of the restored tree-like MBS; and
- the reduction to an ODE system and its resolution with respect to the generalized accelerations $\ddot{u}_u$.

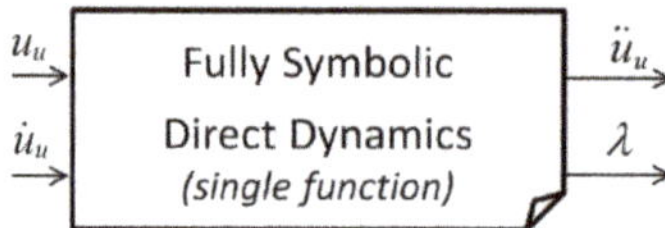

Figure 6. Fully symbolic generation of constrained MBS models.

Nowadays, this represents the most efficient symbolic direct dynamics model that ROBOTRAN can provide, making the most of its symbolic engine capabilities. More details on the above symbolic manipulations and tricks underlying the ROBOTRAN process can be found in details in [8].

4. Hardware Framework

The dynamic models obtained from the above symbolic multibody approach represent the key ingredient of our haptic device framework. Various competing—or even complementary—approaches can be used to reduce the computational complexity of the model, e.g., via deep learning techniques [31] or to make the most of object-oriented languages [32] and parallel programming [33]. Some results are promising but currently remain limited to rather simple MBS [31]. Regarding parallel computation, let us point out that symbolic generation lends itself perfectly to the vectorization of multibody models [8] and represents a promising avenue of exploitation of GPUs or FPGAs architectures in order to further improve model computation performances.

This section will detail the relationship between the multibody model and the human, that are both *in-the-loop*. In particular, we will describe the coupling between ROBOTRAN and the middleware ROS.

4.1. Specifications

First, as the haptic device is designed to be manipulated by a person, the apparatus specifications must meet some requirements of the human body. The human somatosensory system is composed of many sensors present at different levels of perception. According to the authors of [34], the exact contribution of the various mechanoreceptive channels to the formation of haptic perception remains to be established. Moreover, the physical values of perception, that are still the topic of fundamental researches, strongly depend on the application at hand. As regards tactile sensitivity, the latter is hard to characterize in term of frequency sensed [35] and force measurable [36] and needs psychophysical experiments to be able to understand the human tactile perception [37]. Generally speaking, it would appear that an update frequency of 1 kHz is considered as acceptable for human interactions [26,34,35].

In the scope of this paper, only the force feedback is considered, as the physical interface of our devices is exactly the same between the real object and its haptic equivalent, see Sections 5.1 and 5.2. In other words, the above-mentioned tactile sensitivity is intrinsically representative in our applications.

Second, the hardware design of a haptic device is a bidirectional apparatus that contains both motors and sensors, as it is for the human [38]. The latter may inject energy at some point, while receiving energy from the device at another time [39]. This bidirectionality is the most distinguishing feature of haptic interfaces [40].

The sensor resolution must be sufficient, especially when the movement is fast and small [41]. Of course, the sensors must not influence the device behavior, whose dynamics must satisfy the physics laws such as that of Newton [26].

4.2. Human-in-the-Loop Haptics with ROBOTRAN

The multibody approach described in Section 3 can be used to insert a highly dynamic model inside a real-time haptic device, which interacts with a human. Figure 7 shows the relationships between the ROBOTRAN model, the device, and the human body.

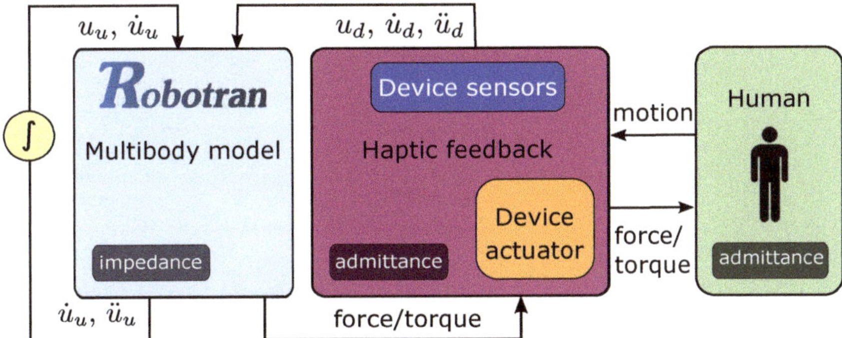

Figure 7. HIL haptic principle with ROBOTRAN. The light blue box, in this case, corresponds to the *Direct Dynamics* of Figure 4.

In haptics, taking the motion as input and the force as output, and vice versa, are the two possibilities, referred as impedance and admittance approaches in [41]. For the flow diagram shown in Figure 7, the device and the human are considered as an admittance operators, while the model can be assimilated to an impedance one. Other scenarios from Figure 4 are obviously possible.

The haptic device contains sensors that are used to measure the motion and to feed the multibody model. Those sensors are critical as they are in-the-loop located. Any error, noise or non-physic behavior will be reflected in the model and ultimately influence the feedback felt by the human who possesses his own somatosensory system, as discussed in Section 4.1.

Let us note of the importance of integration loop on the left in Figure 7, which must be fast, stable and precise enough. Indeed, the loop involving the human and the haptic device is instantaneous, while the integration loop is unavoidably slower, due to a fixed time-step size and thus a constant computation time.

4.3. ROS-ROBOTRAN Coupling

ROS (for "Robot Operating System") is a open-source software framework widely used in robotics. It contains numerous useful tools and libraries that allow for interactions between individual components, both in terms of software and hardware. ROS allows exploiting these tools in real-time to automate and control an electromechanical device.

Many ROS tools have been created by the ROS community to manipulate sensors or actuators and to publish or receive their data on a network. In addition, ROS offers many other benefits such as the ability to distribute the tasks requiring high processor usage across multiple hardware components. Furthermore, it makes possible to easily modify the system hardware without the need to make major changes in the software.

The coupling architecture between ROS and ROBOTRAN is presented in Figure 8.

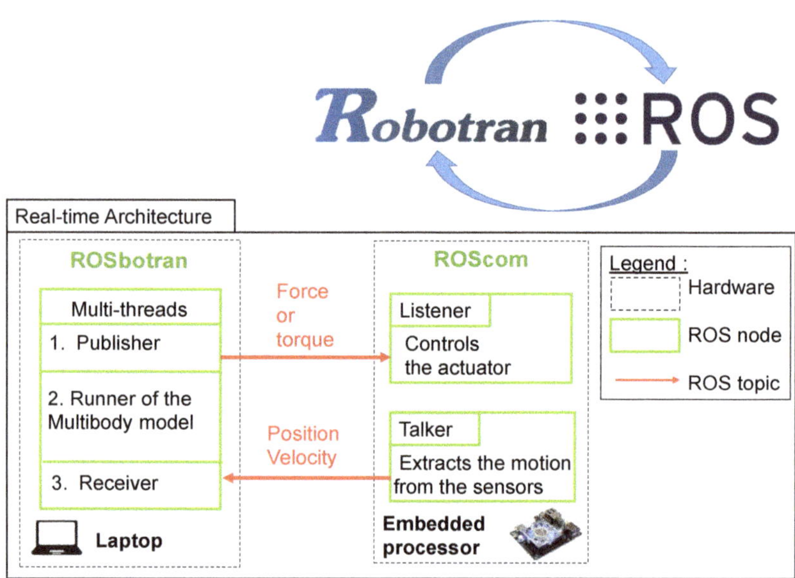

Figure 8. Software architecture with ROS and ROBOTRAN.

Regarding the software, two packages constitute the basis of the coupling. In ROS, a *package* designates a folder containing one or several executable codes, a configuration file for compiling these, as well as other files useful to ROS. A ROS *node* is an executable that is able to process some computation but also to communicate with other nodes thanks to the ROS *messages* and *topics*.

As far as we are concerned, a ROS-ROBOTRAN package—called **ROSbotran**—has been created, that is able to publish/subscribe to the data of a ROS network. It contains the multibody model running in real-time, which allows to simulate the mechanical system and to forecast, analyze, and control its behavior. Inside this package, for our applications of Section 5, this node uses the multi-threads approach to parallelize the calculations.

Another package—called **ROScom**—has simply a listener–receiver role, containing two nodes: *Listener* and *Talker*. Two *topics* are used to exchange kinematic and dynamic information between the ROSbotran and ROScom packages, see Figure 8. The communication is ensured through the SSH protocol and physically via an ethernet cable.

Regarding the hardware implementation, the ROSbotran package is run on a laptop with a high-frequency *Intel Core i7-9700 CPU 3.6 GHz*. This allows to compute in real-time the multibody models, even if they are complex. An embedded processor—the *Odroid UX4*—hosts the ROScom package, which requires less computing power. This processor ensures the interface with the sensors and actuator via appropriate electronics.

This ROS-ROBOTRAN integration is already operational for several applications (see Section 5) but could be easily generalized to all types of models, sensors and actuators. Let us note that the real-time process considered in this work achieves so-called *soft* requirements. Indeed, the Linux OS used does not meet a *hard* real-time implementation. Future works should implement this approach to specific hardware and adapt the software accordingly.

5. Applications

Following the general concepts of Section 4, practical implementations have been designed in our laboratory for rather contrasting applications: a haptic piano key (Section 5.1) and a driving simulator (Section 5.2). Other realizations have also been conceived within the framework of ongoing researches dealing with specific road vehicles.

Several sensors have been utilized in these prototypes, mostly to be ran *in-the-loop*. However, some external sensors presented in this section have also been specifically designed to validate both the multibody model and the corresponding haptic device.

In this section, we take benefits of several sensors to both validate the multibody itself and the ability of the ROS-ROBOTRAN coupling to give a representative feedback to the human.

5.1. Haptic Piano Key

The *touch* of a grand piano mainly results from a dynamic transmission mechanism—called *action*—which propels the hammer up to the string, given the key motion. The multibody model of Figure 9 has been developed at UCLouvain [42].

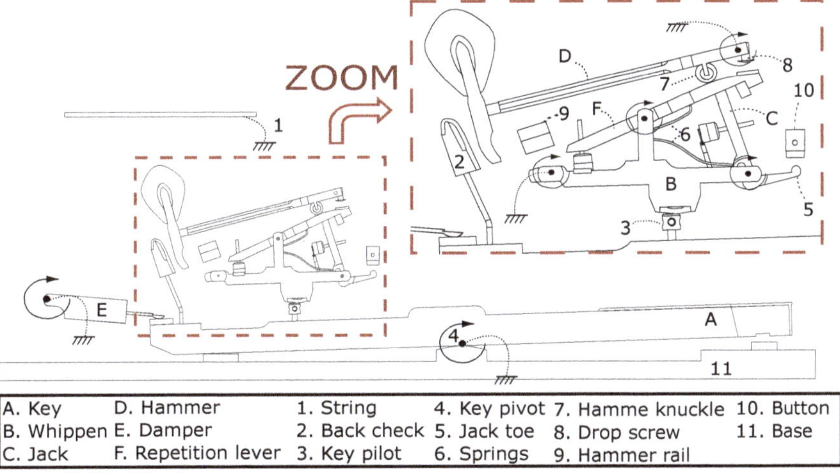

A. Key	D. Hammer	1. String	4. Key pivot	7. Hamme knuckle	10. Button
B. Whippen	E. Damper	2. Back check	5. Jack toe	8. Drop screw	11. Base
C. Jack	F. Repetition lever	3. Key pilot	6. Springs	9. Hammer rail	

Figure 9. Main components of the grand piano action for the MBS model: mobile bodies (resp. other elements) are indicated by letters (resp. numbers). Circular arrows represent the DOFs (adapted from [43]).

This model finely represents the action kinematic and dynamic behavior. A first kinematic validation with an external sensor—a high-speed camera—shows that the hammer motion is well caught by the model, a little less well from the back-check capture occurring at the end of the motion, when the key is fully depressed [42].

5.1.1. Previous Realizations

Several researchers have already worked to reproduce the haptic feeling of a piano keyboard thanks to active devices. In this context, the highly dynamic behavior of a piano action could clearly benefit from a multibody approach. Besides, it is difficult to reproduce its effect with only passive elements, as it is done in current digital piano for which manufacturers try to imitate the touch of a real acoustic piano.

In 1993, Gillespie presented the modeling of the piano action dynamics applied within an electromechanical apparatus [28]. The model is derived analytically and contains strong simplifications, which lack some very important dynamic features as the so-called action *escapement, Timmermans2020*. That being said, almost 30 years ago, using the technologies of that time, it was a first trial for a haptic piano key involving a dynamic model.

In 2002, Oboe developed the MIKEY keyboard [44], which allows one to simulate the type of touch of three instruments: the grand piano, the harpsichord, and the Hammond organ. He improved the grand piano action model from Gillespie's by managing the escapement phase. The considered dynamics still remains simplified compared to the complexity of the piano action. As Gillespie, the experimental prototype uses position sensors and voice coil motors.

Lozada proposed a different approach in 2007 [45], using a magneto-rheological system. He has derived an analytical model of both the physical behavior of the haptic device and the piano action. The key motion is sensed by a contactless photoelectric sensor, and the actuator needs also a force sensing. The final design differs quite significantly from a classic piano key.

With the same purpose to create a haptic piano key, Horvàth proposed in 2014 [46] an experimental fitting of a simplified model with only four constant parameters. He used force and optical sensor to relate the torque acting on the key under different impulses. Unfortunately, no further works has been published by the author to show the practical implementation inside a haptic device.

A broader platform called GENESIS-RT has been introduced by Leonard in 2015 [47] whose goal is to reproduce, among others, the piano action feedback. The piano model is based on a piano-inspired mechanics. The sensors and real-time update rates are quite high, 44 kHz and 4.41 kHz, respectively. No experimental results are shown with the device.

More recently in 2019, Adamou has suggested a two steps approach in [48]. First, an original measurement device has been used to characterize the force feedback according to the key position and velocity. Then, a replication device has been build with an electromagnetic actuator combined with a spring. A laser sensor is placed under the key to measure its position. Again, no validation with the proposed design is proposed.

Those previous attempts have used relatively few sensors, such as optical position sensors or force sensors. The goal was of course to limit the complexity and the cost of such devices. From our perspective, in a purely research context, using multiple sensors of high quality could help in quantifying the haptic feedback quality and prove the approach feasibility. Afterwards, one can imagine to simplify the design complexity for a possible industrial application. For instance, sensors could be integrated in the actuator itself, sensing the position through sensorless methods [49–51].

Using the multibody approach described in Section 3, a haptic piano key has been recently developed in our laboratory, with an experimental validation of its force feedback [30]. This implementation is described in the next section. Some improvements brought to the device, as well as further experimental validation are the subject of the subsequent Sections 5.1.3–5.1.5.

5.1.2. Mechatronic Implementation

For the pianists, the force feedback principle aims at reproducing the same *touch* sensation, i.e., F_{haptic} in Figure 10. An active device reproduces the force F_{act} on the key pilot, based on the reference force F_{mod} computed by the multibody model.

Kinematic sensors measure in real-time the key angular position and velocity, which are inputs of the multibody model, as shown in Figure 7.

Our prototype (Figure 11) consists of a linear actuator rigidly attached to a piano key. An embedded controller computes the model in real-time (i.e., in less than 1 ms), while the electronic boards drive the actuator accordingly. More details about the mechatronic design are given in [30].

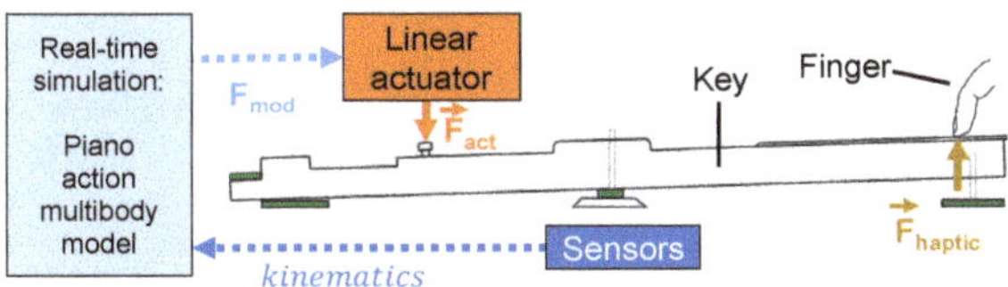

Figure 10. Piano action multibody model force feedback principle, adapted from the work in [30].

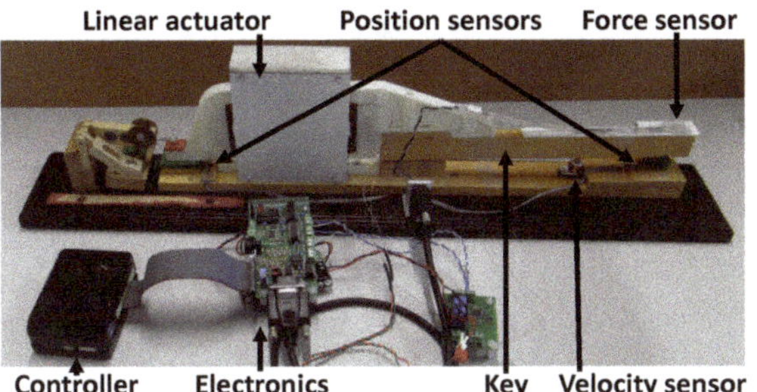

Figure 11. Mechatronic design of one key haptic feedback device, enhanced from the work in [52].

5.1.3. Model Validation

The model output force F_{mod} should render the force at the key pilot that would cause a force F_{haptic} equivalent to that of an acoustic piano keyboard. A custom force sensor has been developed that is based on strain gauges. The sensor sensitivity is 0.765 mV/V, for a measuring range from 0 up to 50 N, with a sampling frequency of 50 kHz, and with a resolution below the milliNewton. The designed load cell of Figure 12 will be inserted instead of the key pilot number 3 in Figure 9 to capture the strain so that the sensor can measure the force applied by the whippen to the key pilot. The latter is exactly the force F_{act} of Figure 10 that the actuator has to apply in the haptic device. This allows us to validate the output force of the multibody model, see Figure 7.

The experimental sensor is inserted in a real piano action demonstrator, a Renner® action (Figure 13). This homemade binocular element contains four strain gauges glued to the sides, connected through a Wheatstone bridge. This setup allows to precisely measure the normal force between the whippen and the key pilot, without influencing the action dynamic behavior.

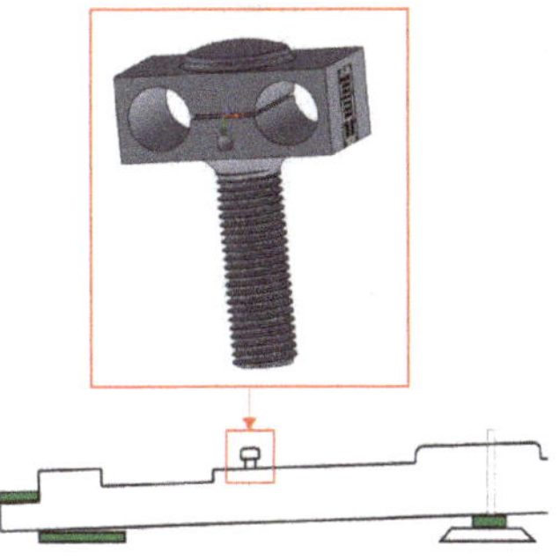

Figure 12. Load cell design.

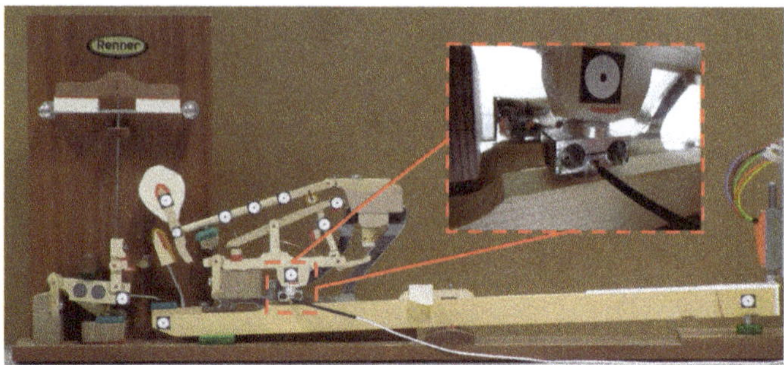

Figure 13. Experimental homemade force sensor set-up.

The key is actuated by an external actuator shown at the right of Figure 13. A high-speed camera at 1000 fps uses the markers to retrieve the action kinematics. The key motion is then supplied as a model input. This allows to validate the model dynamics, considering the measured motion as the reference, keeping in mind the limitations of the sensing method.

In Figure 14, a first experiment shows the hammer behavior, compared to the modeling and the measured angle with the high speed camera. Hammer angles are very close in case of this relatively slow double keystroke.

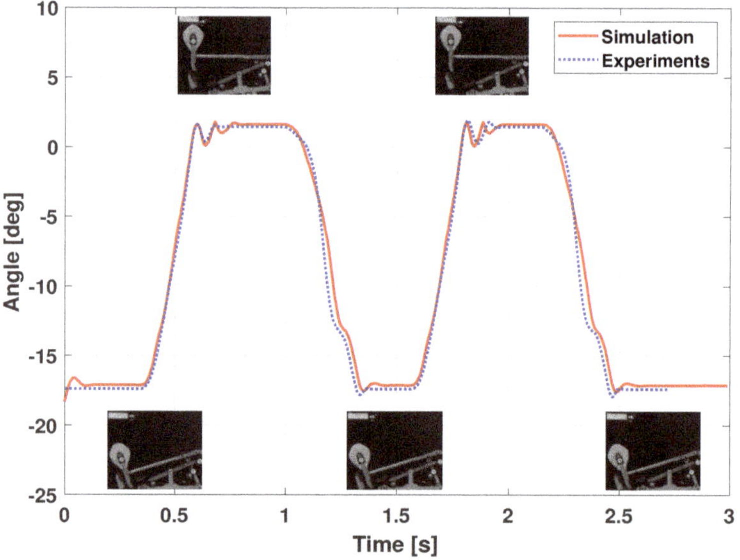

Figure 14. Piano action model dynamic validation: hammer joint position (slow double keystroke).

Results of the measured force versus the offline model-simulated force are shown in Figure 15. A double blow pattern is visible, the first key strike starting around 0.3 s and the second around 1.5 s. In both cases, the force starts by increasing, with oscillations, until a quasi steady-state value close to 3 N. Then, when the key is released—around 0.9 s and 2.1 s—the action returns to its resting position, with forces peaks due to the hammer rebound, whose mass and inertia dynamic contributions are relatively high compared to other elements.

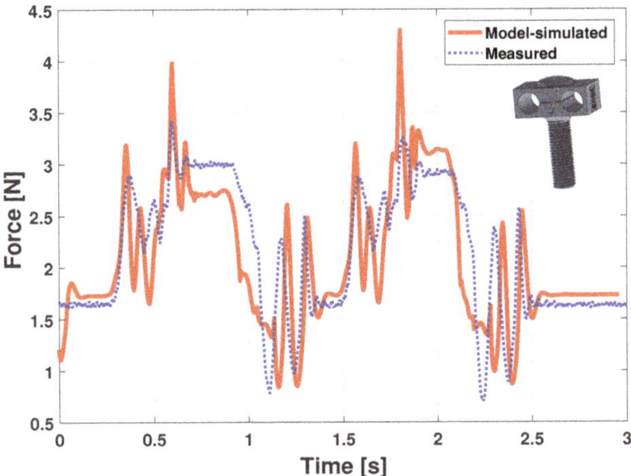

Figure 15. Piano action model dynamic validation: force (slow double keystroke).

In Figure 15, the simulation presents more oscillations all along the curve and shows higher force peaks. Apart from that, the measured force is very similar, which is encouraging for the model quality and its ability to finely reproduce the dynamic behavior of a piano action.

A faster double blow is experimented in Figures 16 and 17.

The hammer behavior is not identical between the simulation and the experiments after the hammer–string contact in Figure 16. In fact, the repetition lever acts differently: in experiments, the lever moves the hammer upwards around [0.25;0.35] s and [0.7;0.8] s, while the simulated behavior is different. Indeed, the repetition lever is difficult to tune both in practice and in simulation.

Despite that, the forces of Figure 17 are very similar, apart from the first part of the strokes, around 0.1 s and 0.5 s. Being a fast double blow, those differences are acceptable, and these results show rather good agreements with experiments, given all the remaining model parameter uncertainties.

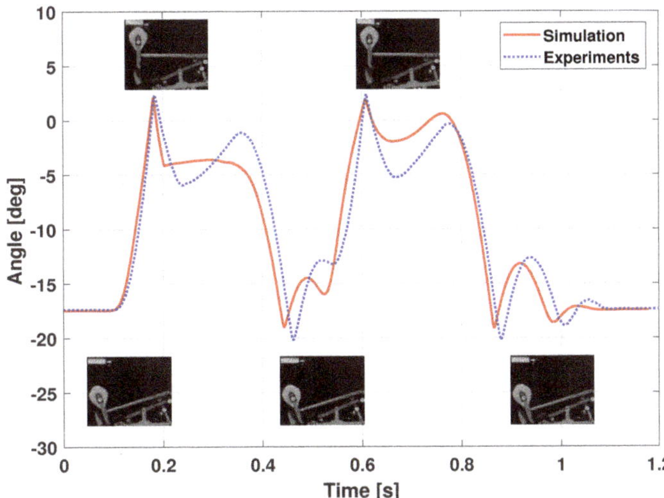

Figure 16. Piano action model dynamic validation: hammer joint position (fast double keystroke).

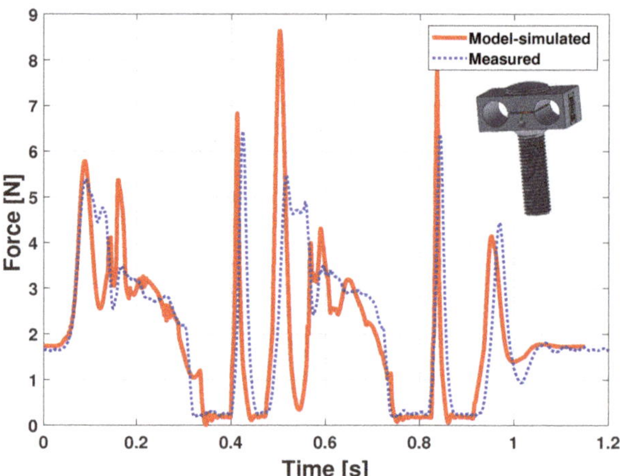

Figure 17. Piano action model dynamic validation: force (fast double keystroke).

5.1.4. Real-Time Sensors Validation

The kinematic input supplied to the multibody model, see Figure 7, is of the utmost importance for the quality of the haptic feedback. Noise, oscillations, or error in the input would unavoidably result in a non-realistic force computed by the multibody model.

In the piano key haptic device of Figure 11, two position sensors and a velocity sensor capture the key kinematics to continuously drive the multibody simulation.

The key angle is retrieved through two Hall effect *Allegro A1301* sensors placed in opposition under the key, one at the front and one at the rear, as presented in Figure 11. These two sensors measure respectively the voltages U_{front} and U_{rear}.

Hall sensors use magnetic field measurements which are not linear versus displacement. However, a chosen combination $U_{combined}$ between U_{front} and U_{rear} allow obtaining experimentally a linear relation between the measurements and the key front height:

$$U_{combined} = k \cdot (U_{front}^2 - U_{rear}^2) + O \tag{17}$$

where k is a proportionality factor determined by calibration and O is a fixed offset, also determined by experiments. The $U_{combined}$ can then be converted to a position measurement and finally to the key angle via simple trigonometry.

The velocity sensor of Figure 11 is an homemade voice coil, with a copper coil fixed on the frame and a moving permanent magnet glued on the key. The motion of the magnet creates a magnetic field variation inside the coil, inducing a varying *back-electromotive* force in the coil. This way, a voltage U_{coil} is created between the coil wire edges, whose value is directly proportional to the speed of the coil:

$$U_{coil} = \kappa \, \dot{z} \tag{18}$$

where κ is a proportionality factor found by experiment and linked to the coil characteristics and $\dot{z}$ is the moving magnet vertical velocity. This velocity is considered as purely vertical because the key angle remains small.

To validate those position and velocity sensors, an external *Polytec OFV-534* vibrometer is used, as presented in Figure 18. This vibrometer independently measures the position and the velocity with a resolution of 175 μm and 4.4 mm/s in our setup.

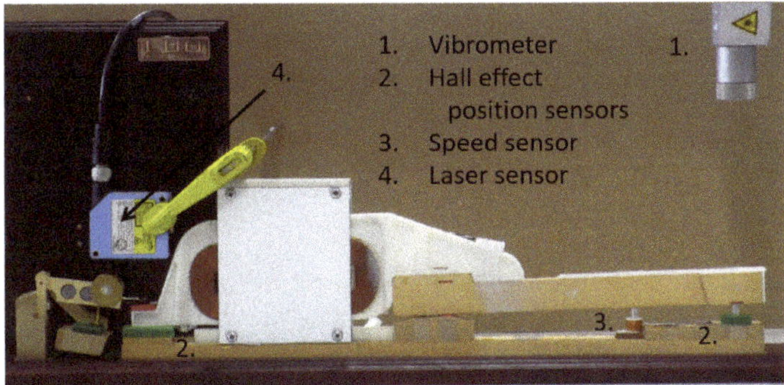

Figure 18. Piano haptic key: kinematics sensing validation.

First, the position estimation is experimented by manually applying up and down movements to the key, as shown in Figure 19.

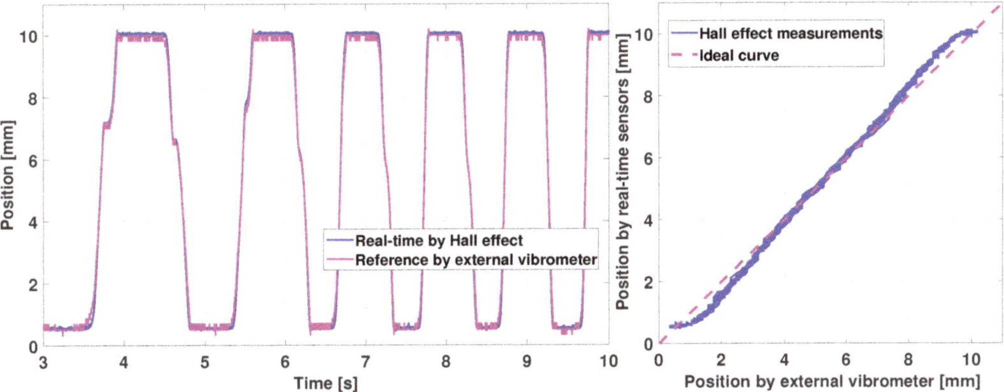

Figure 19. Haptic key validation of position sensors: slow motion.

The Hall effect real-time measured position is close to the reference obtained with the vibrometer. The right graph in Figure 19 shows the relation between measured and reference positions.

The experiment of Figure 19 includes a motion at slow frequency, less than 1 Hz. Performing the exact the process at a higher input frequency, around 5 Hz, provides the results shown in Figure 20. The position sensing is still consistent. However, discrepancies appear that cause the Hall effect measurement to have a maximum error of 10% between the estimation and the reality, which could be enhanced, but is sufficient for the considered application.

Furthermore, the combination of (17) allows to maintain the sensor sensitivity, i.e., the slope of the right curves in Figures 19 and 20. Indeed, the relation between the measured position and the real one of Figure 19 has almost a constant slope, except for a small quadratic deviation at the lower and upper position between [0;1] and [9;10] mm, due to the magnetic field quadratic relation with the position. This does not impact the haptic prototype behavior but it can be improved in the future by enhancing the position sensing via a better calibration.

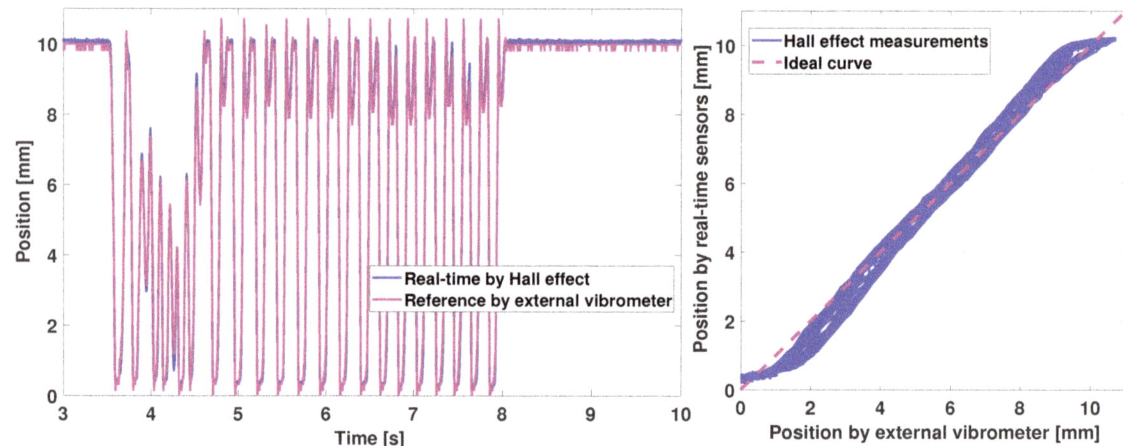

Figure 20. Haptic key validation of position sensors: fast motion.

Regarding the velocity, several up and down motions are again manually applied to the key, and the homemade coil sensor output is compared to the vibrometer measurements in Figure 21.

The homemade sensor captures the velocity amplitude quite well, despite some erroneous values at some points. Indeed, some peaks are present with the homemade but not with the vibrometer, for instance, around 5.5 and 6.7 s in Figure 21, left.

Figure 22 shows the same experiment for *halfway* keystroke, for which the key is not fully depressed but the action is still able to repeat the note, as foreseen by the so-called *double escapement* grand piano action: this feature is fundamental for pianists. The validation of Figure 22 presents our sensor capability to capture this specific movement.

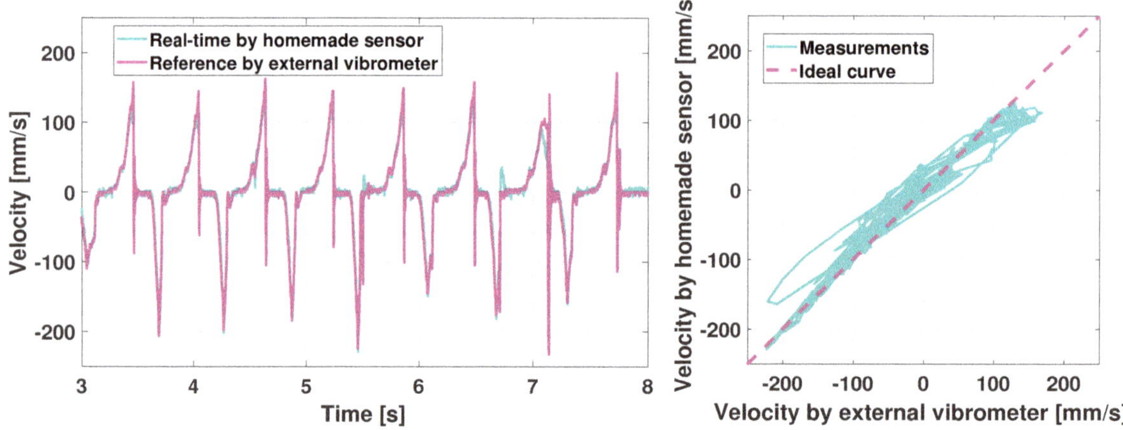

Figure 21. Haptic key validation of the velocity sensor: high amplitude, fast motion.

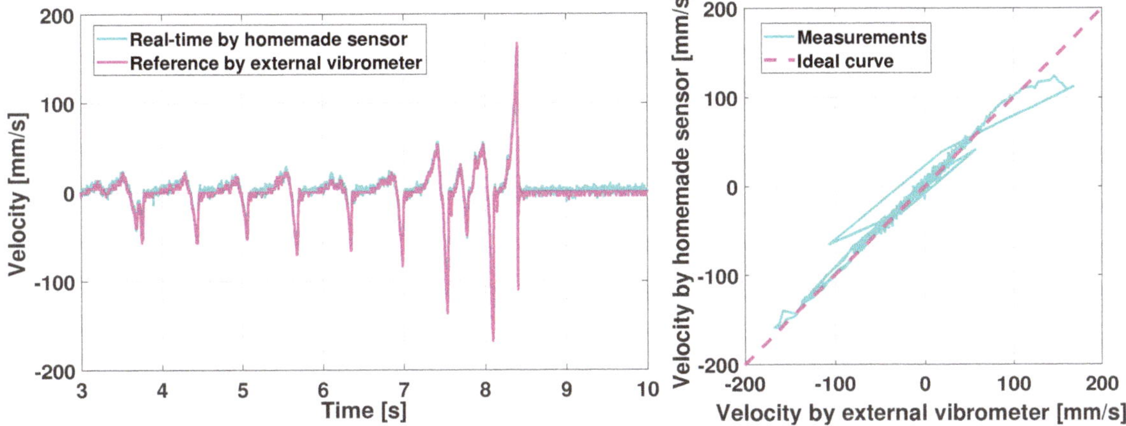

Figure 22. Haptic key validation of the velocity sensor: halfway keystroke.

5.1.5. Haptic Key Results

In [30], the actuator design of our haptic prototype (Figure 11) was exposed and its ability to reproduce the force on the piano key was tested. In the present section, we analyze in further detail the haptic force felt by the pianist in various conditions. For this purpose, we take advantage of the ROS-ROBOTRAN coupling presented above in order to investigate various settings of the piano action like the position of the button height or the mass of the hammer.

The force felt by the pianist is F_{haptic} of Figure 10, which depends of course of F_{mod} and F_{act} but also on the physical components of the prototype. To characterize F_{haptic}, an external actuator *Faulhaber LM1247* has been used to apply the same position-driven profile that corresponds to one full key dip at 10 mm/s. The F_{haptic} force can be deduced from the external actuator current measurement on both the action demonstrator and the prototype, see Figure 23.

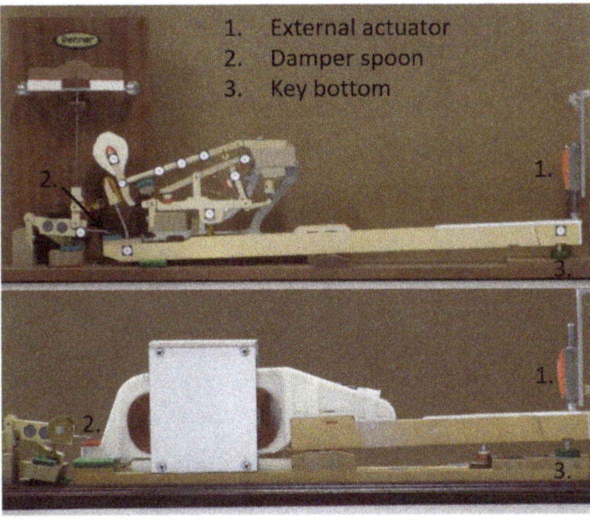

Figure 23. Piano haptic key: measurements of the force felt by the pianist F_{haptic} through an external linear actuator for the Renner® demonstrator (**up**) and the haptic prototype (**bottom**) [30].

We first reproduced and checked the test in [30] with the additional sensors presented and validated above. Figure 24 shows F_{haptic} versus the key vertical position measured at its tip. A key at rest refers to position zero and position 9.5 mm to a fully depressed key, as proposed by [53]. We clearly retrieve the phases $\boxed{A}\!\!-\!\!\boxed{B}\!\!-\!\!\boxed{C}\!\!-\!\!\boxed{D}$ as described in [30].

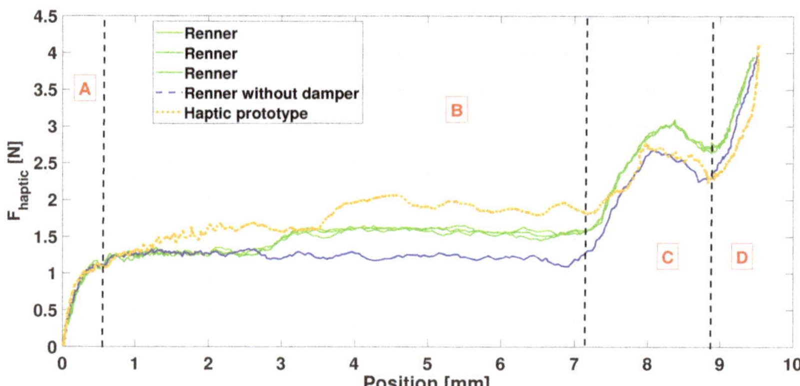

Figure 24. Haptic piano key force validation: prototype comparison with the Renner® demonstrator (results confirming those obtained in [30]).

In addition, similar behaviors can be observed in the profiles of Figure 24, despite some differences due to the physical components that differ between the prototype and the demonstrator. For instance, the blue curve shows the action behavior, without the damper, illustrating its role. The measure on the action demonstrator has been performed three times in a row to highlight the consistency between experiments.

In short, the crucial *escapement* phenomenon occurs during phase $\boxed{C}$ in Figure 24. The difference in the force amplitude between the prototype and the reference may be due to residual inaccuracies in the MBS model.

To illustrate the interest of having a multibody model included in the haptic prototype, variations of the action parameters can be done with the prototype. Their effects on the haptic feedback can be compared with the Renner® demonstrator for which these values have been physically modified.

Figure 25 presents the haptic force for the Renner® (left) and the haptic prototype (right). The parameter is the height of the let-off button [43], see also Figure 9. In the legend, *Normal* means that the setting value is nominal and *Low* (resp. *High*) that the let-off button is lower (resp. higher) than its nominal height, by approximately 0.5 mm.

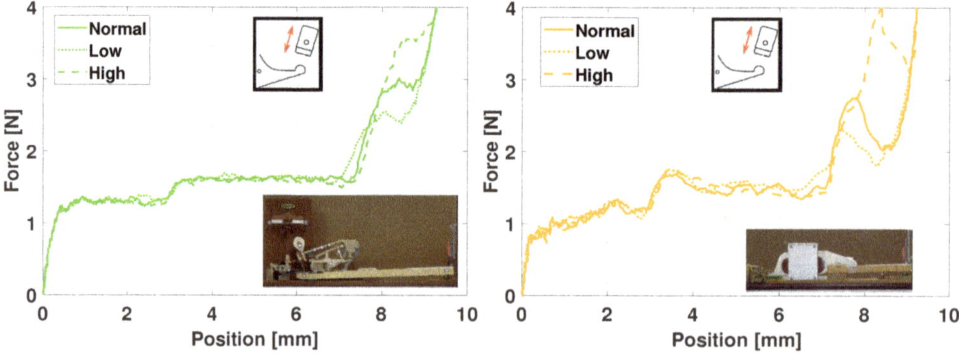

Figure 25. Haptic piano key force validation: variation of the button height, physically on the Renner® demonstrator (**left**) and virtually on the haptic key prototype (**right**).

In Figure 25, the trends are similar between the prototype and the demonstrator: a higher (resp. lower) button causes the escapement phase to have a higher (resp. lower) force value around [8–9] mm. The escapement is even more stressed in the prototype.

Figure 26 shows the effects of the hammer mass variation, i.e., meaning that a punctual mass of 0.003 kg has been added once (resp. twice) for the *High* (resp. *Very high*) case.

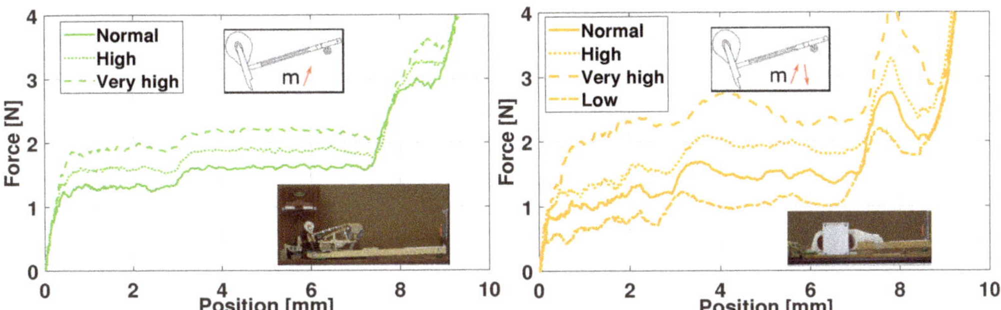

Figure 26. Haptic piano key force validation: variation of the hammer mass, physically on the Renner® demonstrator (**left**) and virtually on the haptic key prototype (**right**).

Again, the impact is similar for the *High* and *Very high* cases, with an increase of the haptic force because the hammer is heavier.

Besides, one advantage of the multibody model is that it can virtually perform many interesting investigations, for instance, the hammer mass can easily be lowered. Doing so in real life would require to manufacture a whole new hammer. This variation is illustrated with the *Low* curve in Figure 26, for the haptic prototype only, resulting with a lower haptic force until the key-bottom contact (phase $\boxed{\text{D}}$ in Figure 24).

Despite some discrepancies, the above results (Figures 24–26) show that the ROS-ROBOTRAN coupling proposed in this paper allows to develop a haptic key for digital pianos able to reproduce the F_{haptic} action dynamic force quite faithfully, i.e., the *gold-standard touch* of a grand piano. Moreover, the symbolic multibody modeling approach makes it very easy to modify any physical parameter of the action or even the action itself in the haptic device, to modulate the haptic force accordingly. This feature was actually appreciated by some pianists and piano tuners who were consulted for this haptic keyboard project.

5.2. Haptic Driving Simulator

Simulators for vehicle driving are nowadays a common tool and used for many applications [54]. Using a real-time multibody model including all the suitable physical parameters, allows to deal with the highly dynamics effects of a vehicle behavior. To analyze the dynamic performances of a vehicle, other approaches consider for example object-oriented programming of autonomous virtual drivers [32], instead of a real human. A prototype has been built that aims at reproducing the handling torque feedback in the steering wheel. Figure 27 shows the experimental set-up with the block schematics. As in all simulators, the driver has a real-time visualisation of the moving environment on a front screen.

In this haptic demonstrator, mainly developed at UCLouvain for educational purpose, a direct drive actuator acts on the steering wheel rotation. It contains an absolute angular encoder *SinCos Hiperface SKM36 Multiturn* which measures the position q and the velocity $\dot{q}$ of the steering wheel. This information is sent through the *data transfer module*—in this case, an embedded processor *Raspberry Pi 3*—to the multibody model of the a full 3D vehicle. Afterwards, the corresponding torque T—computed by the inverse dynamics Equation (13)—is applied by the actuator to the human arms.

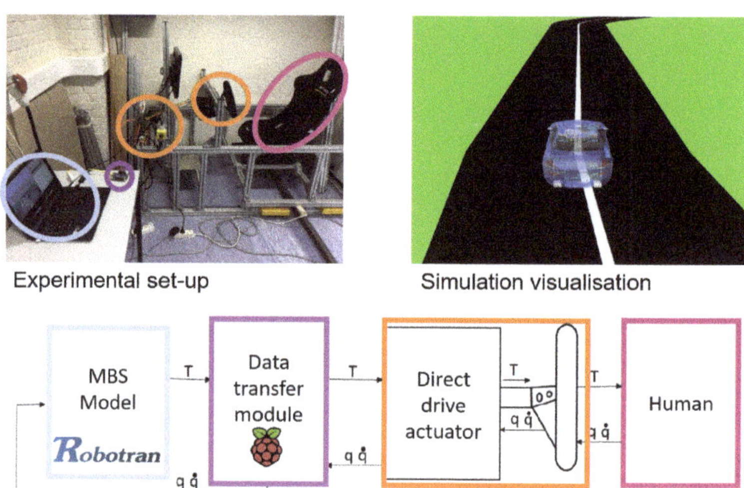

Figure 27. Haptic steering wheel: experimental setup with its real-time visualization and its corresponding block diagram below.

In the following illustrative experiment, in which a driver handles a virtual vehicle, the situation represents an obstacle avoidance while driving on a straight line at constant speed of 75 km/h. During the simulation, an obstacle suddenly and randomly appears, and the driver needs to avoid it by turning on the left and then coming back on the straight line. Meanwhile, for each of the three simulations, the caster distance of the front suspensions is modified without notifying it to the driver.

Figure 28 shows the corresponding lateral versus longitudinal vehicle displacements, as well as the obstacle. Without going into a detailed analysis, one can see that the behavior clearly differs depending on the trials and on the setting of the caster.

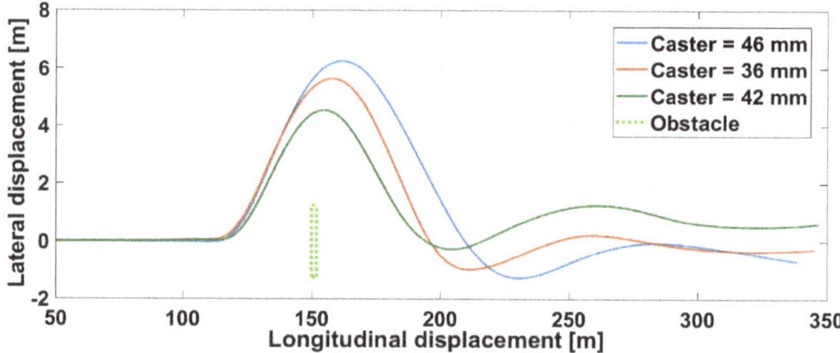

Figure 28. Haptic steering wheel: vehicle displacement on the ground with various caster.

The steering wheel angle, visible in Figure 29, illustrates in a different way the reaction of the driver to the obstacle apparition.

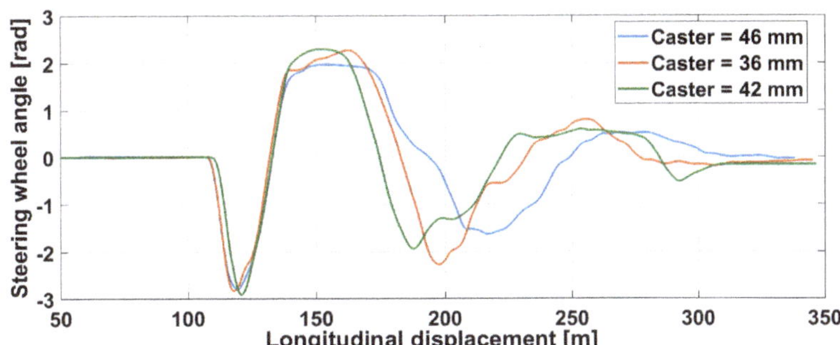

Figure 29. Haptic steering wheel: angle with various caster.

Finally, the torque given as a feedback is shown in Figure 30. Note that this value is taken from the model output, directly in the loop, not measured with an external sensor. For the first negative peak and for a very close angle value, the feedback torque is higher for a higher caster, as expected. After that, the driver counter-steers and tries to stabilize the vehicle towards its initial trajectory.

While a wide range of experiments could be achieved to relate the caster to the driver behavior, this illustrative example clearly highlights the capabilities of a multibody-based haptic steering wheel.

The prototype currently lacks the possibility to control the vehicle velocity. Simulations are done at a constant speed. Current work is ongoing to add *Penny&Giles HLP190* potentiometers to measure the position of the existing brake and accelerator pedals. Adding these sensors would allow to instantaneously adapt the vehicle velocity. Thanks to the ROS-ROBOTRAN coupling described in the previous sections, adding this hardware is quite straightforward in this software environment.

In the future, more experiments can be envisaged with several types of drivers, to analyze their feeling and torque feedback. This way, a wide range of vehicle dynamic parameters can be analyzed, such as the wheel toe-in/toe-out, the tire friction, the roll center height or the anti-roll bar stiffness, among others.

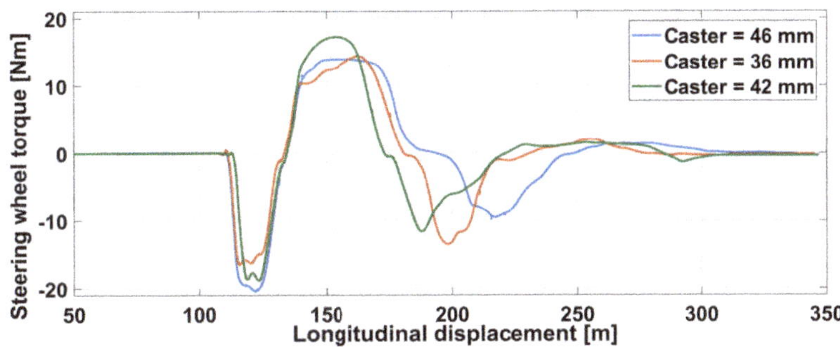

Figure 30. Haptic steering wheel: torque with various caster.

5.3. Other Implementations

The two projects presented before constitute the main current applications of the ROS-ROBOTRAN approach. However, other implementations involving a coupling between multibody models with real-time environments have been developed. In this scope, various sensors have been customized and utilized, showing our growing interest in coupling multibody dynamics with sensors for validation or haptic purposes.

For instance, during the Ph.D. thesis of Docquier [55] at UCLouvain, a small-scale demonstrator of a Narrow Tilting Vehicles (NTV) has been designed. The demonstrator allows to test embedded controllers thanks to inboard sensors that are used in real-time via ROS to update and analyze the vehicle behavior whose multibody model was built in parallel. Figure 31 presents the prototype and zooms on the two main sensors.

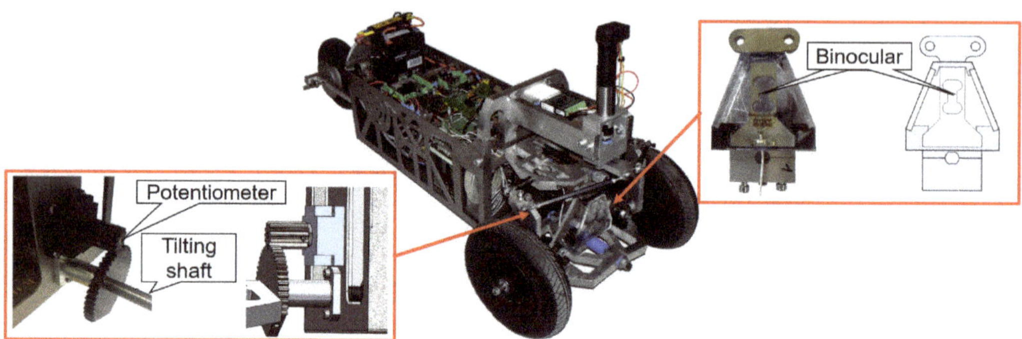

Figure 31. NTV demonstrator and its two main sensors: a rotating potentiometer and a binocular strain gauge for torque measurement.

Besides a classical potentiometer connected to the titling shaft to measure the vehicle tilt, a torque sensor based on a binocular strain gauge provides information associated with the actuated tilting degree of freedom of the vehicle. This information is thereafter used for control and validation purposes. Let us note that the vehicle also carries an IMU on board, to combine the tilting value from both IMU and potentiometer with a Kalman filter.

The second project concerns a so-called "kart" bench designed by and for students, to enable them to visualize and understand the road handling behavior of a four-wheel vehicle (Figure 32), and to compare the real system with its multibody model. The kart is laterally held on a conveyor belt and can be human-driven from a remote steering wheel linked to the front suspension via Bowden-type cables. Real-time multibody model can show live the four lateral forces on the tires, for instance, for different types of driving behaviors and/or kart suspension settings.

Figure 32. Sensors of the kart bench.

Apart from a classical optical sensor measuring the conveyor belt velocity, see Figure 32, a longitudinal and transverse force measurement through a homemade binocular load cell with strain gauges is placed between the frame and the moving kart to quantify the lateral force on the kart. It allows to analyze in real-time the effects on the tire-belt interaction forces and also to compare them with those computed live by the multibody model, while modifying suspension settings such as the toe-in/toe-out.

Let us note that this prototype does not run with ROS. Instead, it exploits *Labview* to communicate with the dedicated electronics and to retrieve the necessary sensors information, as for the above-mentioned applications.

6. Conclusions and Prospects

The main objective of this work was to highlight that the current multibody modeling formulations had reached a sufficient maturity to be exploited within haptic devices applied to mechanical systems involving high dynamics.

In a first step, we presented a multibody formalism in relative coordinates whose constraint equations are eliminated by a proper reduction of the equations. This formalism lends itself perfectly to its programming via the symbolic approach whose capacity of equation manipulation and simplification allows to produce very compact models, i.e., perfect candidates to real-time computation.

The interaction and coupling of these models with the world of sensors is essential, on the one hand, to validate the models themselves with respect to their underlying physics and, on the other hand, to allow a reliable coupling between the model, the sensors, and the actuators of a haptic device. The ROS platform has been chosen as the meta architecture to ensure these couplings successfully: two applications illustrate our developments. The first one, presented in detail, concerns the development of a haptic piano key and demonstrates the capabilities of the approach for a very high dynamic system. The second one refers to a driver simulator under development in our laboratory, whose interest—above all pedagogical—is to show the appeal of multibody models for this kind of application. In particular, it allows students to observe the impact of different parameters on the system dynamics, as taught in the academic courses but through multibody equations that can be a little bit abstruse for some of them.

In terms of perspectives, it would be interesting to continue the investigations on the model side, in particular through the fine-grain parallelization of the symbolic equations. Preliminary tests in ROBOTRAN have indeed shown that recursive equations can be reorganized in very few sequential vectorial steps: this asset could be exploited in order to further reduce the computation time of models to be embedded in devices requiring real-time computation.

Besides, we think that the potentialities of coupling real-time multibody dynamics with sensors and actuators within a middleware platform (e.g., ROS) are very promising for the future. It will enable to develop new concepts of haptic systems that are generic, user-friendly, and efficient. In particular, we see a real pedagogical interest in the use of such an architecture for the fields of multibody modeling, sensor implementation and mechatronic design of haptic systems with a very pronounced dynamic character, such as those presented in the present work.

Author Contributions: Conceptualization, methodology, software N.D. and P.F.; validation, S.T.; formal analysis, investigation, writing N.D., S.T., and P.F.; All authors have read and agreed to the published version of the manuscript.

Funding: Sébastien Timmermans is FRIA Grant Holder of the Fonds de la Recherche Scientifique-FNRS, Belgium.

Institutional Review Board Statement: Not applicable.

Informed Consent Statement: Not applicable.

Data Availability Statement: Not applicable.

Acknowledgments: Authors would like to thank Thierry Daras, Alex Bertholet, Antoine Bietlot, Quentin Docquier and Aubain Verlé from UCLouvain, as well as Théo Tuerlinckx, François Huens, and Sébastien de Longueville for their help. Figure 9 is adapted from the paper published in Mechanism and Machine Theory, Vol 160, Timmermans, S.; Ceulemans, A.E.; Fisette, P., Upright and grand piano actions dynamic performances assessments using a multibody approach, 104296, Copyright Elsevier.

Conflicts of Interest: The authors declare no conflict of interest. The funders had no role in the design of the study; in the collection, analyses, or interpretation of data; in the writing of the manuscript; or in the decision to publish the results.

Abbreviations

The following abbreviations are used in this manuscript:

CPU	Central Processing Unit
DAE	Differential-Algebraic Equation
DOF	Degree of Freedom
FPGA	Field Programmable Gate Area
GPU	Graphics Processing Unit
HIL	Human-in-the-loop
IMU	Inertial Measurement Unit
MBS	Multibody Systems
NTV	Narrow Tilting Vehicles
ODE	Ordinary Differential Equation
OS	Operating System
RAM	Random Access Memory
ROS	Robot Operation System
SSH	Secure Shell
YARP	Yet Another Robot Platform

References

1. Lin, M.C.; Otaduy, M. (Eds). *Haptic Rendering: Foundations, Algorithms, and Applications*, 1st ed.; A K Peters—CRC Press: Boca Raton, FL, USA, 2008. [CrossRef]
2. Sreelakshmi, M.; Subash, T. Haptic Technology: A comprehensive review on its applications and future prospects. *Mater. Today Proc.* **2017**, *4*, 4182–4187. [CrossRef]
3. Wittenburg, J. *Dynamics of Systems of Rigid Bodies*; Teubner Verlag: Berlin, Germany, 1977. [CrossRef]
4. Haug, E. *Computer-Aided Kinematics and Dynamics of Mechanical Systems Volume-I: Basics Methods*; Allyn and Bacon: Boston, MA, USA, 1989; Volume 1, ISBN 0-205-11669-8.
5. Nikravesh, P. *Computer-Aided Analysis of Mechanical Systems*; Prentice-Hall Int.: Hoboken, NJ, USA, 1988; ISBN 0-13-162702-3.
6. Geradin, M.; Cardona, A. *Flexible Multibody Dynamics: A Finite Element Approach*; Wiley–Blackwell: Hoboken, NJ, USA, 2001; ISBN 978-0471489900.
7. Roberson, R.; Schwertassek, R. *Dynamics of Multibody Systems*; Springer: Berlin/Heidelberg, Germany, 1988. [CrossRef]
8. Samin, J.; Fisette, P. *Symbolic Generation of Multibody Systems*; Springer: New York, NY, USA, 2003; ISBN 978-1402016295.
9. Garcia de Jalon, J.; Bayo, E. *Kinematic and Dynamic Simulation of Multibody Systems: The Realtime Challenge*; Springer: New York, NY, USA, 2011; ISBN 978-1461276012.
10. Docquier, N.; Poncelet, A.; Fisette, P. ROBOTRAN: A powerful symbolic gnerator of multibody models. *Mech. Sci.* **2013**, *4*, 199–219. [CrossRef]
11. Pucheta, M.; Cardona, A.; Preidikman, S.; Hecker, R.E. *Multibody Mechatronic Systems (Papers from the MuSMe Conference in 2020)*; Mechanisms and Machine Science Series; Springer International Publishing: Berlin/Heidelberg, Germany, 2021; ISBN 978-3-030-60371-7.
12. Docquier, N.; Lantsoght, O.; Dubois, F.; Brüls, O. Modelling and simulation of coupled multibody systems and granular media using the non-smooth contact dynamics approach. *Multibody Syst. Dyn.* **2020**, *49*, 181–202. [CrossRef]
13. Escalona, J.L.; Urda, P.; Muñoz, S. A Track Geometry Measuring System Based on Multibody Kinematics, Inertial Sensors and Computer Vision. *Sensors* **2021**, *21*, 683. [CrossRef]
14. Risaliti, E.; Tamarozzi, T.; Vermaut, M.; Cornelis, B.; Desmet, W. Multibody model based estimation of multiple loads and strain field on a vehicle suspension system. *Mech. Syst. Signal Process.* **2019**, *123*, 1–25. [CrossRef]
15. Cuadrado, J.; Michaud, F.; Lugrís, U.; Soto, M.P. Using Accelerometer Data to Tune the Parameters of an Extended Kalman Filter for Optical Motion Capture: Preliminary Application to Gait Analysis. *Sensors* **2021**, *21*, 427. [CrossRef]
16. Hashemi, S.; Friedrich, H.; Bobach, L.; Bartel, D. Validation of a thermal elastohydrodynamic multibody dynamics model of the slipper pad by friction force measurement in the axial piston pump. *Tribol. Int.* **2017**, *115*, 319–337. [CrossRef]
17. Zierath, J.; Rachholz, R.; Rosenow, S.E.; Bockhahn, R.; Schulze, A.; Woernle, C. Modal testing on wind turbines for validation of a flexible multibody model. In Proceedings of the ECCOMAS Thematic Conference Multibody Dynamics, Prague, Czech Republic, 19–22 June 2017
18. Torres-Moreno, J.; Blanco-Claraco, J.; Giménez-Fernández, A.; Sanjurjo, E.; Naya, M. Online Kinematic and Dynamic-State Estimation for Constrained Multibody Systems Based on IMUs. *Sensors* **2016**, *16*, 333. [CrossRef]

19. Khadim, Q.; Kaikko, E.P.; Puolatie, E.; Mikkola, A. Targeting the user experience in the development of mobile machinery using real-time multibody simulation. *Adv. Mech. Eng.* **2020**, *12*, 168781402092317. [CrossRef]
20. Rodríguez, A.J.; Pastorino, R.; Carro-Lagoa, Á.; Janssens, K.; Naya, M.Á. Hardware acceleration of multibody simulations for real-time embedded applications. *Multibody Syst. Dyn.* **2020**. [CrossRef]
21. Pastorino, R.; Cosco, F.; Naets, F.; Desmet, W.; Cuadrado, J. Hard real-time multibody simulations using ARM-based embedded systems. *Multibody Syst. Dyn.* **2016**, *37*, 127–143. [CrossRef]
22. Ros, J.; Plaza, A.; Iriarte, X.; Pintor, J.M. Symbolic multibody methods for real-time simulation of railway vehicles. *Multibody Syst. Dyn.* **2017**, 1. [CrossRef]
23. Rivera, Z.B.; Simone, M.C.D.; Guida, D. Unmanned Ground Vehicle Modelling in Gazebo/ROS-Based Environments. *Machines* **2019**, *7*, 42. [CrossRef]
24. Munawar, A.; Wang, Y.; Gondokaryono, R.; Fischer, G.S. A Real-Time Dynamic Simulator and an Associated Front-End Representation Format for Simulating Complex Robots and Environments. In Proceedings of the 2019 IEEE/RSJ International Conference on Intelligent Robots and Systems (IROS), Macau, China, 3–8 November, 2019. [CrossRef]
25. Habra, T.; Dallali, H.; Cardellino, A.; Natale, L.; Tsagarakis, N.; Fisette, P.; Ronsse, R. Robotran-YARP Interface: A Framework for Real-Time Controller Developments Based on Multibody Dynamics Simulations. In *Computational Methods in Applied Sciences*; Springer International Publishing: Berlin/Heidelberg, Germany, 2016; pp. 147–164. [CrossRef]
26. Xia, P. New advances for haptic rendering: State of the art. *Vis. Comput.* **2018**, *34*, 271–287. [CrossRef]
27. Paris, J.N.; Archut, J.L.; Hüsing, M.; Corves, B. Haptic simulation and synthesis of mechanisms. *Mech. Mach. Theory* **2020**, *144*, 103674. [CrossRef]
28. Gillespie, B.; Cutkosky, M. Interactive Dynamics with Haptic Display. In Proceedings of the 2nd annual symposium on Haptic Interfaces for Virtual Environment and Teleoperator Systems, ASME/WAM, New Orleans, LA, USA, November 1993; p. 55-1. Available online: http://www-personal.umich.edu/~brentg/Web/Conference/asme93.pdf (accessed on 12 July 2021)
29. Dialynas, G.; Happee, R.; Schwab, A.L. Design and hardware selection for a bicycle simulator. *Mech. Sci.* **2019**, *10*, 1–10. [CrossRef]
30. Timmermans, S.; Dehez, B.; Fisette, P. Multibody-Based Piano Action: Validation of a Haptic Key. *Machines* **2020**, *8*, 76. [CrossRef]
31. Angeli, A.; Desmet, W.; Naets, F. Deep learning for model order reduction of multibody systems to minimal coordinates. *Comput. Methods Appl. Mech. Eng.* **2021**, *373*. [CrossRef]
32. Perrelli, M.; Cosco, F.; Carbone, G.; Lenzo, B.; Mundo, D. On the Benefits of Using Object-Oriented Programming for the Objective Evaluation of Vehicle Dynamic Performance in Concurrent Simulations. *Machines* **2021**, *9*, 41. [CrossRef]
33. Jahnke, M.D.; Cosco, F.; Novickis, R.; Rastelli, J.P.; Gomez-Garay, V. Efficient Neural Network Implementations on Parallel Embedded Platforms Applied to Real-Time Torque-Vectoring Optimization Using Predictions for Multi-Motor Electric Vehicles. *Electronics* **2019**, *8*, 250. [CrossRef]
34. Hayward, V. A Brief Overview of the Human Somatosensory System. *Music. Haptics* **2018**. [CrossRef]
35. Verrillo, R.T. Vibration sensation in humans. *Music. Percept. Interdiscip. J.* **1992**, *9*, 281–302. [CrossRef]
36. Papetti, S.; Järveläinen, H.; Giordano, B.L.; Schiesser, S.; Fröhlich, M. Vibrotactile sensitivity in active touch: Effect of pressing force. *IEEE Trans. Haptics* **2017**, *10*, 113–122. [CrossRef]
37. Barrea, A.; Delhaye, B.P.; Lefèvre, P.; Thonnard, J.L. Perception of partial slips under tangential loading of the fingertip. *Sci. Rep.* **2018**, *8*. [CrossRef]
38. MacLean, K.E. Haptic interaction design for everyday interfaces. *Rev. Hum. Factors Ergon.* **2008**, *4*, 149–194. [CrossRef]
39. O'Modhrain, S.; Gillespie, R.B. Once More, with Feeling: Revisiting the Role of Touch in Performer-Instrument Interaction. In *Musical Haptics*; Springer: Berlin/Heidelberg, Germany, 2018; pp. 11–28. [CrossRef]
40. Hayward, V.; Astley, O.R.; Cruz-Hernandez, M.; Grant, D.; Robles-De-La-Torre, G. Haptic interfaces and devices. *Sens. Rev.* **2004**, *24*, 16–29. [CrossRef]
41. Hayward, V.; MacLean, K.E. Do it yourself haptics: Part I. *IEEE Robot. Autom. Mag.* **2007**, *14*. [CrossRef]
42. Bokiau, B.; Ceulemans, A.E.; Fisette, P. Historical and dynamical study of piano actions: A multibody modelling approach. *J. Cult. Herit.* **2017**, *27*, S120–S130. [CrossRef]
43. Timmermans, S.; Ceulemans, A.E.; Fisette, P. Upright and grand piano actions dynamic performances assessments using a multibody approach. *Mech. Mach. Theory* **2021**, *160*, 104296. [CrossRef]
44. Oboe, R.; De Poli, G. Multi-instrument virtual keyboard—The MIKEY project. In Proceedings of the Conference on New Instruments for Musical Expres, Dublin, Ireland, 24–26 May 2002.
45. Lozada, J.; Hafez, M.; Boutillon, X. A novel haptic interface for musical keyboards. In Proceedings of the 2007 IEEE/ASME International Conference on Advanced intelligent mechatronics, Zurich, Switzerland, 4–7 September, 2007; pp. 1–6.
46. Horváth, P. Towards to Haptic Keyboard: Modeling the Piano Action. In *Mechatronics 2013*; Springer: Berlin/Heidelberg, Germany, 2014; pp. 49–55.
47. Leonard, J.; Cadoz, C. Physical Modelling Concepts for A Collection of Multisensory Virtual Musical Instruments. In Proceedings of the New Interfaces for Musical Expression 2015, Baton Rouge, LA, USA, 31 May–3 June 2015. Available online: https://hal.archives-ouvertes.fr/hal-01262132 (accessed on 12 July 2021).

48. Adamou, D.; Chin, C.; Rovelli, D.; Szafián, M.; Wood, M.G.; Yanchev, B.; Bailey, N.; Muir, D. Analysis and Reproduction of Keyboard Instrument Touch. In Proceedings of the 8th International Scientific Meeting for the Study of Sound and Musical Instruments–Organological Congress, Belmonte, Portugal, 20–22 September 2019; pp. 20–22.
49. Persson, J.; Blanc, C.; Nguyen, V.; Perriard, Y. Sensorless position estimation of linear voice-coil transducers. In Proceedings of the Conference Record of the 2001 IEEE Industry Applications Conference 36th IAS Annual Meeting (Cat. No.01CH37248), Chicago, IL, USA, 30 September–4 October 2001; Volume 1, pp. 70–74. [CrossRef]
50. Dülk, I.; Kovácsházy, T. A sensorless method for detecting spool position in solenoid actuators. *Carpathian J. Electron. Comput. Eng.* **2013**, *6*, 36.
51. Savioz, G.; Perriard, Y. Towards self-sensed drives in linear haptic systems. In Proceedings of the 2009 International Conference on Electrical Machines and Systems, Tokyo, Japan, 15–18 November 2009; pp. 1–5. [CrossRef]
52. Timmermans, S.; Desclee, Q.; Paillot, G.; Fisette, P.; Dehez, B. Application and Validation of a Linear Electromagnetic Actuator within a Haptic Piano Key. In Proceedings of the 12th International Symposium on Linear Drives for Industry Applications (LDIA), Neuchatel, Switzerland, 1–3 July 2019. [CrossRef]
53. Miedema, W. Active Haptic Feedback within a Musical Keyboard. Master's Thesis, University of Twente, Enschede, The Netherlands, 2016.
54. Wynne, R.A.; Beanland, V.; Salmon, P.M. Systematic review of driving simulator validation studies. *Saf. Sci.* **2019**, *117*, 138–151. [CrossRef]
55. Docquier, Q. Dynamic Analysis and Control of Narrow Track Vehicles via a Multibody Modeling Approach. Ph.D. Thesis, UCLouvain-Université Catholique de Louvain, Ottignies-Louvain-la-Neuve, Belgium, 2020.

*sensors*

MDPI

Article

Estimating the Characteristic Curve of a Directional Control Valve in a Combined Multibody and Hydraulic System Using an Augmented Discrete Extended Kalman Filter

Qasim Khadim [1], Mehran Kiani-Oshtorjani [2],*, Suraj Jaiswal [1], Marko K. Matikainen [1] and Aki Mikkola [1]

[1] Department of Mechanical Engineering, LUT School of Energy Systems, Lappeenranta University of Technology, 53850 Lappeenranta, Finland; qasim.khadim@lut.fi (Q.K.); Suraj.Jaiswal@lut.fi (S.J.); marko.matikainen@lut.fi (M.K.M.); aki.mikkola@lut.fi (A.M.)
[2] Department of Energy Technology, LUT School of Energy Systems, Lappeenranta University of Technology, 53850 Lappeenranta, Finland
* Correspondence: mehran.kiani@lut.fi

Abstract: The estimation of the parameters of a simulation model such that the model's behaviour matches closely with reality can be a cumbersome task. This is due to the fact that a number of model parameters cannot be directly measured, and such parameters might change during the course of operation in a real system. Friction between different machine components is one example of these parameters. This can be due to a number of reasons, such as wear. Nevertheless, if one is able to accurately define all necessary parameters, essential information about the performance of the system machinery can be acquired. This information can be, in turn, utilised for product-specific tuning or predictive maintenance. To estimate parameters, the augmented discrete extended Kalman filter with a curve fitting method can be used, as demonstrated in this paper. In this study, the proposed estimation algorithm is applied to estimate the characteristic curves of a directional control valve in a four-bar mechanism actuated by a fluid power system. The mechanism is modelled by using the double-step semi-recursive multibody formulation, whereas the fluid power system under study is modelled by employing the lumped fluid theory. In practise, the characteristic curves of a directional control valve is described by three to six data control points of a third-order B-spline curve in the augmented discrete extended Kalman filter. The results demonstrate that the highly non-linear unknown characteristic curves can be estimated by using the proposed parameter estimation algorithm. It is also demonstrated that the root mean square error associated with the estimation of the characteristic curve is 0.08% with respect to the real model. In addition, all the errors in the estimated states and parameters of the system are within the 95% confidence interval. The estimation of the characteristic curve in a hydraulic valve can provide essential information for performance monitoring and maintenance applications.

Keywords: parameter estimation; curve fitting method; multibody dynamics; hydraulic system; predictive maintenance; characteristic curve; product life cycle; digital twin

 check for **updates**

Citation: Khadim, Q.; Kiani-Oshtorjani, M.; Jaiswal, S.; Matikainen, M.K.; Mikkola, A. Estimating the Characteristic Curve of a Directional Control Valve in a Combined Multibody and Hydraulic System Using an Augmented Discrete Extended Kalman Filter. *Sensors* **2021**, *21*, 5029. https://doi.org/10.3390/s21155029

Academic Editors: Javier Cuadrado and Miguel Ángel Naya Villaverde

Received: 16 May 2021
Accepted: 20 July 2021
Published: 24 July 2021

1. Introduction

Multibody system dynamics (MBS) approaches enable the creation of the equations of motion that describe a mechanical system and relevant sub-components of complex mechanical systems [1,2]. The use of MBS leads to physics-based models that act as a single source of information [3] and represent the operation of an equivalent physical system in the real world [4]. The data generated by an MBS simulation model can be used to solve real-world problems throughout a product's life cycle [5].

A physical system might have parameters that are difficult to estimate and that could accordingly create uncertainties in MBS models. In the real world, these parameters might be cumbersome or sometimes even impossible to measure directly due to economical

155

limitations and sensor implementation difficulties. In addition, these parameters might change over time due to wear and other factors that come into play during operation. In some cases, parameters can only be interpreted from the manufacturer's catalogues while not manifesting the current state of a product or differences in individual products due to manufacturing tolerances. Estimating these parameters can provide valuable information about the state and working performance of a product [6,7]. Manufacturers can use this information for condition monitoring [8,9], predictive maintenance [10–12], and real-time simulations for digital-twin applications [13,14].

In general, parameter estimation is a discipline that provides the essential tools for the estimation of parameters appearing in the modelling of a system [15]. The most common techniques for parameter estimation are weighted least squares [16,17], Kalman filtering [18,19], orthogonal least squares [20], robust techniques that include clustering [10], and regression diagnostics [21]. Among these algorithms, Kalman filters for parameter estimation have been utilised in a wide variety of engineering studies, ranging from control [22] and mechatronics [23] to heat transfer [24,25], fluid mechanics [26,27], turbulence [28], and others.

In the MBS field, several types of Kalman filter algorithms have been used to estimate system states based on the multibody equations of motion [29–34]. In state estimation, the independent coordinate method was introduced by using the independent positions and velocities of the multibody model as the states of the Kalman filter [30]. Using the independent coordinate method, the MBS formulation offers a general approach for estimating the system coordinates in terms of independent states for open- and closed-loop systems [30,31,34]. Less attention has been given to parameter estimation in MBS systems [35]. This is due to the complexity of the problem. As in many cases, the parameters are not constant and have to be estimated from the measurable variables of the dynamic system. In an-MBS related study, vehicle suspended mass and road friction were estimated in a dual-estimation application by using the extended Kalman filter (EKF) and the unscented Kalman filter (UKF) [36]. The generalised polynomial chaos (gPC) theory was first implemented in the framework of MBS in 2006 to quantify the parametric and external uncertainties [37,38]. However, in [37,38], only constant parameters were estimated.

Contrary to this, in practical systems, the parameters are a function of several system variables and may follow very complicated and unknown non-linear variations during the working cycles [39,40]. For instance, in the case of a hydraulically actuated mobile working machine, the characteristic curve of a hydraulic valve can play a significant role in terms of machine performance [41,42]. The characteristic curve of a hydraulic valve can be expressed as a function of the spool position and the semi-empiric flow rate coefficient [41,42]. The semi-empiric flow rate coefficient relates the discharge coefficient, pressure losses, and flow characteristics that demonstrate the dynamic characteristics of a hydraulic valve [43–45]. Accordingly, the characteristic curve of a hydraulic valve can be used in the condition monitoring and predictive maintenance of hydraulically driven systems [42]. However, in an operating hydraulic system, only the minimum and maximum points on the characteristic curve can be determined from the manufacturer's catalogues with a high level of certainty [41,42]. The characteristic curve of a hydraulic valve remains unclear in a working cycle and varies from one hydraulic valve to another due to manufacturing tolerances and possible wear [41,42]. Applying parameter estimation theories [46,47] in combination with MBS equations of motion can enable the estimation of the characteristic curve of a hydraulically driven physical system in operation by using a limited amount of information.

Generally, unknown parameters are treated as constants in the dynamic equations of motion. The estimation of non-linear parameters typically requires an accurate description of the first derivatives of the corresponding parameters. However, in the real world, the first derivatives of parameters are unclear. The first derivative of a characteristic curve in a hydraulic valve is an example. In the case of a characteristic curve, a vector of data points can be constructed using random points between the minimum and maximum values provided in the manufacturer's catalogues. Through a parameter vector, the characteristic curve of a hydraulic valve in the real world can be estimated by combining parameter estimation algorithms [46,47] and curve-fitting methods [48–52]. Considering parameter estimation constraints, this study proposes the estimation of parameters by combining the augmented discrete extended Kalman filter (ADEKF) with curve-fitting methods.

The objective of this study is to propose a parameter estimation algorithm by combining the ADEKF algorithm with a curve-fitting method in an application for estimating linear and non-linear parameters. To this end, parameters are introduced as vectors in the augmented state vector. Due to the accuracy of the finite difference schemes in the complex plane, as demonstrated in [53,54], an approach to computing the Jacobian of a non-linear system of ordinary differential equations (ODEs) through complex variables in the framework of a parameter estimation algorithm is proposed. Based on the parameter estimation algorithm, the structures of covariance matrices of plant and measurement noises are introduced. The parameter estimation algorithm is applied to estimate the characteristic curve of a directional control valve in a hydraulically driven four-bar mechanism. As reported in [55], the double-step formulation has advantages over Index-3 Augmented Lagrangian formulation due to the use of a coordinate partitioning method [56]. Therefore, the double-step semi-recursive formulation is used to model the four-bar mechanism with relative coordinates. A fluid power system, in turn, is modelled by using the lumped fluid theory. This algorithm is verified by estimating the characteristic curves of the directional control valve using three, four, five, and six vector data control points in the mechanism. The implementation of the parameter estimation algorithm is explained by using MBS simulation models that represent the real model, estimation model, and simulation model. The estimation model considers the actuator position, pump pressure, and the pressure on the piston side as sensor measurements to account for the system responses. Applying the proposed parameter estimation methodology in MBS systems can enable the estimation of parameters of any complex system in a real-world system.

The rest of this paper is organised as follows. In Section 2, the parameter estimation methodology is described. Section 2 details further into the double-step semi-recursive MBS formulation, lumped fluid theory, monolithic approach, the ADEKF with a curve-fitting method, and structure of covariance matrices of plant and measurement noises. The parameter estimation methodology is applied to the case example presented in Section 3. Section 4 demonstrates the results of the parameter estimation algorithm for the case example. Finally, conclusions about parameter estimation are provided in Section 5.

2. Parameter Estimation Methodology

Figure 1 depicts a methodology that can be used to estimate the parameters of a dynamic system by using a simulation model. In this model, an initial covariance matrix $\mathbf{P}_{k-1}^{+} \in \mathbb{R}^{L \times L}$ and an augmented state vector $\hat{\mathbf{x}}_{k-1}^{\prime+} = \begin{bmatrix} \mathbf{x}_{k-1}^{T} & \mathbf{y}_{k-1}^{T} \end{bmatrix}^{T}$ at the time step $k-1$ are introduced. Here, L is the dimension of the augmented state vector, and $\mathbb{R}$ denotes the set of real numbers. $\mathbf{x} \in \mathbb{R}^{L-n_{h_p}}$ and $\mathbf{y} \in \mathbb{R}^{n_{h_p}}$ represent the states and parameters of the system, respectively. Here, n_{h_p} is the number of hydraulic parameters.

In the real world, the sensors shown in the Figure 1 can be replaced by sensor measurements obtained from a physical system, such as a forklift, a tractor, etc. [4,57]. To account for the system response, the sensor measurement vector $\mathbf{o}$ includes the minimum number of measurements required by the ADEKF algorithm to estimate the states and parameters of a real system. In Figure 1, $\mathbf{h}$ corresponds to the sensor measurement function. Note that the parameters should not be included in the measurement vector, i.e., $\mathbf{y} \notin \mathbf{o}$. The parameter estimation algorithm estimates the augmented state vector $\hat{\mathbf{x}}_{k}^{\prime+}$ and covariance matrix $\mathbf{P}_{k}^{+}$ from the minimum information of the real system at time step $k-1$ in the simulation model.

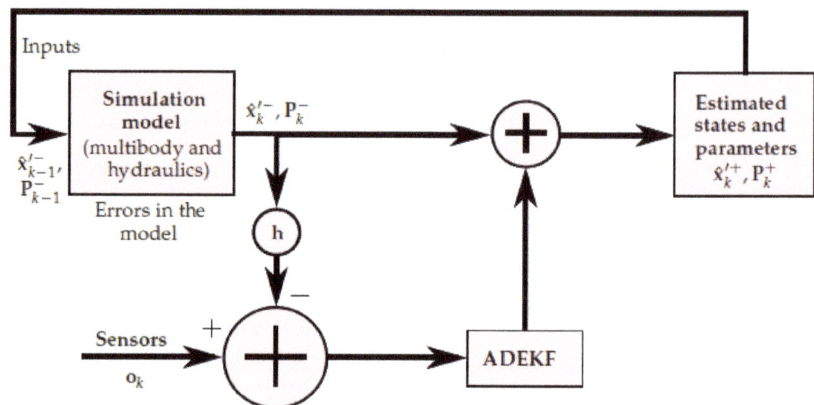

Figure 1. Parameter estimation methodology.

2.1. Multibody Dynamic Formulations

The parameter estimation methodology described in Section 2 is applied to the simulation of a hydraulically driven mechanism. In this study, the hydraulically driven mechanism is modelled using a double-step semi-recursive MBS formulation and the lumped fluid theory. The coupled multibody and hydraulic dynamics are integrated by using a single-step implicit trapezoidal integration scheme in a monolithic coupling approach. As demonstrated in [55], the double-step semi-recursive formulation uses a coordinate partitioning method [58–60] to express the hydraulically driven mechanism in terms of independent coordinates. As a result, the double-step semi-recursive formulation presents an appropriate multibody simulation approach for state and parameter estimation applications.

2.1.1. Double-Step Semi-Recursive Formulation

In the semi-recursive formulation, a body i is defined by the set of six Cartesian velocities as $\mathbf{Z}_i = \begin{bmatrix} \dot{\mathbf{r}}_i^T & \boldsymbol{\omega}_i^T \end{bmatrix}^T$ and six Cartesian accelerations as $\dot{\mathbf{Z}}_i = \begin{bmatrix} \ddot{\mathbf{r}}_i^T & \dot{\boldsymbol{\omega}}_i^T \end{bmatrix}^T$ for a complete description [61,62]. Here, $\dot{\mathbf{r}}_i$, $\ddot{\mathbf{r}}_i$, $\boldsymbol{\omega}_i$ and $\dot{\boldsymbol{\omega}}_i$ are velocities, accelerations, angular velocities, and angular accelerations of the body, respectively. In the relative coordinate system, the position of n_b bodies in a system can be described by using joint coordinates as $\mathbf{z} = \begin{bmatrix} z_1 & z_2 & \dots & z_{n_b} \end{bmatrix}^T$ [61,62]. The absolute velocity $\mathbf{Z}$ and the absolute acceleration $\dot{\mathbf{Z}}$ of the system bodies can be mapped in terms of the relative joint velocity vector $\dot{\mathbf{z}}$ and the relative joint acceleration vector $\ddot{\mathbf{z}}$ by using the velocity transformation matrix as follows [61,62]:

$$\left. \begin{array}{l} \mathbf{Z} = \mathbf{T}\mathbf{R}_d\dot{\mathbf{z}} \\ \dot{\mathbf{Z}} = \mathbf{T}\mathbf{R}_d\ddot{\mathbf{z}} + \mathbf{T}\dot{\mathbf{R}}_d\dot{\mathbf{z}} \end{array} \right\}, \tag{1}$$

where $\mathbf{T} \in \mathbb{R}^{6n_b \times 6n_b}$ is the path matrix that demonstrates the topology of the system, and $\mathbf{R}_d \in \mathbb{R}^{6n_b \times n_b}$ is a block diagonal matrix. The path matrix $\mathbf{T}$ is a lower triangular matrix and contains entries of 6×6 ($\mathbf{I}_6$) unit matrices representing bodies between the body under observation and the root of the system [61]. In Equation (1), the block diagonal matrix $\mathbf{R}_d$ and the product $\dot{\mathbf{R}}_d\dot{\mathbf{z}}$ can be computed with the joint-dependent element of the velocity transformation matrix $\mathbf{b}_i \in \mathbb{R}^{6 \times 1}$ and the joint-dependent element of the acceleration transformation vector $\mathbf{d}_i \in \mathbb{R}^{6 \times 1}$, respectively [61,62]. The semi-recursive formulation is described hereafter, but the interested reader is referred to [61,62] for further details of $\mathbf{T}$, $\mathbf{b}_i$, and $\mathbf{d}_i$. The composite mass matrix $\overline{\mathbf{M}}_i \in \mathbb{R}^{6\times6}$ and the composite force vector $\overline{\mathbf{Q}}_i \in \mathbb{R}^{6\times1}$ of the ith body can be described using absolute coordinates as [61,62]:

$$\overline{\mathbf{M}}_i = \begin{bmatrix} m_i\mathbf{I}_3 & -m_i\tilde{\mathbf{g}}_i \\ m_i\tilde{\mathbf{g}}_i & \mathbf{J}_i - m_i\tilde{\mathbf{g}}_i\tilde{\mathbf{g}}_i \end{bmatrix}, \tag{2}$$

$$\overline{\mathbf{Q}}_i = \begin{bmatrix} \overline{\mathbf{f}}_i - \tilde{\omega}_i(\tilde{\omega}_i m_i \mathbf{g}_i) \\ \mathbf{n}_i - \tilde{\omega}_i\mathbf{J}_i\omega_i + \tilde{\mathbf{g}}_i(\mathbf{f}_i - \tilde{\omega}_i(\tilde{\omega}_i m_i\mathbf{g}_i)) \end{bmatrix}, \tag{3}$$

where m_i is the mass of the i body, $\overline{\mathbf{f}}_i \in \mathbb{R}^{3\times1}$ is a vector of external forces, $\tilde{\omega}_i$ is the skew-symmetric matrix of the angular velocity vector, $\mathbf{n}_i \in \mathbb{R}^{3\times1}$ is the vector of external moments, $\mathbf{I}_3$ is a 3×3 unit matrix, and $\tilde{\mathbf{g}}_i \in \mathbb{R}^{3\times3}$ is the skew-symmetric matrix of the position vector of the centre of mass of the body in the global coordinate system. In Equation (2), $\mathbf{J}_i$ is the inertia tensor of the ith body, which can be computed as described in [61]. Applying the principle of virtual work and using Equation (1) yields the equation of motion for an open-loop system in the simplified form [61,62]:

$$\mathbf{R}_d{}^T\mathbf{T}^T\overline{\mathbf{M}}\mathbf{T}\mathbf{R}_d\ddot{\mathbf{z}} = \mathbf{R}_d{}^T\mathbf{T}^T(\overline{\mathbf{Q}} - \overline{\mathbf{M}}\mathbf{T}\dot{\mathbf{R}}_d\dot{\mathbf{z}}), \tag{4}$$

where $\overline{\mathbf{M}} \in \mathbb{R}^{6n_b \times 6n_b}$ is the block diagonal matrix consisting of the composite mass matrices of the bodies. The force vector $\overline{\mathbf{Q}} \in \mathbb{R}^{6n_b \times 1}$ is the column matrix of composite forces. To incorporate closed-loop systems, the double-step semi-recursive formulation is used in this study [62]. In this method, a set of m constraint equations $\boldsymbol{\Phi}(\mathbf{z}) = \mathbf{0}$ are introduced for closure of an open-loop system. This method employs Gaussian elimination with a full pivoting approach to identify the independent and dependent columns of the Jacobian matrix $\boldsymbol{\Phi}_\mathbf{z}$ [62–64]. Through this formulation, relative joint-independent coordinates can be used to define the dynamics of a system, i.e., the relative joint-dependent coordinates can be computed in terms of the relative joint-independent coordinates. Hence, this provides an appropriate option for the state and parameter estimation applications. The relative joint velocity vector $\dot{\mathbf{z}}$ can be described using the coordinate partitioning method [59]:

$$\begin{bmatrix} \dot{\mathbf{z}}^\mathrm{d} \\ \dot{\mathbf{z}}^\mathrm{i} \end{bmatrix} = \begin{bmatrix} -\left(\boldsymbol{\Phi}_\mathbf{z}^\mathrm{d}\right)^{-1}\boldsymbol{\Phi}_\mathbf{z}^\mathrm{i} \\ \mathbf{I} \end{bmatrix}\dot{\mathbf{z}}^\mathrm{i} \equiv \mathbf{R}_\mathbf{z}\dot{\mathbf{z}}^\mathrm{i}, \tag{5}$$

where $\dot{\mathbf{z}}^\mathrm{d} \in \mathbb{R}^m$ are the relative joint-dependent velocities, $\dot{\mathbf{z}}^\mathrm{i} \in \mathbb{R}^{n_f}$ are the relative joint-independent velocities, $\mathbf{R}_\mathbf{z} \in \mathbb{R}^{n_b \times n_f}$ is the velocity transformation matrix, $\boldsymbol{\Phi}_\mathbf{z}^\mathrm{d} \in \mathbb{R}^{m\times m}$, and $\boldsymbol{\Phi}_\mathbf{z}^\mathrm{i} \in \mathbb{R}^{m\times n_f}$ are the Jacobian matrices of the constraint equations with respect to the dependent and independent relative joint positions, respectively. In Equation (5), it is assumed that neither singular configurations nor redundant constraints exist; as a consequence, the inverse of matrix $\boldsymbol{\Phi}_\mathbf{z}^\mathrm{d}$ can be obtained [55,59]. The relative joint acceleration vector can be expressed by differentiating Equation (5) [59]:

$$\ddot{\mathbf{z}} = \mathbf{R}_\mathbf{z}\ddot{\mathbf{z}}^\mathrm{i} + \dot{\mathbf{R}}_\mathbf{z}\dot{\mathbf{z}}^\mathrm{i}, \tag{6}$$

where $\ddot{\mathbf{z}}^\mathrm{i}$ are the relative joint-independent accelerations, and $\dot{\mathbf{R}}_\mathbf{z}$ is the derivative of the velocity transformation matrix. Substituting Equation (6) into Equation (4) results in an equation of motion for a closed-loop system in a simplified form [55,56,58,63,64]:

$$\mathbf{R}_\mathbf{z}^T\mathbf{R}_d^T\mathbf{T}^T\overline{\mathbf{M}}\mathbf{T}\mathbf{R}_d\mathbf{R}_\mathbf{z}\ddot{\mathbf{z}}^\mathrm{i} = \mathbf{R}_\mathbf{z}^T\mathbf{R}_d^T\left(\mathbf{T}^T\overline{\mathbf{Q}} - \mathbf{T}^T\overline{\mathbf{M}}\mathbf{D}\right), \tag{7}$$

where $\mathbf{D} = \mathbf{T}\mathbf{R}_d\begin{bmatrix} -\left(\boldsymbol{\Phi}_\mathbf{z}^\mathrm{d}\right)^{-1}\left(\dot{\boldsymbol{\Phi}}_\mathbf{z}\dot{\mathbf{z}}\right) \\ \mathbf{0} \end{bmatrix} + \mathbf{T}\dot{\mathbf{R}}_d\dot{\mathbf{z}}$ represent the absolute accelerations when the vector $\ddot{\mathbf{z}}$ is zero in Equation (6). Equation (7) can be further simplified using the accumulated mass matrix $\mathbf{M}^{\Sigma} = \mathbf{R}_z^T \mathbf{R}_d^T \mathbf{T}^T \overline{\mathbf{M}} \mathbf{T} \mathbf{R}_d \mathbf{R}_z$ and the accumulated force matrix $\mathbf{Q}^{\Sigma} = \mathbf{R}_z^T \mathbf{R}_d^T \left(\mathbf{T}^T \overline{\mathbf{Q}} - \mathbf{T}^T \overline{\mathbf{M}} \mathbf{D} \right)$.

2.1.2. Hydraulic Lumped Fluid Theory

The lumped fluid theory can be used to compute pressures within a hydraulic circuit [65]. Using this approach, a hydraulic circuit is assumed to be composed of discrete volumes. The pressures inside the volumes are equally distributed, with the acoustic waves having a negligible effect on the pressures [41,65]. In any hydraulic volume V_h, the differential pressure $\dot{p}_h$ can be computed [41,65] as

$$\dot{p}_h = \frac{k_p + p_h k_0}{V_h} \sum_{h=1}^{n_f} Q_h, \tag{8}$$

where Q_h is the sum of incoming and outgoing volume flow rates, k_0 is the flow gain, k_p is the pressure flow coefficient, and n_f is the total amount of volume flows. By employing a semi-empirical method, the volume flow rate Q_R through a throttle valve can be described as [66]

$$Q_R = C_R \operatorname{sgn}(\Delta p) \sqrt{|\Delta p|}, \tag{9}$$

where Δp is the pressure difference and $C_R = C_d A_R \sqrt{\frac{2}{\rho}}$ is the semi-empirical flow rate coefficient of the throttle valve. Here, C_d is the flow discharge coefficient and ρ is the fluid density. In Equation (9), A_R is the cross-sectional area of the pressure relief valve. Similarly, the volume flow rate Q_d through a directional control valve can be computed as [67,68]

$$Q_d = C_v U \operatorname{sgn}(\Delta p) \sqrt{|\Delta p|}, \tag{10}$$

where C_v is the semi-empiric flow rate coefficient, and U is the relative position of the spool. Equation (10) is complemented by the following first-order differential equation:

$$\dot{U} = \frac{U_{ref} - U}{\tau}, \tag{11}$$

where U_{ref} is the reference voltage signal, and τ is the time constant. The incoming flow rate Q_{in} and outgoing flow rate Q_{out} in the hydraulic cylinder can be described as

$$Q_{in} = \dot{s} A_1, \quad Q_{out} = \dot{s} A_2, \tag{12}$$

where $\dot{s}$ is the piston velocity, and A_1 and A_2 are the areas on the piston and piston-rod side of the cylinder, respectively. The force F_h produced by the cylinder can be written as

$$F_h = p_1 A_1 - p_2 A_2 - F_\mu, \tag{13}$$

where p_1 and p_2 are, respectively, the pressure on the piston and piston-rod side, which can be calculated by using Equation (8). F_μ is the total friction force in the hydraulic cylinder caused by the hydraulic sealing. As proposed in [69], this friction force can be calculated by employing the Brown and McPhee model [70], which is valid for both positive and negative tangential velocity. The actuator velocity dependent friction force can be written in the vector form as

$$\mathbf{F}_\mu = \left(F_c \tanh\left(4 \frac{\|\dot{s}\|}{v_s} \right) + (F_s - F_c) \frac{\frac{\|\dot{s}\|}{v_s}}{\left(\frac{1}{4} \left(\frac{\|\dot{s}\|}{v_s} \right)^2 + \frac{3}{4} \right)^2} \right) \operatorname{sgn}(\dot{s}) + \sigma_2 \dot{s} \tanh(4), \tag{14}$$

where F_c is the Coulomb friction, v_s is the Stribeck velocity, F_s is the static friction, σ_2 is the coefficient of viscous friction, and $\dot{s}$ is the actuator velocity vector.

2.1.3. Monolithic Approach: Coupling MBS and Hydraulic Dynamic Systems

The MBS formulation can be combined with the fluid power system solver to form a unified set of non-linear differential equations in a monolithic approach:

$$\left.\begin{array}{c} \mathbf{M}^{\Sigma}(\mathbf{z})\ddot{\mathbf{z}}^{\mathrm{i}} = \mathbf{Q}^{\Sigma}(\mathbf{z}, \dot{\mathbf{z}}, \mathbf{p}) \\ \dot{\mathbf{p}} = \mathbf{v}(\mathbf{z}, \dot{\mathbf{z}}, \mathbf{p}, \mathbf{y}, U) \end{array}\right\}, \tag{15}$$

where $\mathbf{p}$ is the pressure vector, and $\mathbf{y}$ is the vector of hydraulic parameters. Equation (15) is a set of non-linear equations that can be represented as $\mathbf{f}(\bar{\mathbf{x}}, U) = 0$. Here, the vector $\bar{\mathbf{x}} = \left[\mathbf{z}^T \ \dot{\mathbf{z}}^T \ \mathbf{p}^T \ \mathbf{y}^T\right]^T$. The solution of the non-linear equations described in Equation (15) is stiff. A stiff equation can be solved by using single-step implicit trapezoidal integration scheme [55,71–73]. In this integration scheme, the relative joint-independent positions and the pressures are initially predicted as $\mathbf{z}_{k+1}^{\mathrm{i}} = \mathbf{z}_k^{\mathrm{i}} + \dot{\mathbf{z}}_k^{\mathrm{i}}\Delta k + \frac{1}{2}\ddot{\mathbf{z}}_k^{\mathrm{i}}\Delta k^2$ and $\mathbf{p}_{k+1} = \mathbf{p}_k + \dot{\mathbf{p}}_k\Delta k$, respectively [73]. The derivatives of $\mathbf{z}_{k+1}^{\mathrm{i}}$ and $\mathbf{p}_{k+1}$ can be predicted as

$$\left.\begin{array}{c} \dot{\mathbf{z}}_{k+1}^{\mathrm{i}} = \dfrac{2}{\Delta k}\mathbf{z}_{k+1}^{\mathrm{i}} + \check{\mathbf{z}}_k^{\mathrm{i}} \\[2mm] \ddot{\mathbf{z}}_{k+1}^{\mathrm{i}} = \dfrac{4}{\Delta k^2}\mathbf{z}_{k+1}^{\mathrm{i}} + \check{\ddot{\mathbf{z}}}_k^{\mathrm{i}} \\[2mm] \dot{\mathbf{p}}_{k+1} = \dfrac{2}{\Delta k}\mathbf{p}_{k+1} + \check{\mathbf{p}}_k \end{array}\right\}, \tag{16}$$

where $\check{\mathbf{z}}_k^{\mathrm{i}} = -(\frac{2}{\Delta k}\mathbf{z}_k^{\mathrm{i}} + \dot{\mathbf{z}}_k^{\mathrm{i}})$, $\check{\ddot{\mathbf{z}}}_k^{\mathrm{i}} = -(\frac{4}{\Delta k^2}\mathbf{z}_k^{\mathrm{i}} + \frac{4}{\Delta k}\dot{\mathbf{z}}_k^{\mathrm{i}} + \ddot{\mathbf{z}}_k^{\mathrm{i}})$ and $\check{\mathbf{p}}_k = -(\frac{2}{\Delta k}\mathbf{p}_k + \dot{\mathbf{p}}_k)$. Note that the relative joint-dependent positions $\mathbf{z}_{k+1}^{\mathrm{d}}$ are obtained from $\mathbf{z}_{k+1}^{\mathrm{i}}$ and the previous step $\mathbf{z}_k^{\mathrm{d}}$ by solving the position problem $\mathbf{\Phi}(\mathbf{z}) = 0$ [59–61]. The non-linear constraint equations are solved iteratively with the Newton–Raphson method [59–61]. The derivatives of the relative joint-dependent positions $\mathbf{z}_{k+1}^{\mathrm{d}}$ are computed from Equations (5) and (6) at the velocity and acceleration levels, respectively [55]. Substituting Equation (16) into Equation (15) leads to a set of dynamic equilibrium equations as $\overline{\mathbf{F}}(\chi_{k+1}) = 0$ at the time step $k+1$ as

$$\left.\begin{array}{c} \mathbf{M}^{\Sigma}\mathbf{z}_{k+1}^{\mathrm{i}} - \dfrac{\Delta k^2}{4}\mathbf{Q}_{k+1}^{\Sigma} + \dfrac{\Delta k^2}{4}\mathbf{M}^{\Sigma}\check{\ddot{\mathbf{z}}}_k^{\mathrm{i}} = 0 \\[2mm] \dfrac{\Delta k}{2}\mathbf{p}_{k+1} - \dfrac{\Delta k^2}{4}\mathbf{v}_{k+1} + \dfrac{\Delta k^2}{4}\check{\mathbf{p}}_k = 0 \end{array}\right\}, \tag{17}$$

where $\chi_{k+1} = \left[(\mathbf{z}^{\mathrm{i}})_{k+1}^T \ \mathbf{p}_{k+1}^T\right]^T$ is unknown. The Newton–Raphson method is employed on the non-linear Equation (17) to iteratively compute the unknown variables [73,74]:

$$\left[\dfrac{\partial\overline{\mathbf{F}}(\chi)}{\partial\chi}\right]_{k+1}^{(h)} \Delta\chi_{k+1}^{(h)} = -\left[\overline{\mathbf{F}}(\chi)\right]_{k+1}^{(h)}, \tag{18}$$

where $\left\|\Delta\mathbf{z}_{k+1}^{\mathrm{i}}\right\| < 1 \times 10^{-10}$ rad and $\left\|\Delta\mathbf{p}_{k+1}\right\| < 1 \times 10^{-2}$ Pa are the error tolerances in the relative joint independent positions and pressures provided in the Newton–Raphson method. In Equation (18), $\left[\overline{\mathbf{F}}(\chi)\right]_{k+1}^{(h)}$ is the residual vector, which can be computed as

$$\left[\overline{\mathbf{F}}(\chi)\right]_{k+1}^{(h)} = \dfrac{\Delta k^2}{4}\left[\begin{array}{c} \mathbf{M}^{\Sigma}\ddot{\mathbf{z}}^{\mathrm{i}} - \mathbf{Q}^{\Sigma} \\ \dot{\mathbf{p}} - \mathbf{v} \end{array}\right]_{k+1}^{(h)}. \tag{19}$$

In Equation (18), $\left[\frac{\partial \overline{\mathbf{F}}(\chi)}{\partial \chi}\right]_{k+1}^{(h)}$ is the tangent matrix, which can be numerically approximated at a point χ_0 by using a forward finite differences, as demonstrated in the literature [72,75]. Now, by computing $\Delta \chi_{k+1}^{(h+1)}$ from Equation (18), the next iteration $\chi_{k+1}^{(h+1)}$ can be calculated.

2.2. Estimation Algorithm: ADEKF with a Curve-Fitting Method

In this section, the ADEKF parameter estimation algorithm is introduced in the framework of a B-spline curve-fitting method. It is important to note, however, that the introduced procedure can be easily modified for applications of other curve-fitting methods, as mentioned in [48,49]. Parameter estimation through the ADEKF comprises prediction and update stages. At the prediction stage, in the case of the coupled multibody and hydraulic systems, the augmented state vector can be described as $\mathbf{x}' = \left[(\mathbf{z}^i)^T \quad (\dot{\mathbf{z}}^i)^T \quad \mathbf{p}^T \quad \mathbf{y}^T\right]^T$. At this step, $\hat{\mathbf{x}}_k'^-$ is calculated in the time integration of a dynamic model described as [46]

$$\hat{\mathbf{x}}_k'^- = \mathbf{f}(\hat{\mathbf{x}}'^+_{k-1}, U_k), \tag{20}$$

To account for unknown parameters, the proposed parameter estimation algorithm employs the curve-fitting method. Through this method, a B-spline curve is constructed with the knot vector $\mathbf{u}$ for non-uniform open splines [48,49] at the current time step:

$$C(\mathbf{u}) = \sum_{i=0}^{n} B^{i,d}(\mathbf{u})\mathbf{N}^i, \tag{21}$$

where n is the number of control points, d is the degree, $B^{i,d}(\mathbf{u})$ are the dth order of B-spline basis functions, and $\mathbf{N}^i$ is the control point vector. The control point vector can be expressed in terms of the system parameters $\mathbf{y}$. For instance, in the case of the characteristic curve, the control point vector can be written in terms of the spool position and semi-empiric flow rate coefficient as $\mathbf{N} = \begin{bmatrix} U_{\min} & U_1 & \cdots & U_n & U_{\max} \\ C_{v_{\min}} & C_{v_1} & \cdots & C_{v_n} & C_{v_{\max}} \end{bmatrix}$. Here, $U_{\min}, U_1$, and U_n represent spool positions, and $C_{v_{\min}}, C_{v_1}$, and $C_{v_{\max}}$ are the semi-empiric flow rate coefficients of a hydraulic valve. $B^{i,d}(\mathbf{u})$ can be defined by using the Cox–de Boor recursion formula [48,49]:

$$B^{i,0}(\mathbf{u}) = \begin{cases} 1 & \mathbf{u}^i \le \mathbf{u} < \mathbf{u}^{i+1} \\ 0, & \text{otherwise} \end{cases}, \tag{22}$$

$$B^{i,j}(\mathbf{u}) = \frac{\mathbf{u} - \mathbf{u}^i}{\mathbf{u}^{i+j} - \mathbf{u}^i} B^{i,j-1}(\mathbf{u}) + \frac{\mathbf{u}^{i+j+1} - \mathbf{u}}{\mathbf{u}^{i+j+1} - \mathbf{u}^{i+1}} B^{i+1,j-1}(\mathbf{u}), \tag{23}$$

where $\mathbf{u}^i$ is the ith element of the knot vector for non-uniform open splines. Next, the numerical values of parameters, which are scalar, should be evaluated by using Equation (21) at time step k to be incorporated in Equation (20). The calculation of Equation (20) at the desired input signal is straightforward. However, the numerical computation of the Jacobian matrix $\mathbf{f}_{\mathbf{x}'_{k-1}}$ could be challenging when using a curve-fitting method. Each term of the Jacobian matrix can be approximated by using complex variables to reduce the truncation error [53,54] for very small increments. For instance, in the case of a multi-variable function, the Jacobian column of Equation (20) with respect to the rth term of the augmented state vector $\hat{\mathbf{x}}'_{k-1}$ can be written in the partial derivative form as

$$\frac{\partial \mathbf{f}(\hat{x}'_{k-1,1}, \hat{x}'_{k-1,2}, \dots, \hat{x}'_{k-1,L})}{\partial \hat{x}'_{k-1,r}} = \frac{Im(\mathbf{f}(\hat{x}'^+_{k-1,1}, \hat{x}'^+_{k-1,2}, \dots, \hat{x}'_{k-1,r} + i\delta, \dots, \hat{x}'^+_{k-1,L}))}{\delta} + O(\delta^2), \tag{24}$$

where $r \in [1, L]$, and $i\delta$ represents a very small increment in the complex plane. $\delta = L\varepsilon$ is a real value. Epsilon ε is a very small number and depends on the machine's precision. The rth term of the state vector $\hat{x}'_{k-1,r} + i\delta$ is expanded by using the Taylor series [53,54].

The evaluation of $Im(f(\hat{x}'^{+}_{k-1,1}, \hat{x}'^{+}_{k-1,2}, ..., \hat{x}'_{k-1,r} + i\delta, ..., \hat{x}'^{+}_{k-1,L}))$ for the parameter vector u_k requires the evaluation of $C(u_{k-1})$ as complex numbers. However, the knot vector cannot contain any complex numbers [48,49]. Therefore, a criterion $|t - t^{i}_{k-1}| < \Xi$ is introduced, where Ξ can be a small real number, such as 0.1, which implies that the knot-point distance between t^{i}_{k-1} and the complex input argument t should be less than Ξ. Using this criterion enables the knot points to be evaluated with the curve-fitting method through complex input. With the Jacobian of the dynamic system $f_{x'_{k-1}}$, the covariance matrix P^{-}_{k} is approximated as [46]

$$P^{-}_{k} = f_{x'_{k-1}} P^{+}_{k-1} f^{T}_{x'_{k-1}} + Q_k, \tag{25}$$

where Q_k is the covariance matrix of the plant noise. In the update stage, the sensor measurements are taken into account to improve the estimated augmented state vector $\hat{x}'^{-}_{k}$. The innovation Δ_k is calculated as [46]

$$\Delta_k = o_k - h(\hat{x}'^{-}_{k}), \tag{26}$$

where o_k are the sensor measurements at the k time step, and $h(x'^{-}_{k})$ is the sensor measurement function. With the Jacobian of the sensor measurement function $h_{x'}$, the innovation in the covariance matrix S_k and the Kalman gain K_k can be calculated as [46]

$$\left.\begin{aligned} S_k &= h_{x'_k} P^{-}_{k} h^{T}_{x'_k} + R_k \\ K_k &= P^{-}_{k} h^{T}_{x'_k} S^{-1}_{k} \end{aligned}\right\}, \tag{27}$$

where R_k is the covariance matrix of the measurement noise. Finally, the augmented state vector $\hat{x}'^{+}_{k}$ and covariance matrix P^{+}_{k} are updated at the time step k, which will be used for the next time step as [46]

$$\left.\begin{aligned} \hat{x}'^{+}_{k} &= \hat{x}'^{-}_{k} + K_k \Delta_k \\ P^{+}_{k} &= (I_L - K_k h_{x'_k}) P^{-}_{k} \end{aligned}\right\}, \tag{28}$$

where I_L is the identity matrix of dimension L.

Covariance Matrices of Process and Measurement Noises

It is well known that when applying Kalman filters, the tuning of the filter parameters is crucial, especially the covariance matrices of the plant and measurement noises. Furthermore, it was established in [30,31] that in dealing with non-linear systems, the improper tuning/setting of these covariance matrices can make the algorithm unstable. In this study, the properties of measurement noise are precisely known because the measurements are built from a dynamic model (providing ground truth) with an addition of white Gaussian noise to replicate real sensors. Thus, the covariance matrix of measurement noise can be obtained. For instance, when using position and pressure sensors, the covariance matrix of the measurement noise, R, would then take the form [30,31,34]

$$R = \begin{bmatrix} \left(\sigma'_s\right)^2 I_n & 0_{n \times n_p} \\ 0_{n_p \times n} & \left(\sigma'_p\right)^2 I_{n_p} \end{bmatrix}, \tag{29}$$

where σ'_s and σ'_p are the standard deviations of measurement noises at the position and pressure levels, respectively. In Equation (29), n is the number of actuator sensors and n_p is the number of pressure sensors. I_n, I_{n_p}, $0_{n \times n_p}$, and $0_{n_p \times n}$ are the identity and zero matrices of corresponding orders, respectively. In the case of a multibody model along with positions and velocities as the state vector, the structure of the plant noise in the discrete-time frame was well established in [30,31] and can be written as

$$
\mathbf{Q} = \begin{bmatrix} \dfrac{\sigma_{\ddot{\mathbf{x}}}^2 \Delta t^3 \mathbf{I}_{n_f}}{3} & \dfrac{\sigma_{\ddot{\mathbf{x}}}^2 \Delta t^2 \mathbf{I}_{n_f}}{2} \\[2ex] \dfrac{\sigma_{\ddot{\mathbf{x}}}^2 \Delta t^2 \mathbf{I}_{n_f}}{2} & \sigma_{\ddot{\mathbf{x}}}^2 \Delta t \mathbf{I}_{n_f} \end{bmatrix}, \tag{30}
$$

where Δt is the size of the integration time step and n_f is the number of degrees of freedom of the system. It should be noted that Equation (30) includes the variance at the acceleration level, $\sigma_{\ddot{\mathbf{x}}}$, because of the acceleration errors arising from the inaccurate description of forces and mass distribution. Furthermore, the state vector in this study also includes the hydraulic pressures and the hydraulic parameters, along with the positions and velocities, and errors can occur at the pressure and parameter levels as well. Therefore, inspired by [34], the variance of hydraulic pressures, $\sigma_{p,D}$, and the variance of hydraulic parameters, $\sigma_{h_p,D}$, can be directly incorporated as the diagonal elements in Equation (30). Accordingly, the structure of the covariance matrix of the plant noise, $\mathbf{Q}$, in the parameter estimation can be written as

$$
\mathbf{Q} = \begin{bmatrix} \dfrac{\sigma_{\ddot{\mathbf{x}}}^2 \Delta t^3 \mathbf{I}_{n_f}}{3} & \dfrac{\sigma_{\ddot{\mathbf{x}}}^2 \Delta t^2 \mathbf{I}_{n_f}}{2} & \mathbf{0}_{n_f \times n_p + n_{h_p}} & \mathbf{0}_{n_f \times n_p + n_{h_p}} \\[2ex] \dfrac{\sigma_{\ddot{\mathbf{x}}}^2 \Delta t^2 \mathbf{I}_{n_f}}{2} & \sigma_{\ddot{\mathbf{x}}}^2 \Delta t \mathbf{I}_{n_f} & \mathbf{0}_{n_f \times n_p + n_{h_p}} & \mathbf{0}_{n_f \times n_p + n_{h_p}} \\[2ex] \mathbf{0}_{n_p + n_{h_p} \times n_f} & \mathbf{0}_{n_p + n_{h_p} \times n_f} & \sigma_{p,D}^2 & \mathbf{0}_{n_f \times n_p + n_{h_p}} \\[2ex] \mathbf{0}_{n_p + n_{h_p} \times n_f} & \mathbf{0}_{n_p + n_{h_p} \times n_f} & \mathbf{0}_{n_p + n_{h_p} \times n_f} & \sigma_{h_p,D}^2 \end{bmatrix}. \tag{31}
$$

In this study, the integration errors are assumed to be negligible in comparison to the acceleration, pressure, and parameter errors.

3. Case Example: Hydraulically Actuated System

The parameter estimation methodology described in Section 2 is applied to estimate the characteristic curves at the a, b, c, and d ports of the 4/3 directional control valve shown in Figure 2. A four-bar mechanism actuated by a hydraulic circuit is presented in Figure 2. The dynamics of the mechanism are modelled using the semi-recursive and hydraulic lumped fluid theories, as described below.

3.1. Dynamic Model of the System

The bodies of the mechanism are assumed to be rectangular beams whose lengths are $L_1 = 2$ m, $L_2 = 8$ m, and $L_3 = 5$ m and whose masses are $m_1 = 100$ kg, $m_2 = 400$ kg, and $m_3 = 250$ kg, respectively. The position vector at point D is $\mathbf{r}_D = \begin{bmatrix} -\frac{L_1}{2} & 0 & 0 \end{bmatrix}^T$. The point G is located at the centre of mass of body 1. The double-step semi-recursive formulation described in Section 2.1.1 is used to model the four-bar mechanism.

The four-bar mechanism is actuated using the sinusoidal reference input signal, which is taken as $U_{ref} = 10 \sin(0.4\pi k)$, where k is the simulation run time. The simulations are performed for 5 s. The hydraulic circuit consists of a double-acting hydraulic cylinder, connecting hoses 1 and 2, a 4/3 directional control valve, a pressure relief valve, a connecting hose of volume V_p, a differential pump of pressure p_p, and a tank with a constant pressure source p_T.

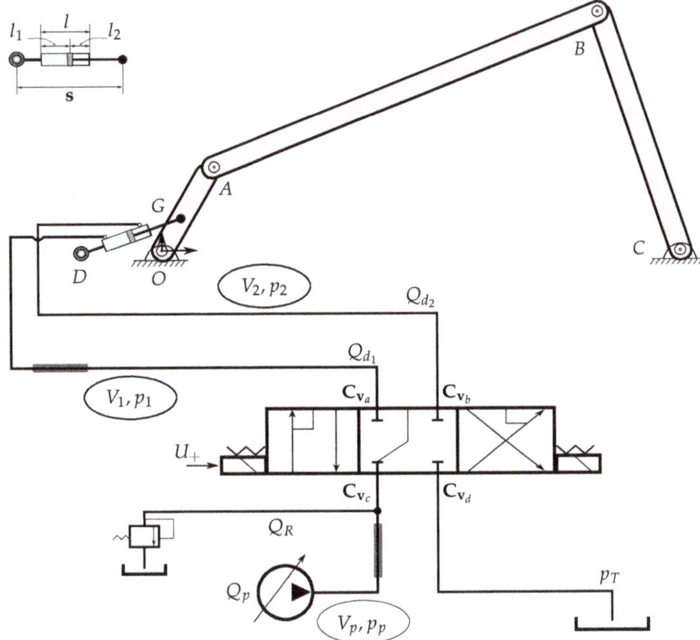

Figure 2. Hydraulically actuated four-bar mechanism. The mechanism is actuated by a differential pressure pump. $\mathbf{C_{v_a}},\mathbf{C_{v_b}},\mathbf{C_{v_c}}$, and $\mathbf{C_{v_d}}$ represent the semi-empiric flow rate coefficients at the a, b, c, and d ports of the 4/3 directional control valve. Grey rectangles indicate the pressure sensors on the control volumes V_p and V_1.

The lumped fluid theory described in Section 2.1.2 is used to compute the pressures within the hydraulic circuit. In the application of the lumped fluid theory, the hydraulic circuit can be divided into three control volumes V_p, V_1, and V_2. The pressure derivatives $\dot{p}_p$, $\dot{p}_1$, and $\dot{p}_2$ through these volumes can be computed as

$$\left.\begin{aligned}
\dot{p}_p &= \frac{k_p + p_p k_0}{V_p}(Q_p - Q_R - Q_{d_1}) \\
\dot{p}_1 &= \frac{k_p + p_1 k_0}{V_1}(Q_{d_1} - A_1 \dot{s}) \\
\dot{p}_2 &= \frac{k_p + p_2 k_0}{V_2}(A_2 \dot{s} - Q_{d_2})
\end{aligned}\right\} , \qquad (32)$$

where Q_{d_1} and Q_{d_2} are the flow rates in the control volumes 1 and 2. In Equation (32), Q_p and Q_R are the pump flow rate and flow rate through the pressure relief valve, respectively. The flow rates Q_R, Q_{d_1}, and Q_{d_2} can be computed by employing Equations (9) and (10), respectively. The constant hydraulic parameters are tabulated in Table 1. In Equation (32), $\dot{s}$ is the actuator velocity, which can be determined from the actuator position vector $\mathbf{s}$. Following Figure 2, the vector $\mathbf{s}$ can be calculated from the position vectors $\mathbf{r}_G$ and $\mathbf{r}_D$ as

$$\left.\begin{aligned}
\mathbf{s} &= \mathbf{r}_G - \mathbf{r}_D \\
\dot{s} &= \frac{d\,|\mathbf{s}|}{\Delta t} = \dot{\mathbf{s}} \cdot \frac{\mathbf{s}}{|\mathbf{s}|} = \dot{\mathbf{r}}_G \cdot \frac{\mathbf{s}}{|\mathbf{s}|}
\end{aligned}\right\} , \qquad (33)$$

where $\dot{r}_G$ is the velocity vector of point G. The control volumes V_1 and V_2 appearing in Equation (32) can be calculated as follows:

$$\left.\begin{aligned} V_1 &= V_{h_1} + A_1 l_1 \\ V_2 &= V_{h_2} + A_2 l_2 \end{aligned}\right\},\tag{34}$$

where V_{h_1}, V_{h_2}, and V_p are the control volumes of the respective hoses, as described in Table 1. In Equation (34), l_1 and l_2 are the lengths of the piston side and the piston-rod side chambers, respectively. l_1 and l_2 can be calculated with the vector $\mathbf{s}$ as

$$\left.\begin{aligned} l_1 &= l_{1_0} - |\mathbf{s}| + s_0 \\ l_2 &= l_{2_0} + |\mathbf{s}| - s_0 \end{aligned}\right\},\tag{35}$$

where l_{1_0} and l_{2_0} are the initial piston side length and the initial piston-rod side length, respectively. l_{1_0} and l_{2_0} are computed from the length of cylinder l, which is given in Table 1.

Table 1. Parameters of the hydraulic circuit.

Parameter	Symbol	Value
Pump flow rate	Q_p	$0.001\ \mathrm{m^3/s}$
Tank pressure	p_T	$0.1\ \mathrm{MPa}$
Volume of the hose p	V_p	$3.42 \times 10^{-3}\ \mathrm{m^3}$
Volume of the hose 1	V_{h_1}	$3.42 \times 10^{-1}\ \mathrm{m^3}$
Volume of the hose 2	V_{h_2}	$3.42 \times 10^{-1}\ \mathrm{m^3}$
Oil density	ρ	$869\ \mathrm{kg/m^3}$
Hydraulic parameter	k_p	$1600\ \mathrm{MPa}$
Hydraulic parameter	k_0	0.5
Area of the piston	A_1	$2 \times 10^{-3}\ \mathrm{m^2}$
Area of the piston-rod	A_2	$1.8 \times 10^{-3}\ \mathrm{m^2}$
Length of the cylinder/piston	l	$\sqrt{3}\ \mathrm{m}$
Area of pressure relief valve	A_r	$2.24 \times 10^{-12}\ \mathrm{m^2}$
Area of directional control valve	A_d	$1.96 \times 10^{-6}\ \mathrm{m^2}$
Coulomb friction force	F_c	$210\ \mathrm{N}$
Static friction force	F_s	$830\ \mathrm{N}$
Stribeck velocity	v_s	$1.25 \times 10^{-2}\ \mathrm{m/s}$
Coefficient of viscous friction	σ	$330\ \mathrm{Ns/m}$
Discharge coefficient	C_d	0.5
Area of throttle	A_R	$2.24 \times 10^{-12}\ \mathrm{m^2}$

Using the vector $\mathbf{s}$, the hydraulic force $\mathbf{F}_h$ produced by the double-acting cylinder can be calculated as

$$\mathbf{F}_h = \begin{bmatrix} \frac{s_X}{|\mathbf{s}|} F_h & \frac{s_Y}{|\mathbf{s}|} F_h & \frac{s_Z}{|\mathbf{h}|} F_h \end{bmatrix}^T,\tag{36}$$

where F_h is computed from Equation (13). The hydraulic force vector $\mathbf{F}_h$ is combined with the external force vector $\mathbf{f}_i$ to calculate $\overline{\mathbf{Q}}$ in Equation (3). The resultant equations of motion (15) are formulated for the hydraulically driven four-bar mechanism. Equations (15) are solved by using an implicit single-step trapezoidal integration scheme in a monolithic approach, which was described in Section 2.1.3.

3.1.1. Real and Estimation Models

In this study, three dynamic versions of the mechanism are used to demonstrate the implementation of the parameter estimation algorithm. One of the models is the real model. The sensor measurements $\mathbf{o}$ are taken from the real model. The modelling errors are introduced in the force model of the estimation model with respect to the real model. The properties of the estimation model and the simulation model are the same. In Table 2, the properties of the real model, the estimation model, and the simulation model are provided.

Note that the simulation model is used in this study to demonstrate the differences between the simulated world and the real world.

Table 2. Properties of the real model, the estimation model, and the simulation model. Errors in the simulation model and the estimation model are given in comparison to the real model. s_{1_0}, p_{p_0}, and p_{1_0} represent the initial actuator position, the initial pump pressure, and the initial pressure on the piston side as the system states. The system parameters $\mathbf{C_{v_a}}$, $\mathbf{C_{v_b}}$, $\mathbf{C_{v_c}}$, $\mathbf{C_{v_d}}$, k_0, and k_p represent the semi-empiric flow rate coefficient at the a, b, c, and d ports of the directional control valve, the flow gain, and the pressure flow coefficients, respectively.

Errors	Symbol	Real Model	Estimation Model	Simulation Model
State	s_{1_0}	$\sqrt{3}$ m	1.62 m	1.62 m
State	p_{p_0}	7.6 MPa	5.6 MPa	5.6 MPa
State	p_{1_0}	1 MPa	2 MPa	2 MPa
Parameter	$\mathbf{C_{v_a}}$	Non-linear	Linear	Linear
Parameter	$\mathbf{C_{v_b}}$	Non-linear	Linear	Linear
Parameter	$\mathbf{C_{v_c}}$	Non-linear	Linear	Linear
Parameter	$\mathbf{C_{v_d}}$	Non-linear	Linear	Linear
Parameter	k_0	0.5	0.4	0.4
Parameter	k_p	1600 MPa	1500 MPa	1500 MPa

As in practise, the minimum and maximum points on the characteristic curves of a directional control valve can be determined from the manufacturer's catalogues. Using this limited information, the characteristic curves are defined linearly at all ports of the directional control valve in the cases of the estimation model and the simulation model. The linear characteristic curves are implemented by using the minimum and maximum values of the semi-empiric flow rate coefficients $\mathbf{C_{v_a}}$, $\mathbf{C_{v_b}}$, $\mathbf{C_{v_c}}$, and $\mathbf{C_{v_d}}$ at the valve closing and the valve opening positions, respectively. The linear characteristic curves of the directional control valve affect the dynamics of the estimation model throughout the simulation runtime. In the case of the real model, the characteristic curves of the directional control valve are unclear and can be non-linear. With Equation (21), the non-linear characteristic curves of the directional control valve are implemented using $\mathbf{C_{v_a}}$, $\mathbf{C_{v_b}}$, $\mathbf{C_{v_c}}$, and $\mathbf{C_{v_d}}$ in the hydraulic circuit of the real model. Similarly, the initial actuator positions s_{1_0} of the real model and the estimation model are different. Note that the initial relative joint coordinates of the bodies in the system can be found from s_{1_0} and $\dot{s}_{1_0}$ by using geometrical relationships. To avoid instabilities in the integration process, the simulations are started in the static equilibrium position, the details of the mechanism of which can be found in [34].

3.1.2. Sensor Measurements

In this study, the measurable observations $\mathbf{o} = \begin{bmatrix} s & p_p & p_1 \end{bmatrix}^T$ are taken from the real model. In the real model, the actuator position sensor measures the actuator position s [76]. Gauge pressure sensors are used for the pressure measurements p_p and p_1 [34]. These pressure sensors measure the pressure with respect to the atmospheric pressure. The pressure sensors are installed on their respective volumes, as also shown in Figure 2. The numerical values of the standard deviation, as mentioned in Equation (29), are taken as $\sigma'_s = 1.12 \times 10^{-3}$ m and $\sigma'_p = 1.5 \times 10^5$ Pa for the actuator and pressure sensors, respectively.

3.2. Parameter Estimation Algorithm

In the parameter estimation algorithm, the augmented state vector $\hat{\mathbf{x}}'$ is defined as

$$\hat{\mathbf{x}}' = \begin{bmatrix} \underbrace{s \quad \dot{s} \quad p_p \quad p_1 \quad p_2}_{\hat{x}} & \underbrace{k_p \quad k_0 \quad \mathbf{C_v}}_{\hat{y}} \end{bmatrix}^T, \qquad (37)$$

where s is the actuator position, $\dot{s}$ is the actuator velocity, p_p, p_1, and p_2 are the pressures, k_p is the pressure flow coefficient, k_0 is the flow gain, and $\mathbf{C_v} = \begin{bmatrix} \mathbf{C_{v_a}} & \mathbf{C_{v_b}} & \mathbf{C_{v_c}} & \mathbf{C_{v_d}} \end{bmatrix}$ are the semi-empiric flow rate coefficients at the corresponding ports of the directional control valve. In Equation (37), $\hat{\mathbf{x}}$ and $\hat{\mathbf{y}}$ present the states and the parameters of the hydraulically driven four-bar mechanism, respectively. Equations (20)–(28) are implemented to estimate the augmented state vector $\hat{\mathbf{x}}'$ and the characteristic curves of the directional control valve. In this application, the third-order B-spline interpolation method is combined with the ADEKF. For the case example, three, four, five, and six control points are used in the parameter estimation algorithm to compute Equations (20) and (24). As mentioned earlier, the first, third, and fourth state variables are measured. Therefore, the sensor measurement function $\mathbf{h}(\hat{\mathbf{x}}'^{-}_{k})$ and its Jacobian $\mathbf{h_{x'}}$ can be written as

$$
\left.
\begin{aligned}
\mathbf{h}(\hat{\mathbf{x}}'^{-}_{k}) &= \begin{bmatrix} \hat{\mathbf{x}}'^{-}_{k,1} & \hat{\mathbf{x}}'^{-}_{k,3} & \hat{\mathbf{x}}'^{-}_{k,4} \end{bmatrix}^{T} \\[6pt]
\mathbf{h_{x'}} &= \begin{bmatrix}
1 & 0 & 0 & 0 & 0 & 0 & 0 & \dfrac{\partial h}{\partial \mathbf{C_{v_a}}} & \dfrac{\partial h}{\partial \mathbf{C_{v_b}}} & \dfrac{\partial h}{\partial \mathbf{C_{v_c}}} & \dfrac{\partial h}{\partial \mathbf{C_{v_d}}} \\[10pt]
0 & 0 & 1 & 0 & 0 & 0 & 0 & \dfrac{\partial h}{\partial \mathbf{C_{v_a}}} & \dfrac{\partial h}{\partial \mathbf{C_{v_b}}} & \dfrac{\partial h}{\partial \mathbf{C_{v_c}}} & \dfrac{\partial h}{\partial \mathbf{C_{v_d}}} \\[10pt]
0 & 0 & 0 & 1 & 0 & 0 & 0 & \dfrac{\partial h}{\partial \mathbf{C_{v_a}}} & \dfrac{\partial h}{\partial \mathbf{C_{v_b}}} & \dfrac{\partial h}{\partial \mathbf{C_{v_c}}} & \dfrac{\partial h}{\partial \mathbf{C_{v_d}}}
\end{bmatrix}
\end{aligned}
\right\} , \tag{38}
$$

where $\mathbf{C_v} = \begin{bmatrix} \mathbf{C_{v_a}} & \mathbf{C_{v_b}} & \mathbf{C_{v_c}} & \mathbf{C_{v_d}} \end{bmatrix}$. $\mathbf{h}(\hat{\mathbf{x}}'^{-}_{k})$ and $\mathbf{h_{x'}}$ are used in Equations (26) and (27), respectively, in the parameter estimation algorithm.

4. Results and Discussion

In this section, the results of the simulation of the estimation model of the parameter estimation algorithm are presented. The results of the estimation model are compared to those of the real model and the simulation model. The initial covariance $\mathbf{P}_0$ used in the augmented state estimator includes $\sigma_s^2 = 1 \times 10^{-4}$ m^2 for the actuator position, $\sigma_{\dot{s}}^2 = 1 \times 10^{-4}$ m^2/s^2 for the actuator velocity, and three pressure terms of $\sigma_p^2 = 22.50 \times 10^7$ Pa^2 in the diagonal. For the hydraulic parameters, the initial covariance values $\sigma_{k_p}^2 = 1 \times 10^{14}$ Pa^2, $\sigma_{k_0}^2 = 1 \times 10^2$ and $\sigma_{\mathbf{C_v}}^2 = 9 \times 10^2 \; \frac{m^6}{s^2 Pa}$ are used in the diagonal. The numerical values of the plant noise $\sigma_{\ddot{x}}^2 = 0.8$ m^2/s^4 and $\sigma_p^2 = 259.81 \times 10^7$ Pa^2 for Equation (31) are obtained through trial and error. All models are run with a time step of 1 ms and provide sensor data to the parameter estimation algorithm at 1000 Hz.

4.1. Estimating the Characteristic Curve of the Valve

In the real model, as only the minimum point c_{min} and the maximum point c_{max} on the characteristic curves are known at the a, b, c, and d ports of the directional control valve, the characteristic curves are generally unclear in the working cycles of the real model. The characteristic curves may vary from one valve to another and can be highly non-linear in the working cycle. In Figure 3, *Spline* 1 and *Spline* 2, which are in the cyan colour, are used to demonstrate the non-linear behaviour of the directional control valve in the real model.

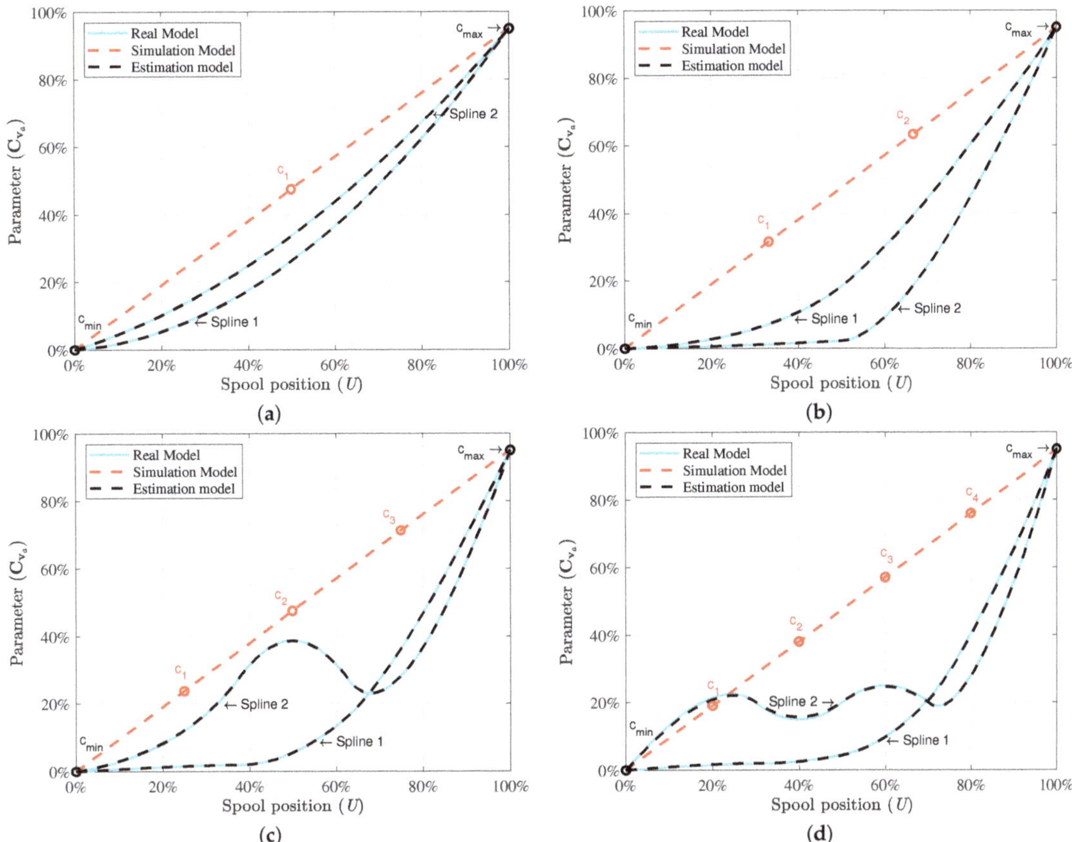

Figure 3. The estimation of the characteristic curves of the directional control valve by using the ADEKF with third-order B-spline interpolation. (**a**) Three-point B-spline estimation. (**b**) Four-point B-spline estimation. (**c**) Five-point B-spline estimation. (**d**) Six-point B-spline estimation.

The proposed parameter estimation algorithm can be used to estimate the characteristic curves of the real model with this limited information. To this end, the semi-empiric flow rate coefficients C_{v_a}, C_{v_b}, C_{v_c}, and C_{v_d} are defined with the data control points between c_{min} and c_{max} in Equation (37). Equation (37) is further used in terms of the control point vector $\mathbf{N}$ in Equations (20)–(28) to estimate the characteristic curves.

For instance, in the case of Figure 4, the control point vector $\mathbf{N}_a$ at port a of the directional control valve can be defined in terms of c_1, c_2, c_3, and c_4 as

$$\mathbf{N}_a = \begin{bmatrix} c_{min} & c_1 & c_2 & c_3 & c_4 & c_{max} \end{bmatrix}^T. \tag{39}$$

To estimate the characteristic curve, three, four, five, and six control points are used in the control point vector $\mathbf{N}_a$. As an example, these data control points for *Spline* 1 in each case are presented in Table 3.

The abscissa of vector $\mathbf{N}_a$ represents the spool position U, whereas the ordinate of vector $\mathbf{N}_a$ indicates the semi-empiric flow rate coefficients C_{v_a} at port a of the directional control valve. The results of the second-order B-spline are described in Appendix A. The results of the third-order B-spline demonstrate the characteristic curve of the directional control valve relatively better in a working cycle, as shown in Figure 3. As can be seen, *Spline*

1 and *Spline* 2 of the third-order B-spline are drawn in each data control point estimation case. The dashed red-coloured line indicates the characteristic curve of the simulation model. The dashed black-coloured line demonstrates the estimation model. In Figure 3a, three points, c_{min}, c_1, and c_{max}, are used to estimate *Spline* 1 and *Spline* 2 of the real model. The characteristic curve of the estimation model precisely follows the real model in the case of three points. Further, the percentages of the root mean square error (RMSE) are described in the Table 3 for *Spline* 1 and *Spline* 2 to verify the observations.

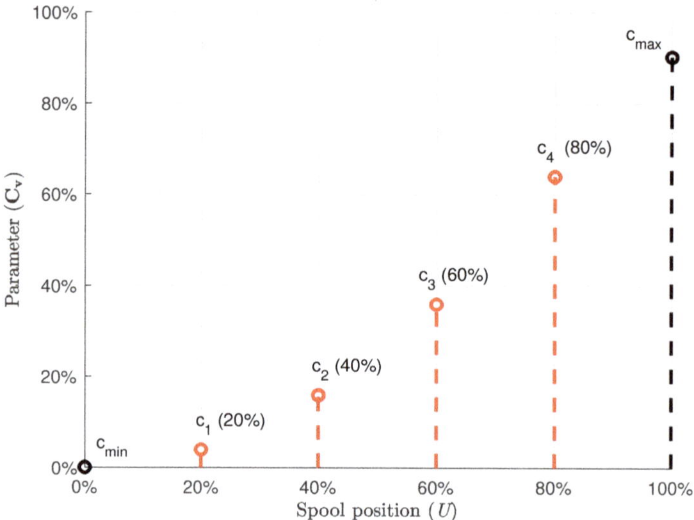

Figure 4. Data control points between c_{min} and c_{max} on the characteristic curve of the directional control valve.

Table 3. Root mean square error in the estimation of the characteristic curve. The third and fourth columns represent the root mean square errors in *Spline* 1 and *Spline* 2, respectively.

Control Points	Control Point Vector N_a	RMSE	RMSE
Three points	$\begin{bmatrix} 0 & 5 & 10 \\ 0 & 47.5 & 95 \end{bmatrix}$	0.04%	0.03%
Four points	$\begin{bmatrix} 0 & 3.3 & 6.6 & 10 \\ 0 & 31.6 & 63.3 & 95 \end{bmatrix}$	0.05%	0.01%
Five points	$\begin{bmatrix} 0 & 2.5 & 5 & 7.5 & 10 \\ 0 & 23.7 & 47.5 & 71.2 & 95 \end{bmatrix}$	0.06%	0.07%
Six points	$\begin{bmatrix} 0 & 2 & 4 & 6 & 8 & 10 \\ 0 & 19 & 38 & 57 & 76 & 95 \end{bmatrix}$	0.07%	0.08%

Figure 3b shows the estimation of the characteristic curve when using four points c_{min}, c_1, c_2, and c_{max}. As can be seen in Figure 3b, the semi-empiric flow rate coefficient $\mathbf{C_{v_a}}$ for *Spline* 2 changes with small increments until 52% opening of the spool as compared to *Spline* 1 in the real model. After this point, the parameter $\mathbf{C_{v_a}}$ increases sharply towards the maximum point c_{max}. The difference of the estimated curve from the real model's curve is indistinguishable. The RMSEs of these curves are given in Table 3. The relatively complicated non-linear behaviours of the directional control valve can be estimated by using five control points and six control points. This can be seen in Figure 3c,d. By using the estimated characteristic curves, the working conditions of the directional control valve can be predicted.

4.2. Convergence of the Vector Data Control Points

The convergence rate of the data control points in the parameter vector $\mathbf{C}_{v_a}$ is further explained in Figure 5 to describe the estimation process. These plots demonstrate the convergence rate of data control points in the case of *Spline* 2, as presented in Figure 3. For instance (see Figure 5b), c_1 and c_2 converge towards the corresponding point on the curve of the real model at 0.22 s. However, during the estimation process, c_1 briefly becomes negative, and shortly thereafter converges smoothly to the real model.

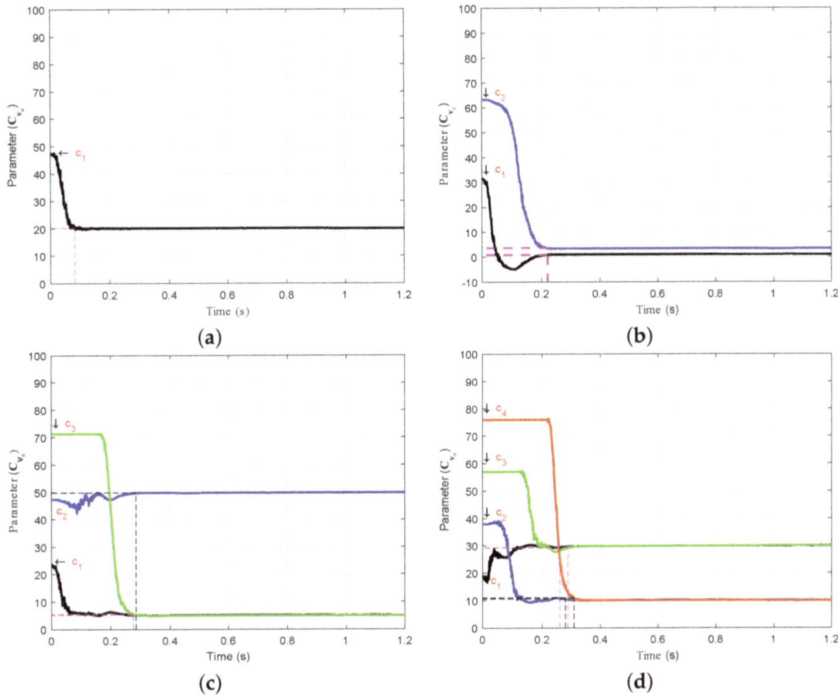

Figure 5. Convergence of the control points in the vector $\mathbf{C}_{v_a}$ in the case of *Spline* 2. (**a**) Convergence of c_1 in the three-point estimation process. (**b**) Convergence of c_1 and c_2 in the four-point estimation process. (**c**) Convergence of c_1, c_2, and c_3 in the five-point estimation process. (**d**) Convergence of c_1, c_2, c_3, and c_4 in the six-point estimation process.

The curves of *Spline* 2 change into an S-shape during the working cycle, as shown in Figure 3c,d. In these cases, c_2, c_3, and c_4 converge at different simulation times according to the corresponding order in the vector $\mathbf{C}_{v_a}$. Through the ADEKF algorithm, unknown curves start converging within a range of $0 < t \leq 0.3$ s when using the three-, four-, five-, and six-point estimation techniques.

4.3. Accuracy Requirements of State Estimations

The successful application of the parameter estimation algorithm requires the accurate estimation of the system states **x**. To demonstrate this requirement, the errors in the estimated actuator position s, estimated actuator velocity $\dot{s}$, estimated pump pressure p_p, estimated piston side pressure p_1, estimated piston-rod side pressure p_2, and estimated parameter $\mathbf{C}_{v_a}$ in the case of *Spline* 2 (described in Figure 3d) are shown in Figure 6. The errors in the estimated parameters k_p and k_0 are presented in Appendix B. The average of the parameter vector $\mathbf{C}_{\mathbf{v}_a}$ at each time step is considered in Figure 6.

The errors are computed from $\pm 1.96\sigma$. Here, σ is the standard deviation calculated from the covariance matrix $\mathbf{P}_k^+$ at each time step. These plots demonstrate the requirement of an accurate estimation of the system's states to estimate the system's parameters. As can be seen in Figure 6, the 95% confidence interval (CI) is used by the system states in the 5 s simulation period. The errors in s, $\dot{s}$, p_p, p_1, and $\dot{s}$ fluctuate in the confidence interval. As indicated earlier, s, $\dot{s}$, and p_p are measured in this example. The errors in the parameters $\mathbf{C}_{\mathbf{v}_a}$, k_p, and k_0 are also in the CI, as can be seen in the corresponding plots. The key to the parameter estimation is that the estimated system states should be in the 95% CI during the working cycle.

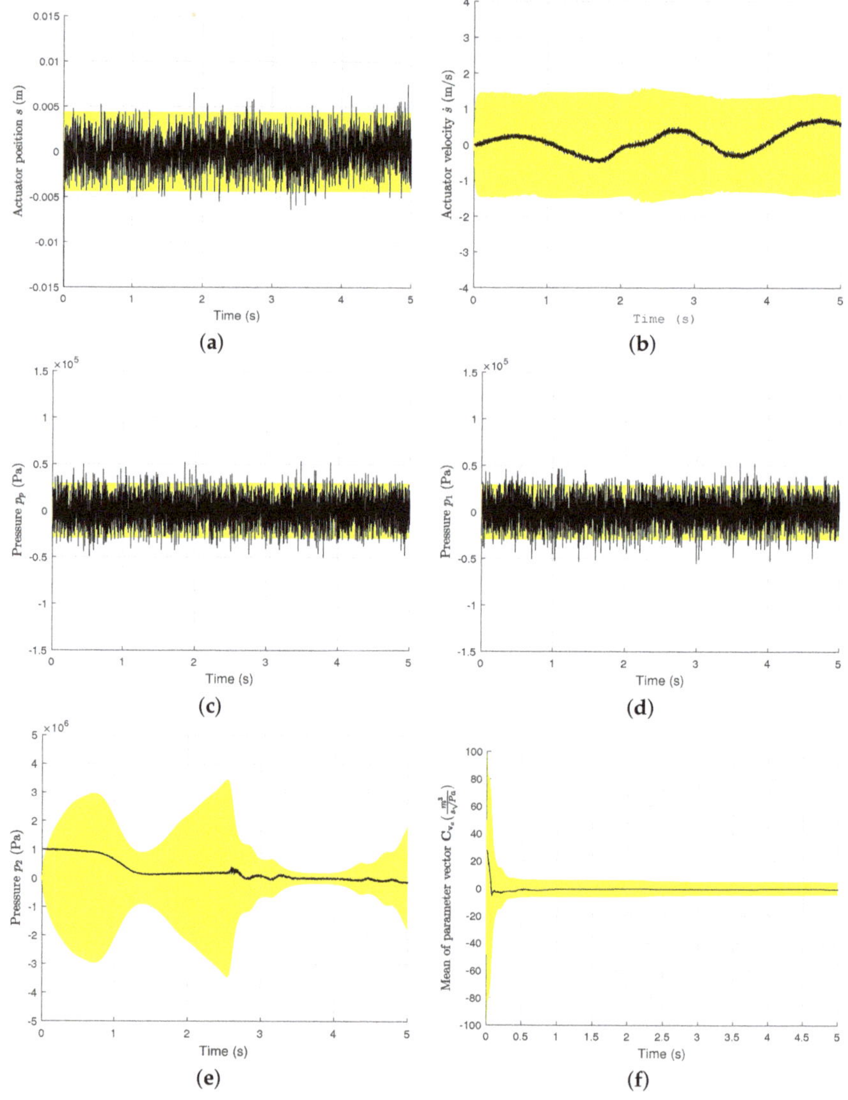

Figure 6. Requirements for the accuracy in the system states for parameter estimation. (**a**) Error in s with 95% CI. (**b**) Error in $\dot{s}$ with 95% CI. (**c**) Error in p_p with 95% CI. (**d**) Error in p_1 with 95% CI. (**e**) Error in p_2 with 95% CI. (**f**) Error in parameter $\mathbf{C}_\mathbf{v}$ with 95% CI.

5. Conclusions

This work proposes the estimation of the parameters of a system by combining parameter estimation theories and curve-fitting methods. The ADEKF algorithm is introduced in the framework of a B-spline curve-fitting method. Using the proposed algorithm, the parameters can be defined as a vector containing a set of data control points. This algorithm is applied on a hydraulically driven four-bar mechanism to estimate the characteristic curves of a directional control valve. The double-step semi-recursive formulation and lumped fluid theories are used to model the four-bar mechanism and the hydraulic system, respectively. The measurements taken from the real system include the actuator position, pump pressure, and piston side pressure. The semi-empiric flow rate coefficient vector $\mathbf{C_{v_a}}$ is defined with three to six data control points in order to define the characteristic curve of the directional control valve.

The unknown non-linear nature of the characteristic curves of the directional control valve are precisely estimated. The maximum RMSE observed in the estimation of the characteristic curves is 0.08%. This implies that the characteristic curves are accurately estimated. The data control points in the parameter vector $\mathbf{C_{v_a}}$ converge in the range of $0 < t \leq 0.3$ s in these estimation cases. To account for the system's response, the estimation of the system's state vector variables should be located in the 95% confidence interval. By using the estimated characteristic curves, important information about the discharge coefficient, pressure losses, and flow characteristics of the directional control valve can be interpreted. With this valuable information, manufacturers and users can monitor the condition of a system and make decisions about the repair and maintenance of hydraulically driven systems.

Applying the parameter estimation algorithm in the real world by using a multibody-based estimation model can enable the estimation of important parameters. This can be challenging, as the estimation model might not be as accurate as the real world necessitates. However, despite implementation challenges, the application of this parameter estimation algorithm will provide an interesting area for manufacturers and researchers. Manufacturers can use these parameters in condition monitoring, repair and maintenance, and the anticipation of product life cycles. With a product's application history, important design changes can be introduced in future designs of the product. This will ultimately lead to more efficient MBS-based digital-twin applications through the use of real-time simulations and more sustainable future products.

Author Contributions: Conceptualization, M.K.-O. and A.M.; Funding acquisition, A.M.; Investigation, Q.K. and M.K.-O.; Methodology, Q.K. and M.K.-O.; Resources, A.M.; Software, Q.K. and M.K.-O.; Supervision, M.K.M. and A.M.; Visualization, Q.K. and M.K.-O.; Writing—original draft, Q.K.; Writing—review & editing, Q.K., M.K.-O., S.J., M.K.M. and A.M. All authors have read and agreed to the published version of the manuscript.

Funding: This work was supported in the Research Project of DigiBuzz at Lappeenranta University of Technology, Lappeenranta, Finland.

Institutional Review Board Statement: Not applicable.

Informed Consent Statement: Not applicable.

Data Availability Statement: Not applicable.

Conflicts of Interest: The authors declare no conflict of interest.

Appendix A. The Estimation of the Curve Using Second-Order (Linear) B-Spline Interpolation

When using second-order (linear) B-spline interpolation, the characteristic curves of the directional control valve are not continuous, as shown in the Figure A1. Therefore, to demonstrate the real-world application, it is recommended to use the third-order B-spline or above.

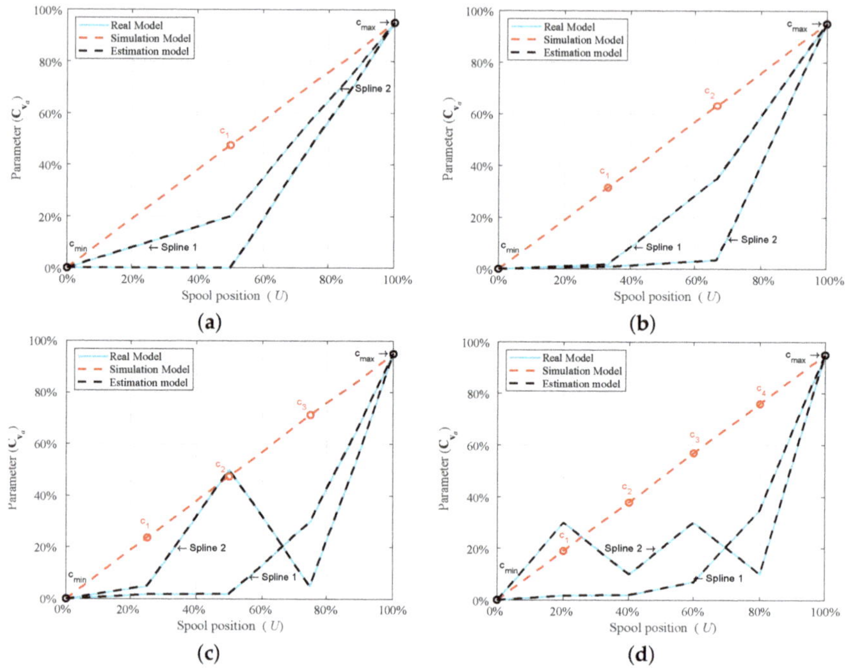

Figure A1. The estimation of the characteristic curve of the directional control valve by using the ADEKF algorithm with second-order (linear) B-spline interpolation. (**a**) Three-point B-spline estimation. (**b**) Four-point B-spline estimation. (**c**) Five-point B-spline estimation. (**d**) Six-point B-spline estimation.

Appendix B. Estimation of the Pressure Flow Coefficient and the Flow Gain

The estimation of the pressure flow coefficient k_p and the flow gain k_0 in the case example is represented below.

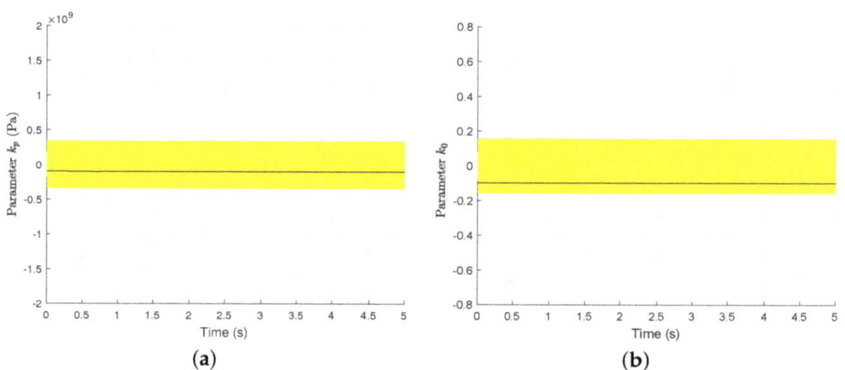

Figure A2. Requirements for the accuracy in the system states for parameter estimation. (**a**) Error in k_p with 95% CI. (**b**) Error in k_0 with 95% CI.

References

1. Schiehlen, W. *Multibody Systems Handbook*; Springer: Berlin/Heidelberg, Germany, 1990; Volume 6.
2. Schiehlen, W. Multibody System Dynamics: Roots and Perspectives. *Multibody Syst. Dyn.* **1997**, *1*, 149–188. [CrossRef]
3. Khadim, Q.; Hannola, L.; Donoghue, I.; Mikkola, A.; Kaikko, E.P.; Hukkataival, T. Integrating the user experience throughout the product lifecycle with real-time simulation-based digital twins. In *Real-Time Simulation for Sustainable Production: Enhancing User Experience and Creating Business Value*; Routledge: London, UK, 2021
4. Khadim, Q.; Kaikko, E.P.; Puolatie, E.; Mikkola, A. Targeting the user experience in the development of mobile machinery using real-time multibody simulation. *Adv. Mech. Eng.* **2020**, *12*, 1687814020923176. [CrossRef]
5. Ukko, J.; Saunila, M.; Heikkinen, J.; Semken, R.S.; Mikkola, A. *Real-Time Simulation for Sustainable Production: Enhancing User Experience and Creating Business Value*; Routledge: London, UK, 2021.
6. Boschert, S.; Rosen, R. Digital Twin—The Simulation Aspect. In *Mechatronic Futures*; Springer: Berlin/Heidelberg, Germany, 2016; pp. 59–74.
7. Lim, K.Y.H.; Zheng, P.; Chen, C.H. A state-of-the-art survey of Digital Twin: techniques, engineering product lifecycle management and business innovation perspectives. *J. Intell. Manuf.* **2019**, 1–25. [CrossRef]
8. Son, J.; Zhou, S.; Sankavaram, C.; Du, X.; Zhang, Y. Remaining useful life prediction based on noisy condition monitoring signals using constrained Kalman filter. *Reliab. Eng. Syst. Saf.* **2016**, *152*, 38–50. [CrossRef]
9. Beebe, R.S.; Beebe, R.S. *Predictive Maintenance of Pumps Using Condition Monitoring*; Elsevier: Amsterdam, The Netherlands, 2004.
10. Yang, M.S.; Su, C.F. On parameter estimation for normal mixtures based on fuzzy clustering algorithms. *Fuzzy Sets Syst.* **1994**, *68*, 13–28. [CrossRef]
11. Yang, S.; Liu, T. State estimation for predictive maintenance using Kalman filter. *Reliab. Eng. Syst. Saf.* **1999**, *66*, 29–39. [CrossRef]
12. Yang, S. An experiment of state estimation for predictive maintenance using Kalman filter on a DC motor. *Reliab. Eng. Syst. Saf.* **2002**, *75*, 103–111. [CrossRef]
13. Kiani-Oshtorjani, M.; Mikkola, A.; Jalali, P. Numerical Treatment of Singularity in Hydraulic Circuits Using Singular Perturbation Theory. *IEEE/ASME Trans. Mechatron.* **2018**, *24*, 144–153. [CrossRef]
14. Kiani-Oshtorjani, M.; Ustinov, S.; Handroos, H.; Jalali, P.; Mikkola, A. Real-Time Simulation of Fluid Power Systems Containing Small Oil Volumes, Using the Method of Multiple Scales. *IEEE Access* **2020**, *8*, 196940–196950. [CrossRef]
15. Beck, J.V.; Arnold, K.J. *Parameter Estimation in Engineering and Science*; John Wiley and Sons Inc.: Hoboken, NJ, USA, 1977.
16. Asparouhov, T.; Muthén, B. Weighted Least Squares Estimation with Missing Data. *Mplus Tech. Append.* **2010**, *2010*, 1–10.
17. Lee, C.; Tan, O.T. A weighted-least-squares parameter estimator for synchronous machines. *IEEE Trans. Power Appar. Syst.* **1977**, *96*, 97–101. [CrossRef]
18. Haykin, S. *Kalman Filtering and Neural Networks*; John Wiley & Sons: Hoboken, NJ, USA, 2004; Volume 47.
19. Moradkhani, H.; Sorooshian, S.; Gupta, H.V.; Houser, P.R. Dual state–parameter estimation of hydrological models using ensemble Kalman filter. *Adv. Water Resour.* **2005**, *28*, 135–147. [CrossRef]
20. Billings, S.; Jones, G. Orthogonal least-squares parameter estimation algorithms for non-linear stochastic systems. *Int. J. Syst. Sci.* **1992**, *23*, 1019–1032. [CrossRef]
21. Zhang, Z. Parameter estimation techniques: A tutorial with application to conic fitting. *Image Vis. Comput.* **1997**, *15*, 59–76. [CrossRef]
22. Zhang, Y.; Xu, Q. Adaptive Sliding Mode Control With Parameter Estimation and Kalman Filter for Precision Motion Control of a Piezo-Driven Microgripper. *IEEE Trans. Control. Syst. Technol.* **2016**, *25*, 728–735. [CrossRef]
23. Soltani, M.; Bozorg, M.; Zakerzadeh, M.R. Parameter estimation of an SMA actuator model using an extended Kalman filter. *Mechatronics* **2018**, *50*, 148–159. [CrossRef]
24. Radecki, P.; Hencey, B. Online building thermal parameter estimation via Unscented Kalman Filtering. In *2012 American Control Conference (ACC)*; IEEE: Montreal, QC, Canada, 2012; pp. 3056–3062.
25. Oka, Y.; Ohno, M. Parameter estimation for heat transfer analysis during casting processes based on ensemble Kalman filter. *Int. J. Heat Mass Transf.* **2020**, *149*, 119232. [CrossRef]
26. Nonomura, T.; Shibata, H.; Takaki, R. Dynamic mode decomposition using a Kalman filter for parameter estimation. *AIP Adv.* **2018**, *8*, 105106. [CrossRef]
27. Asl, R.M.; Hagh, Y.S.; Simani, S.; Handroos, H. Adaptive square-root unscented Kalman filter: An experimental study of hydraulic actuator state estimation. *Mech. Syst. Signal Process.* **2019**, *132*, 670–691.
28. Pan, Z.; Zhang, Y.; Gustavsson, J.P.; Hickey, J.P.; Cattafesta, L.N., III. Unscented Kalman filter (UKF)-based nonlinear parameter estimation for a turbulent boundary layer: A data assimilation framework. *Meas. Sci. Technol.* **2020**, *31*, 094011. [CrossRef]
29. Cuadrado, J.; Dopico, D.; Barreiro, A.; Delgado, E. Real-time state observers based on multibody models and the extended Kalman filter. *J. Mech. Sci. Technol.* **2009**, *23*, 894–900. [CrossRef]
30. Sanjurjo, E.; Naya, M.Á.; Blanco-Claraco, J.L.; Torres-Moreno, J.L.; Giménez-Fernández, A. Accuracy and efficiency comparison of various nonlinear Kalman filters applied to multibody models. *Nonlinear Dyn.* **2017**, *88*, 1935–1951. [CrossRef]
31. Sanjurjo, E.; Dopico, D.; Luaces, A.; Naya, M.Á. State and force observers based on multibody models and the indirect Kalman filter. *Mech. Syst. Signal Process.* **2018**, *106*, 210–228. [CrossRef]
32. Cuadrado, J.; Dopico, D.; Perez, J.A.; Pastorino, R. Automotive observers based on multibody models and the extended Kalman filter. *Multibody Syst. Dyn.* **2012**, *27*, 3–19. [CrossRef]

33. Naets, F.; Pastorino, R.; Cuadrado, J.; Desmet, W. Online state and input force estimation for multibody models employing extended Kalman filtering. *Multibody Syst. Dyn.* **2014**, *32*, 317–336. [CrossRef]
34. Jaiswal, S.; Sanjurjo, E.; Cuadrado, J.; Sopanen, J.; Mikkola, A. State Estimator Based on an Indirect Kalman filter for a Hydraulically Actuated Multibody System. *Multibody Syst. Dyn.* **2021**, in review.
35. Blanchard, E.D.; Sandu, A.; Sandu, C. A Polynomial Chaos-Based Kalman Filter Approach for Parameter Estimation of Mechanical Systems. *J. Dyn. Syst. Meas. Control* **2010**, *132*. [CrossRef]
36. Rodríguez, A.J.; Sanjurjo, E.; Pastorino, R.; Naya, M.Á. State, parameter and input observers based on multibody models and Kalman filters for vehicle dynamics. *Mech. Syst. Signal Process.* **2021**, *155*, 107544. [CrossRef]
37. Sandu, A.; Sandu, C.; Ahmadian, M. Modeling Multibody Systems with Uncertainties. Part I: Theoretical and Computational Aspects. *Multibody Syst. Dyn.* **2006**, *15*, 369–391. [CrossRef]
38. Sandu, C.; Sandu, A.; Ahmadian, M. Modeling multibody systems with uncertainties. Part II: Numerical applications. *Multibody Syst. Dyn.* **2006**, *15*, 241–262. [CrossRef]
39. Kolansky, J.; Sandu, C. Enhanced Polynomial Chaos-Based Extended Kalman Filter Technique for Parameter Estimation. *J. Comput. Nonlinear Dyn.* **2018**, *13*, 021012. [CrossRef]
40. Pence, B.L.; Fathy, H.K.; Stein, J.L. Recursive Estimation for Reduced-Order State-Space Models Using Polynomial Chaos Theory Applied to Vehicle Mass Estimation. *IEEE Trans. Control. Syst. Technol.* **2013**, *22*, 224–229. [CrossRef]
41. Manring, N.D.; Fales, R.C. *Hydraulic Control Systems*; John Wiley & Sons: Hoboken, NJ, USA, 2019.
42. Skousen, P.L. *Valve Handbook*; McGraw-Hill Education: New York, NY, USA, 2011.
43. Wu, D.; Li, S.; Wu, P. CFD simulation of flow-pressure characteristics of a pressure control valve for automotive fuel supply system. *Energy Convers. Manag.* **2015**, *101*, 658–665. [CrossRef]
44. Lisowski, E.; Filo, G.; Rajda, J. Pressure compensation using flow forces in a multi-section proportional directional control valve. *Energy Convers. Manag.* **2015**, *103*, 1052–1064. [CrossRef]
45. Lisowski, E.; Rajda, J. CFD analysis of pressure loss during flow by hydraulic directional control valve constructed from logic valves. *Energy Convers. Manag.* **2013**, *65*, 285–291. [CrossRef]
46. Simon, D. *Optimal State Estimation: Kalman, H Infinity, and Nonlinear Approaches*; John Wiley & Sons: Hoboken, NJ, USA, 2006.
47. Grewal, M.S.; Andrews, A.P. *Kalman Filtering: Theory and Practice Using MATLAB*; John Wiley & Sons: Hoboken, NJ, USA, 2014.
48. De Boor, C.; De Boor, C. *A Practical Guide to Splines*; Springer: New York, NY, USA, 1978; Volume 27,
49. De Boor, C. On calculating with B-splines. *J. Approx. Theory* **1972**, *6*, 50–62. [CrossRef]
50. Jauch, J.; Bleimund, F.; Rhode, S.; Gauterin, F. Recursive B-spline approximation using the Kalman filter. *Eng. Sci. Technol. Int. J.* **2017**, *20*, 28–34. [CrossRef]
51. Lim, K.H.; Seng, K.P.; Ang, L.M. River Flow Lane Detection and Kalman Filtering-Based B-spline Lane Tracking. *Int. J. Veh. Technol.* **2012**, *2012*, 465819. [CrossRef]
52. Charles, G.; Goodall, R.; Dixon, R. Wheel-Rail Profile Estimation. In Proceedings of the 2006 IET International Conference on Railway Condition Monitoring, Birmingham, UK, 29–30 November 2006; pp. 32–37.
53. Squire, W.; Trapp, G. Using Complex Variables to Estimate Derivatives of Real Functions. *Siam Rev.* **1998**, *40*, 110–112. [CrossRef]
54. Lai, K.L.; Crassidis, J.; Cheng, Y.; Kim, J. New complex-step derivative approximations with application to second-order kalman filtering. In Proceedings of the AIAA Guidance, Navigation, and Control Conference and Exhibit, San Francisco, CA, USA, 15–18 August 2005; p. 5944.
55. Jaiswal, S.; Rahikainen, J.; Khadim, Q.; Sopanen, J.; Mikkola, A. Comparing double-step and penalty-based semirecursive formulations for hydraulically actuated multibody systems in a monolithic approach. *Multibody Syst. Dyn.* **2021**, *52*, 1–23. [CrossRef]
56. Rodríguez, J.I.; Jiménez, J.M.; Funes, F.J.; de Jalón, J.G. Recursive and Residual Algorithms for the Efficient Numerical Integration of Multi-Body Systems. *Multibody Syst. Dyn.* **2004**, *11*, 295–320. [CrossRef]
57. Jaiswal, S.; Korkealaakso, P.; Åman, R.; Sopanen, J.; Mikkola, A. Deformable Terrain Model for the Real-Time Multibody Simulation of a Tractor With a Hydraulically Driven Front-Loader. *IEEE Access* **2019**, *7*, 172694–172708. [CrossRef]
58. Callejo, A.; Pan, Y.; Ricon, J.L.; Kövecses, J.; Garcia de Jalon, J. Comparison of Semirecursive and Subsystem Synthesis Algorithms for the Efficient Simulation of Multibody Systems. *J. Comput. Nonlinear Dyn.* **2017**, *12*, 011020. [CrossRef]
59. De Jalon, J.G.; Bayo, E. *Kinematic and Dynamic Simulation of Multibody Systems: The Real-Time Challenge*; Springer Science & Business Media: Berlin/Heidelberg, Germany, 2012.
60. Flores, P.; Ambrósio, J.; Claro, J.P.; Lankarani, H.M. *Kinematics and Dynamics of Multibody Systems with Imperfect Joints: Models and Case Studies*; Springer Science & Business Media: Berlin/Heidelberg, Germany, 2008; Volume 34.
61. Cuadrado, J.; Dopico, D.; Naya, M.A.; Gonzalez, M. Real-Time Multibody Dynamics and Applications. In *Simulation Techniques for Applied Dynamics*; Springer: Vienna, Austria, 2008; pp. 247–311.
62. de García Jalón, J.; Álvarez, E.; De Ribera, F.; Rodríguez, I.; Funes, F. A Fast and Simple Semi-Recursive Formulation for Multi-Rigid-Body Systems. In *Advances in Computational Multibody Systems*; Springer: Dordrecht, The Netherlands, 2005; Volume 2, pp. 1–23.
63. Hidalgo, A.F.; Garcia de Jalon, J. Real-Time Dynamic Simulations of Large Road Vehicles Using Dense, Sparse, and Parallelization Techniques. *J. Comput. Nonlinear Dyn.* **2015**, *10*, 031005. [CrossRef]

64. Pan, Y.; He, Y.; Mikkola, A. Accurate real-time truck simulation via semirecursive formulation and Adams–Bashforth–Moulton algorithm. *Acta Mech. Sin.* **2019**, *35*, 641–652. [CrossRef]
65. Watton, J. *Fluid Power Systems: Modeling, Simulation, Analog, and Microcomputer Control*; Prentice Hall: Hoboken, NJ, USA, 1989.
66. Handroos, H.; Vilenius, M. Flexible Semi-Empirical Models for Hydraulic Flow Control Valves. *J. Mech. Des.* **1991**, *113*, 232–238. [CrossRef]
67. Lu, Q.; Tiainen, J.; Kiani-Oshtorjani, M.; Ruan, J. Radial Flow Force at the Annular Orifice of a Two-Dimensional Hydraulic Servo Valve. *IEEE Access* **2020**, *8*, 207938–207946. [CrossRef]
68. Mohammadi, M.; Kiani-Oshtorjani, M.; Mikkola, A. The Effects of Oil Entrained Air on the Dynamic Performance of a Hydraulically Driven Multibody System. *Int. Rev. Model. Simul. (IREMOS)* **2020**, *13*. [CrossRef]
69. Jaiswal, S.; Sopanen, J.; Mikkola, A. Efficiency Comparison of Vrious Fiction models of a Hydraulic Cylinder in the Framework of Multibody System Dynamics. *Nonlinear Dyn.* **2021**, *104*, 3497–3515. [CrossRef]
70. Brown, P.; McPhee, J. A Continuous Velocity-Based Friction Model for Dynamics and Control With Physically Meaningful Parameters. *J. Comput. Nonlinear Dyn.* **2016**, *11*, 054502. [CrossRef]
71. Naya, M.A.; Cuadrado, J.; Dopico, D.; Lugris, U. An efficient unified method for the combined simulation of multibody and hydraulic dynamics: comparison with simplified and co-integration approaches. *Arch. Mech. Eng.* **2011**, *58*, 223–243. [CrossRef]
72. Rahikainen, J.; Kiani-Oshtorjani, M.; Sopanen, J.; Jalali, P.; Mikkola, A. Computationally efficient approach for simulation of multibody and hydraulic dynamics. *Mech. Mach. Theory* **2018**, *130*, 435–446. [CrossRef]
73. Wanner, G.; Hairer, E. *Solving Ordinary Differential Equations II*; Springer: Berlin/Heidelberg, Germany, 1996; Volume 375.
74. Eich-Soellner, E.; Führer, C. *Numerical Methods in Multibody Dynamics*; Springer: Berlin/Heidelberg, Germany, 1998; Volume 375.
75. Brenan, K.E.; Campbell, S.L.; Petzold, L.R. *Numerical Solution of Initial-Value Problems in Differential-Algebraic Equations*; SIAM: University City, PA, USA, 1995.
76. SMC. Actuator Position Sensor. Available online: http://ca01.smcworld.com/catalog/New-products-en/mpv/es20-257-d-mp/data/es20-257-d-mp.pdf (accessed on 20 July 2021).

Article

Multibody-Based Input and State Observers Using Adaptive Extended Kalman Filter

Antonio J. Rodríguez [1,*], Emilio Sanjurjo [1], Roland Pastorino [2] and Miguel Á. Naya [1]

[1] Laboratorio de Ingeniería Mecánica, University of A Coruna, Escuela Politécnica Superior, Mendizábal s/n, 15403 Ferrol, Spain; emilio.sanjurjo@udc.es (E.S.); miguel.naya@udc.es (M.Á.N.)

[2] Test Division, Siemens Digital Industries Software, Interleuvenlaan 68, B-3001 Leuven, Belgium; roland.pastorino@siemens.com

* Correspondence: antonio.rodriguez.gonzalez@udc.es

Abstract: The aim of this work is to explore the suitability of adaptive methods for state estimators based on multibody dynamics, which present severe non-linearities. The performance of a Kalman filter relies on the knowledge of the noise covariance matrices, which are difficult to obtain. This challenge can be overcome by the use of adaptive techniques. Based on an error-extended Kalman filter with force estimation (errorEKF-FE), the adaptive method known as maximum likelihood is adjusted to fulfill the multibody requirements. This new filter is called adaptive error-extended Kalman filter (AerrorEKF-FE). In order to present a general approach, the method is tested on two different mechanisms in a simulation environment. In addition, different sensor configurations are also studied. Results show that, in spite of the maneuver conditions and initial statistics, the AerrorEKF-FE provides estimations with accuracy and robustness. The AerrorEKF-FE proves that adaptive techniques can be applied to multibody-based state estimators, increasing, therefore, their fields of application.

Keywords: adaptive Kalman filter; multibody dynamics; nonlinear models; virtual sensing; multibody based observers

Citation: Rodríguez, A.J.; Sanjurjo, E.; Pastorino, R.; Naya, M.Á. Multibody-Based Input and State Observers Using Adaptive Extended Kalman Filter. *Sensors* **2021**, *21*, 5241. https://doi.org/10.3390/s21155241

Academic Editor: Stefano Mariani

Received: 31 May 2021
Accepted: 29 July 2021
Published: 3 August 2021

1. Introduction

In many applications, it is required to acquire data in order to analyze the performance of a particular machine, to make decisions through control strategies or to gather data for durability purposes. Most of the time, it is possible to get all the data through sensors. However, there are particular cases where a sensor cannot be installed due to economical or technical issues. In these situations, data acquisition relies on estimations.

Kalman filter [1] stands out as one of the most used estimation techniques. It combines a model describing the system with sensor measurements. The estimations are based on the statistical properties of the desired variables conditioned on the sensors measurements [2]. Although the Kalman filter was designed for linear systems, several approaches have been developed in the literature extending its use to non-linear models. Such is the case of the extended Kalman filter (EKF) or unscented Kalman filter (UKF) [2]. These approaches introduced the Kalman filter to real problems, which are predominantly non-linear.

Similarly, the interest for mulibody models has grown in the recent years in several fields of the industry, such as automotive, biomechanics or machinery. Through multibody dynamics, a system can be modeled with high accuracy [3], at expenses of a moderate computational cost. Recently, multibody models have started to be used in combination with Kalman filters for estimation. Such is the case of virtual sensing in automotive applications [4]. However, the assembly between a multibody model and a Kalman filter is not trivial due to the high non-linearities of multibody models.

The first approach presented in the literature is [5], where an EKF was combined with a simple mechanism. However, the complexity of the implementation and the computa-

179

tional cost limits its application. Several approaches have been presented later on [6–9]. In [7], an UKF based on a multibody model is proposed, reducing the complexity of the implementation. However, it entails high computational cost: the estimations are based on a set of sample points propagated through the model. Hence, it requires to simulate as many multibody models as sample points each time step. From the work presented in [8], a new Kalman filter known as error-state extended Kalman filter (errorEKF) shows accurate results while reducing the computational cost. In addition, the filter entails low coupling with the model. Thus, the implementation of any model becomes simpler. The errorEKF has been improved in more recent works, extending its estimations to inputs [9] and parameters [10].

The Kalman filter, besides being developed for linear systems, assumes that the noise of the measurements and the process (or system) are zero-mean white Gaussian with known covariance matrices [1]. These assumptions are hard to satisfy in real scenarios. The use of wrong statistics in the design of the filter can lead to large estimation errors or even a divergence in the filter [11]. In practice, these matrices are estimated through a tedious trial-and-error procedure. This is especially critical for the process noise covariance matrix (PNCM), since it is not possible to determine the accuracy of the model. Regarding the measurement noise covariance matrix (MNCM), each sensor is characterized and, hence, an initial value can be approximated.

However, even though a proper value of the noise covariance matrices is found, they remain constant reducing the robustness of the filter [10,12]. In some scenarios, the system can experiment a noticeable change during a particular maneuver. This change can affect the model accuracy and, thus, invalidate the initial noise covariances. This explains the increasing interest on developing methods for noise covariance matrices estimation [13]. These methods are usually known as Adaptive Kalman filters (AKF).

Adaptive Kalman filters can be divided in two major approaches: multiple-model based adaptive estimation (MMAE) and innovation-based adaptive estimation (IAE). Both approaches share the same concept of using the new information of the innovation sequence of the filter (difference between real and estimated measurements) [14]. MMAE methods implement a bank of several different models and compute the probability for each model to be the true system model given the measurement sequence and under the assumption that one of the models in the bank is the correct one. IAE performs the adaptation based directly on the innovation sequence, using the fact that the innovation sequence is white noise if the filter has the right values of the covariance matrices [15].

Multibody models can be computationally expensive and achieving real-time performance in practical applications is a challenge [16]. Hence, executing several models in parallel following the MMAE approach would compromise real-time performance and, thus, the viability of the solution.

Focusing on the IAE, there are many alternatives presented in the literature. In [14], an approach based on the maximum likelihood criteria (ML) is proposed to estimate both the PNCM and MNCM for inertial navigation, where Kalman filters are employed to overcome the lack of GPS signals in certain areas. The proposed approach is also compared against conventional Kalman filters. The ML aims to maximize the probability that a set of data (i.e., innovation) is observed for a given a parameter (i.e., noise covariance matrices). The ML depends on a window of past innovation data. If the sample data grows without bound, the estimate converges to the true value of the covariance matrices. As a counterpart, the ML could be biased for small sample sizes (over-trained). This issue is addressed in [17] by combining the ML with a fuzzy control in order to select only the viable satellites for the estimation, removing outlier data from the innovation sequence.

In [18], an IAE adaptive Kalman filter is also proposed for sensor fusion in INS/GPS navigation. The algorithm is equivalent to the maximum likelihood criteria. Combining the GPS with an IMU, the position can be accurately tracked, even in environments without GPS signal.

For spacecraft navigation, in [12], an AKF based on ML combined with fuzzy logic is proposed. The fuzzy logic is based on the discrepancy between the estimated covariance and the theoretical covariance bounds. The use of fuzzy logic helped to reduce the computational cost. However, the use of fuzzy logic incremented the estimation errors.

In [19], a Sage-Husa (SH) filter is combined with a Kalman filter to obtain the angular velocities of a vehicle from an inertial navigation system which uses only accelerometers. The SH algorithm is equivalent to those resulting from maximum likelihood estimation [20]. It is based on the maximum a posterior probability (MAP), which maximizes the probability of a parameter for a given set of data. It means that the likelihood is weighted by the prior information. In practice, it can be seen as applying weights in the ML estimates, giving more relevance to the most recent estimations. Sage-Husa algorithm is an alternative to the ML when there are not enough data available. However, it cannot estimate the PNCM and MNCM simultaneously without causing divergence in the estimations [21].

The SH algorithm is also used with modifications in [22] to estimate the MNCM of an UKF for state estimation in a vehicle. A similar approach is followed in [21], where a robust UKF is complemented with the SH algorithm for estimating the PNCM for autonomous underwater vehicle acoustic navigation.

Another approach for adaptive Kalman filtering is known as Variational Bayesian (VB). While ML and MAP give a fixed value for the noise covariance matrices (point estimators), the VB gives a probability density function. The advantage of this method is that it mitigates the effect of over-trained filters [23]. As an example, the VB is applied in [24] in order to improve the measurements of inertial navigation systems. It shows robustness against anomalous measurements. As a counterpart, the VB assumes that the PNCM is accurately known. Thus, it can be only used for estimating the states and the MNCM [25]. This is an important drawback since, as reflected in [26,27], the filter performance will decrease with the uncertainties of the PNCM. The VB approach is extended in [26,28] including the calculation of the predicted error covariance matrix instead of the process covariance matrix seeking for higher accuracy. In the same way, in [29], the VB approach is enhanced by its combination with a Sage-Husa algorithm for the PNCM estimation achieving satisfactory results.

The aim of this work is to explore the use of adaptive algorithms with multibody based Kalman filters, solving the actual limitation of the unknown statistics of the noise covariance matrices. For that purpose, a Kalman filter based on multibody models has to be combined with an adaptive method. This implies to study how to combine the different equations involved, performing the required modifications. Most of the research made in adaptive Kalman filtering is related to navigation applications. However, multibody models present strong non-linearities that can affect the performance of the adaptive filters. To the authors' knowledge, adaptive Kalman filtering is a technique that has never been applied to multibody-based estimation and, at the state of the art of Kalman filtering in this field, it is the next natural step to be taken.

As novelty, this work presents a state and input estimator for multibody models, which includes an adaptive methodology in order to estimate the noise covariance matrices. The lack of knowledge on the statistics of the system is an actual limitation for industrial applications of multibody-based estimation. That is the case, for example, of automotive applications, where the maneuvers are continuously changing and adjusting the noise covariance matrices following a trial-and-error procedure is not viable. Following the presented approach, this limitation can be overcome, and the use of estimation based on multibody models can be increased in real industrial applications.

2. Materials and Methods

This work continues the development of an accurate and efficient Kalman filter for multibody models. Starting from the work presented in [9], the errorEKF with force estimation is extended with adaptive methods improving its accuracy and robustness. The errorEKF with force estimation is one of the best options in terms of accuracy and computational efficiency for multibody-based Kalman filtering. Regarding the adaptive method, whereas the Variational Bayesian can only be used for estimating the measurement noise covariance matrix, the Sage-Husa and maximum likelihood can estimate the process and measurement noise covariance matrices. The former has shown lack of stability if both matrices are estimated simultaneously. Hence, the filter of this work combines the maximum likelihood method with the errorEKF with force estimation.

The adaptive techniques used in this work are evaluated when applied to mechanisms described by multibody models. In order to present a fair comparison with the previous version of the filter, the mechanisms employed in this work are the same as in [9]: a four bar linkage (Figure 1a) and a five bar linkage (Figure 1b). The dimensions of both mechanisms are summarized in Tables 1 and 2.

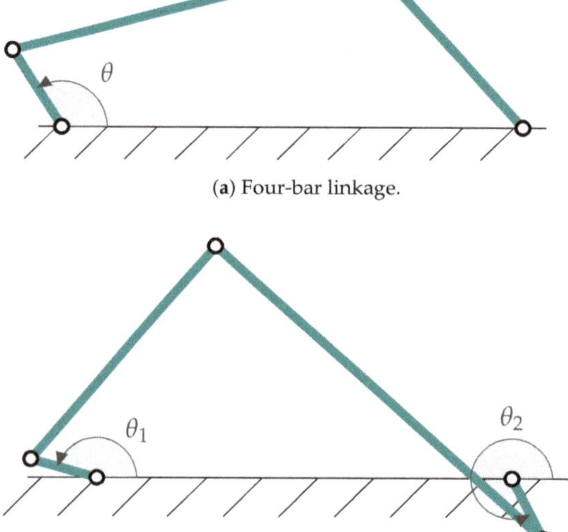

(**a**) Four-bar linkage.

(**b**) Five-bar linkage.

Figure 1. Mechanisms employed in this work.

Table 1. Properties of the four-bar linkage.

	Crank	Coupler	Rocker	Ground Element
Mass (kg)	2	8	5	-
Length (m)	2	8	5	10

Table 2. Properties of the five-bar linkage.

	Left Crank	Left Coupler	Right Coupler	Right Crank	Ground Element
Mass (kg)	3	1	2	3	-
Length (m)	0.5	2.062	3.202	0.5	3

2.1. Multibody Modeling

The input and state observer that is used in this work, the errorEKF with force estimation, can be applied to any multibody formulation and integrator [8]. This is one of its main advantages, since it reduces the complexity of implementing estimators based on multibody models (MBS). For example, in [9], an augmented Lagrangian formulation of index 3 (ALI3P) using natural coordinates is employed. Meanwhile, in [10], the multibody model is based on a semi-recursive method using relative coordinates.

In this work, the ALI3P formulation is selected following the work in [9]. First presented in [30], this formulation is widely used in multibody applications due to its efficiency and robustness [16,31–34]. The formulation is briefly described next. The reader is referred to [30,31,35] for further details. The multibody models are simulated using MBDE, a MATLAB® Open Source toolbox (See https://github.com/MBDS/mbde-matlab (accessed on 12 May 2021)).

The dynamics of a multibody system can be described by an index-3 augmented Lagrangian formulation as,

$$\mathbf{M\ddot{q}} + \mathbf{\Phi_q}^{\mathrm{T}}\alpha\mathbf{\Phi} + \mathbf{\Phi_q}^{\mathrm{T}}\lambda^* = \mathbf{Q} \tag{1}$$

being $\mathbf{M}$ the mass matrix, $\mathbf{\ddot{q}}$ the accelerations of the natural coordinates, $\mathbf{\Phi}$ the constraints vector, $\mathbf{\Phi_q}$ the Jacobian matrix of the constraint equations with respect to the generalized coordinates $\mathbf{q}$, α the penalty factor, $\mathbf{Q}$ the vector of generalized forces, and λ^* the vector of Lagrange multipliers obtained through the following iteration process [35],

$$\lambda_{i+1}^* = \lambda_i^* + \alpha\mathbf{\Phi}_{i+1}, \qquad i = 0,1,2\ldots \tag{2}$$

where i is the index of the iteration.

It should be noted that this method directly includes the constraint equations at position level as a dynamical system, penalized by a large factor, α. The larger the penalty factor, the better the constraints will be achieved. In order to avoid numerical ill conditioning, a factor between 10^7 and 10^9 (depending on the mass of the system) gives excellent results for double arithmetic [3].

In order to integrate Equation (1), the implicit single-step trapezoidal rule has been adopted. The corresponding difference equations in velocities and accelerations can be derived as,

$$\mathbf{\dot{q}}_{n+1} = \frac{2}{\Delta t}\mathbf{q}_{n+1} + \mathbf{\dot{q}}_n^*; \quad \mathbf{\dot{q}}_n^* = -\left(\frac{2}{\Delta t}\mathbf{q}_n + \mathbf{\dot{q}}_n\right) \tag{3}$$

$$\mathbf{\ddot{q}}_{n+1} = \frac{4}{\Delta t^2}\mathbf{q}_{n+1} + \mathbf{\hat{q}}_n^*; \quad \mathbf{\hat{q}}_n^* = -\left(\frac{4}{\Delta t^2}\mathbf{q}_n + \frac{4}{\Delta t}\mathbf{\dot{q}}_n + \mathbf{\ddot{q}}_n\right) \tag{4}$$

where Δt is the time-step. Introducing the previous equations in Equation (1) leads to a nonlinear system of equations, $\mathbf{f}$, in which the positions at time-step $n + 1$ are the unknowns. Hence,

$$\mathbf{f}(\mathbf{q}_{n+1}) = \mathbf{0} \tag{5}$$

Since Equation (5) is a nonlinear system of algebraic equations, the Newton–Raphson iteration can be used to find a solution. The residual vector is,

$$\mathbf{f}(\mathbf{q}) = \frac{\Delta t^2}{4}(\mathbf{M\ddot{q}} + \mathbf{\Phi_q}^{\mathrm{T}}\alpha\mathbf{\Phi} + \mathbf{\Phi_q}^{\mathrm{T}}\lambda^* - \mathbf{Q}) \tag{6}$$

and the approximated tangent matrix,

$$\frac{\partial\mathbf{f}(\mathbf{q})}{\partial\mathbf{q}} = \mathbf{M} + \frac{\Delta t}{2}\mathbf{C} + \frac{\Delta t^2}{4}\left(\mathbf{\Phi_q}^{\mathrm{T}}\alpha\mathbf{\Phi_q} + \mathbf{\bar{K}}\right) \tag{7}$$

where $\mathbf{\bar{K}}$ and $\mathbf{C}$ represent the contribution of damping and elastic forces of the system.

After converging to a solution, the positions satisfy the equations of motion and the constraint conditions, $\mathbf{\Phi} = \mathbf{0}$. However, there is no guarantee on satisfying the constraint equations at velocity, $\dot{\boldsymbol{\Phi}} = \mathbf{0}$, and acceleration level, $\ddot{\boldsymbol{\Phi}} = \mathbf{0}$, since they were not imposed. To overcome this issue, the velocities and accelerations are projected,

$$\frac{\partial \mathbf{f(q)}}{\partial \mathbf{q}} \dot{\mathbf{q}} = \left[\mathbf{M} + \frac{\Delta t}{2} \mathbf{C} + \frac{\Delta t^2}{4} \bar{\mathbf{K}} \right] \tilde{\dot{\mathbf{q}}} - \frac{\Delta t^2}{4} \boldsymbol{\Phi}_{\mathbf{q}}^{\mathsf{T}} \alpha \boldsymbol{\Phi}_t \tag{8}$$

$$\frac{\partial \mathbf{f(q)}}{\partial \mathbf{q}} \ddot{\mathbf{q}} = \left[\mathbf{M} + \frac{\Delta t}{2} \mathbf{C} + \frac{\Delta t^2}{4} \bar{\mathbf{K}} \right] \tilde{\ddot{\mathbf{q}}} - \frac{\Delta t^2}{4} \boldsymbol{\Phi}_{\mathbf{q}}^{\mathsf{T}} \alpha \left(\dot{\boldsymbol{\Phi}}_{\mathbf{q}} \dot{\mathbf{q}} + \dot{\boldsymbol{\Phi}}_t \right) \tag{9}$$

where $\tilde{\dot{\mathbf{q}}}$ and $\tilde{\ddot{\mathbf{q}}}$ are the velocities and accelerations obtained from the integration process, i.e., non-compliant with the constraints equations.

2.2. Adaptive Error-State Extended Kalman Filter with Force Estimation (AerrorEKF-FE)

The errorEKF with force estimation is based on indirect estimation, also known as error-state estimation. It combines the efficiency of an EKF filter with the easy integration of the dynamical system in the UKF filters [36]. The coupling between the filter and model is limited to share information regarding the states, which are the error in the kinematics of the model. The multibody model can be used without modifications and, thus, any formulation or integrator can be employed.

The diagram of Figure 2 illustrates the procedure followed by the adaptive errorEKF with force estimation for a time step k. The method starts integrating the multibody equations, obtaining the predicted coordinates $\mathbf{q}_{MBS}$, velocities $\dot{\mathbf{q}}_{MBS}$ and accelerations $\ddot{\mathbf{q}}_{MBS}$. These coordinates are later used to obtain the *virtual* measurements $\mathbf{h}(\mathbf{q}_{MBS}, \dot{\mathbf{q}}_{MBS}, \ddot{\mathbf{q}}_{MBS})$. Later, the equations of the filter are used to perform the estimation.

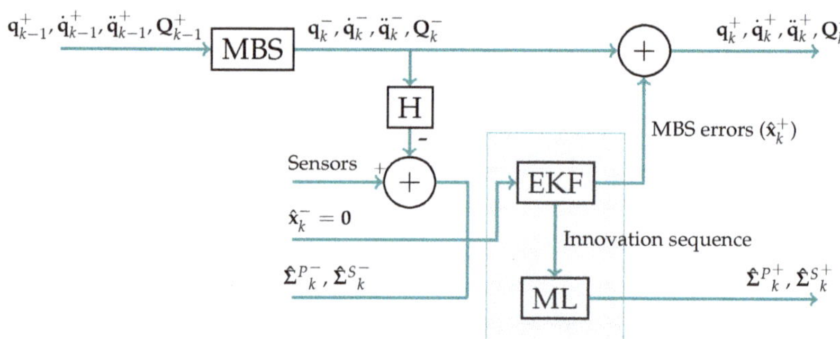

Figure 2. Simplified flow diagram of a time step of the adaptive errorEKF with force estimation. MBS refers to the multibody system, whose output is the predicted position, velocities and accelerations of the system. H is the observation matrix, employed to obtain the *virtual* measurements, which are compared with the measurements from the sensors. $\hat{\boldsymbol{\Sigma}}^P$ and $\hat{\boldsymbol{\Sigma}}^S$ are the PNCM and MNCM, respectively, where the superscripts $^-$ and $^+$ refer to the a priori and *a posteriori* estimations. EKF represents the application of the extended Kalman filter equations, whose outputs are the estimated errors in the predicted state vector (MBS errors). ML refers to maximum likelihood and is where the noise covariance matrices are estimated.

2.2.1. State and Input Estimation

For estimating the states and inputs of the system, the state vector consists in the *error* in position, velocity and acceleration of the degrees of freedom of the mechanism, ensuring a certain level of independence within the states. Hence,

$$\mathbf{x}^{\mathsf{T}} = \left[\Delta \mathbf{z}^{\mathsf{T}}, \Delta \dot{\mathbf{z}}^{\mathsf{T}}, \Delta \ddot{\mathbf{z}}^{\mathsf{T}} \right] \tag{10}$$

where $\Delta\mathbf{z}$, $\Delta\dot{\mathbf{z}}$, $\Delta\ddot{\mathbf{z}}$ are the errors in position, velocity and acceleration of the independent coordinates (coincident with the degrees of freedom), respectively, such that,

$$\mathbf{z} = \mathbf{z}_{MBS} + \Delta\mathbf{z} \tag{11}$$

$$\dot{\mathbf{z}} = \dot{\mathbf{z}}_{MBS} + \Delta\dot{\mathbf{z}} \tag{12}$$

$$\ddot{\mathbf{z}} = \ddot{\mathbf{z}}_{MBS} + \Delta\ddot{\mathbf{z}} \tag{13}$$

being $\mathbf{z}_{MBS}$, $\dot{\mathbf{z}}_{MBS}$, $\ddot{\mathbf{z}}_{MBS}$ the *virtual* positions, velocities and accelerations of the degrees of freedom (i.e., the coordinates predicted by the multibody model), and $\mathbf{z}$, $\dot{\mathbf{z}}$, $\ddot{\mathbf{z}}$ the *real* positions, velocities and accelerations of the degrees of freedom of the mechanism. It should be noted that estimating the accelerations is closely related with the estimation of the forces, as can be seen from Equations (22)–(24).

Combining the information from the *real* and *virtual* measurements into the Kalman filter, the state of the multibody model is corrected. Thus, the expected error for the next time step is null. As a consequence, the propagation phase takes the form of,

$$\hat{\mathbf{x}}_k^- = \mathbf{0} \tag{14}$$

$$\mathbf{P}_k^- = \mathbf{f}_{\mathbf{x}k-1}\mathbf{P}_{k-1}^+ \mathbf{f}_{\mathbf{x}k-1}^{\mathrm{T}} + \hat{\mathbf{\Sigma}}^{P+}_{k-1} \tag{15}$$

where $\mathbf{P}$ is the covariance matrix of the estimation error, $\mathbf{f}_{\mathbf{x}}$ is the Jacobian matrix of the transition model and $\hat{\mathbf{\Sigma}}^P$ is the estimated covariance matrix of the plant or process (i.e., the multibody model), referred as PNCM. Note that since the filter is adaptive, the value of $\hat{\mathbf{\Sigma}}^P$ is estimated each time step. For the propagation phase, the best estimation of the $\hat{\mathbf{\Sigma}}^P$ is the one obtained in the previous time step.

Equations (14) and (15) are equivalent to the equations employed in the discrete extended Kalman filter (DEKF) [2,37]. Following [9], the Jacobian matrix of the transition model can be derived through a forward Euler integrator of the states and a random walk for modeling the acceleration error,

$$\mathbf{f}_{\mathbf{x}} = \left[\begin{array}{c:c:c} \mathbf{I} + \frac{1}{2}\frac{\partial\Delta\ddot{\mathbf{z}}}{\partial\mathbf{z}}\Delta t^2 & \mathbf{I}\Delta t + \frac{1}{2}\frac{\partial\Delta\ddot{\mathbf{z}}}{\partial\dot{\mathbf{z}}}\Delta t^2 & \frac{1}{2}\mathbf{I}\Delta t^2 \\ \hdashline \frac{\partial\Delta\ddot{\mathbf{z}}}{\partial\mathbf{z}}\Delta t & \mathbf{I} + \frac{\partial\Delta\ddot{\mathbf{z}}}{\partial\dot{\mathbf{z}}}\Delta t & \mathbf{I}\Delta t \\ \hdashline \mathbf{0} & \mathbf{0} & \mathbf{I} \end{array} \right] \tag{16}$$

where the terms $\frac{\partial\Delta\ddot{\mathbf{z}}}{\partial\mathbf{z}}$ and $\frac{\partial\Delta\ddot{\mathbf{z}}}{\partial\dot{\mathbf{z}}}$ reflect the effect of the acceleration error through the force models of the multibody model. Deriving expressions for these terms is not straightforward and is out of the scope of this work. The reader is referred to [9,38] for a comprehensive explanation on how to obtain these terms.

Following the equations defined for the DEKF [2,37], the corrections can be propagated through the states obtaining the *a posteriori* state vector, $\hat{\mathbf{x}}^+$, yielding,

$$\mathbf{y}_k = \mathbf{o}_k - \mathbf{h}(\mathbf{q}_{MBS}, \dot{\mathbf{q}}_{MBS}, \ddot{\mathbf{q}}_{MBS}) \tag{17}$$

$$\mathbf{\Sigma}_k = \mathbf{h}_{\mathbf{x}k}\mathbf{P}_k^-\mathbf{h}_{\mathbf{x}k}^{\mathrm{T}} + \hat{\mathbf{\Sigma}}^{S+}_{k-1} \tag{18}$$

$$\mathbf{K}_k = \mathbf{P}_k^-\mathbf{h}_{\mathbf{x}k}^{\mathrm{T}}\mathbf{\Sigma}_k^{-1} \tag{19}$$

$$\hat{\mathbf{x}}_k^+ = \mathbf{0} + \mathbf{K}_k\mathbf{y}_k \tag{20}$$

$$\mathbf{P}_k^+ = (\mathbf{I} - \mathbf{K}_k\mathbf{h}_{\mathbf{x}k})\mathbf{P}_k^- \tag{21}$$

being $\mathbf{y}_k$ the innovation sequence, $\mathbf{o}_k$ the measurements, $\mathbf{\Sigma}_k$ the innovation covariance matrix, $\mathbf{K}_k$ the Kalman gain, and $\hat{\mathbf{\Sigma}}^S$ the covariance matrix of the sensors or measurements (MNCM). Similarly to $\hat{\mathbf{\Sigma}}^P$ in Equation (15), $\hat{\mathbf{\Sigma}}^S$ can vary within time steps. Hence, $\hat{\mathbf{\Sigma}}^{S+}_{k-1}$ is the more accurate value for propagating the corrections. The state vector and covariance matrix are estimated in Equations (20) and (21), respectively.

From the *a posteriori* state vector, $\hat{\mathbf{x}}^+$, and through Equation (13), the estimated position $\hat{\mathbf{z}}$, velocities $\dot{\hat{\mathbf{z}}}$ and accelerations $\ddot{\hat{\mathbf{z}}}$ for the time step k are obtained. Since the model relies on the set of dependent coordinates, it is required to compute the errors in these coordinates. Projecting the independent coordinates over the constraints manifold, the dependent coordinates can be obtained as stated in [9].

After correcting the states of the model, the forces (or inputs) can be estimated. The error in forces can be seen as the forces required to avoid the error in accelerations, which is referred as $\Delta\mathbf{Q}$. Since (in general) there are more dependent coordinates than degrees of freedom, there are infinite force combinations producing the same effect. Hence, depending on the application, different criteria should be considered in order to find a proper force distribution [10]. A common solution, valid for simple mechanisms, is to assume that the unknown forces are applied to the degrees of freedom. Therefore,

$$\Delta\mathbf{Q}_k^{\mathbf{q}} = \mathbf{0} \tag{22}$$

$$\Delta\mathbf{Q}_k^{\mathbf{z}} = \Delta\mathbf{Q}_{k-1}^{\mathbf{z}} + \bar{\mathbf{M}}\Delta\ddot{\hat{\mathbf{z}}} \tag{23}$$

$$\Delta\mathbf{Q} = \left[(\Delta\mathbf{Q}^{\mathbf{z}})^{\mathrm{T}} \quad (\Delta\mathbf{Q}^{\mathbf{q}})^{\mathrm{T}} \right]^{\mathrm{T}} \tag{24}$$

being $\Delta\mathbf{Q}^{\mathbf{q}}$ and $\Delta\mathbf{Q}^{\mathbf{z}}$ the force correction acting on the dependent and independent coordinates (equivalent to the degrees of freedom), respectively.

2.2.2. Process and Measurement Covariance Matrices Estimation

As reflected in Section 1, knowing the covariance matrices of the process and measurements noises is critical for a proper behavior of the filter in terms of accuracy and robustness. For the particular case of multibody models, the PNCM is the most critical. The PNCM is completely unknown and can vary depending on the maneuver due to, for example, external perturbations which are not modeled. On the other hand, the MNCM can be approximated by the information provided by the sensor manufacturer. In addition, fluctuations of this noise is not expected within maneuvers. In order to address this issue, in this work, the previous errorEKF with force estimation is extended with adaptive strategies for the PNCM. The effect of a proper estimation of the MNCM is also addressed.

Following the work presented in [14], through the maximum likelihood technique, the PNCM and MNCM can be estimated. The ML has the attractive property of uniqueness and consistency of the solution. This ensures, in a probabilistic sense, a convergence to the true value of the covariance matrices. However, the ML relies on a sliding window of the innovation sequence, whose size directly affects its performance: small sample sizes would lead to biased estimations.

In addition, the solution proposed in [14] is developed for navigation purposes. It assumes that the filter transition matrix, $\mathbf{f_x}$, and the design matrix, $\mathbf{h}$, are time invariant. This is not true when using the errorEKF. Thus, the algorithm should be adapted for the use-case of this work.

The detailed procedure for estimating the PNCM and MNCM through the ML method is presented in [14]. Before presenting the method, it should be noted the difference between probability and likelihood. Probability refers to the chances of an event to happen based on the statistical distribution of some data. Likelihood corresponds to finding the distribution that fits best to a given set of data.

The ML estimation aims to solve the question of, given the observed data, which are the model parameters that maximize the likelihood of the observed data occurring. The likelihood function can be defined as $L(\alpha|\mathbf{Z})$, where α are the parameters and $\mathbf{Z}$ the observed data. In innovation-based adaptive Kalman filtering, the parameters are the noise covariance matrices, and the observed data corresponds to the innovation sequence. In plain words, the ML estimates the noise covariance matrices searching for the values which give the maximum probability of obtaining a particular innovation sequence.

In order to calculate the likelihood, it is expressed as the probability of observing a certain dataset under an unknown statistical distribution, that is $P(\mathbf{Z}|\boldsymbol{\alpha})$. Hence, the target is to find the value of $\boldsymbol{\alpha}$, which maximizes $P(\mathbf{Z}|\boldsymbol{\alpha})$. If the sets of observed data, N, are independent, the likelihood of observing the data can be expressed as the product of the probability of observing each data point individually yielding,

$$L(\boldsymbol{\alpha}|\mathbf{Z}) = \prod_{j=1}^{N} P(\mathbf{Z}_j|\boldsymbol{\alpha}) \tag{25}$$

where $\boldsymbol{\alpha}$, for this particular case, corresponds to the elements of the noise covariance matrices, and $\mathbf{Z}$ are the innovations of the errorEKF with force estimation, $\mathbf{y}$. Regarding $\prod_{j=1}^{N}(\cdot)$, it implies the product of the full set of innovations. However, storing the entire set of past data is not viable for online estimation. Thus, a slide window is defined so that the filter only processes a fixed number of past events.

Assuming that the innovation sequence is Gaussian distributed, the likelihood function can be expressed as,

$$L(\boldsymbol{\alpha}|\mathbf{y}) = \prod_{j=1}^{N} \frac{1}{\sqrt{(2\pi)^m |\boldsymbol{\Sigma}_j|}} \exp^{-\frac{1}{2}\mathbf{y}_j^T \boldsymbol{\Sigma}_j^{-1} \mathbf{y}_j} \tag{26}$$

being $\mathbf{y}$ the innovation sequence, $\boldsymbol{\Sigma}$ the covariance matrix, m the number of measurements per time step, $|\cdot|$ the determinant operator, and exp the natural base.

The value of $\boldsymbol{\alpha}$ maximizing the likelihood can be obtained by deriving the likelihood function (Equation (25)) with respect to $\boldsymbol{\alpha}$ and setting it to zero. However, differentiate Equation (25) is difficult. As an alternative, Equation (25) is simplified by taking the natural logarithm of the equation yielding,

$$L(\boldsymbol{\alpha}|\mathbf{y}) = -\frac{1}{2} \sum_{j=j_0}^{k} \left[ln|\boldsymbol{\Sigma}_j| + \mathbf{y}_j^T \boldsymbol{\Sigma}_j^{-1} \mathbf{y}_j + c_j \right] \tag{27}$$

where c_j is a constant term independent of the adaptive parameters, $\boldsymbol{\alpha}$. Maximizing the previous equation is equivalent to minimizing its negative version. Neglecting the constant terms, the maximum likelihood condition can be derived,

$$\sum_{j=j_0}^{k} ln|\boldsymbol{\Sigma}_j| + \sum_{j=j_0}^{k} \mathbf{y}_j^T \boldsymbol{\Sigma}_j^{-1} \mathbf{y}_j = min \tag{28}$$

where k is the time step in which the covariance matrix is estimated and j is the counter inside the estimating window. Equation (28) aims at finding the covariance matrix, which results in the smallest error norm, being complementary to the state estimation.

As stated before, the solution for Equation (28) can be constructed by taking its derivative with respect to the adaptation parameters and setting the results to zero. This yields,

$$\sum_{j=j_0}^{k} \left[tr\left\{ \boldsymbol{\Sigma}_j^{-1} \frac{\partial \boldsymbol{\Sigma}_j^S}{\partial \alpha_k} \right\} - \mathbf{y}_j \boldsymbol{\Sigma}_j^{-1} \frac{\partial \boldsymbol{\Sigma}_j^S}{\partial \alpha_k} \boldsymbol{\Sigma}_j^{-1} \mathbf{y}_j^T \right] = 0 \tag{29}$$

In addition, using Equations (15) and (18), the covariance matrix, $\boldsymbol{\Sigma}_k$, can be replaced by $\boldsymbol{\Sigma}^P$ and $\boldsymbol{\Sigma}^S$. After some manipulation [14], the equation that maximizes the likelihood function for the adaptation parameters can be expressed as,

$$\sum_{j=j_0}^{k} tr\left\{ \left[\boldsymbol{\Sigma}_j^{-1} - \boldsymbol{\Sigma}_j^{-1} \mathbf{y}_j \mathbf{y}_j^T \boldsymbol{\Sigma}_j^{-1} \right] \left[\frac{\partial \boldsymbol{\Sigma}_j^S}{\partial \alpha_k} + \mathbf{h}_{xj} \frac{\partial \boldsymbol{\Sigma}_{j-1}^P}{\partial \alpha_k} \mathbf{h}_{xj} \right] \right\} = 0 \tag{30}$$

From Equation (30), expressions for $\mathbf{\Sigma}^P$ and $\mathbf{\Sigma}^S$ can be obtained independently. For deriving the expression of $\mathbf{\Sigma}^P$, it is assumed that $\mathbf{\Sigma}^S$ is known and independent of α (i.e., $\frac{\partial \mathbf{\Sigma}_{j-1}^P}{\partial \alpha_k} = \mathbf{0}$). Under this assumption, the adaptation relies on $\mathbf{\Sigma}^P$. This means that the covariance noises of the process are the adaptive parameters (i.e., $\alpha = \sigma^2$). Considering $\mathbf{\Sigma}^P$ as a diagonal matrix leads to

$$\sum_{j=j_0}^{k} tr\left\{ \left[\mathbf{\Sigma}_j^{-1} - \mathbf{\Sigma}_j^{-1} \mathbf{y}_j \mathbf{y}_j^T \mathbf{\Sigma}_j^{-1} \right] \left[0 + \mathbf{h}_{\mathbf{x}j} \mathbf{I} \mathbf{h}_{\mathbf{x}j} \right] \right\} = 0 \tag{31}$$

which can be reduced to

$$\sum_{j=j_0}^{k} \mathbf{h}_{\mathbf{x}j}^T \left[\mathbf{\Sigma}_j^{-1} - \mathbf{\Sigma}_j^{-1} \mathbf{y}_j \mathbf{y}_j^T \mathbf{\Sigma}_j^{-1} \right] \mathbf{h}_{\mathbf{x}j} = 0 \tag{32}$$

Introducing now the expressions of the errorEKF, the covariance matrix $\mathbf{\Sigma}$ can be replaced by $\mathbf{\Sigma}^P$, leading to

$$\hat{\mathbf{\Sigma}}^P{}_k = \frac{1}{N} \sum_{j=j_0}^{k} \left[\Delta \mathbf{x}_j \Delta \mathbf{x}_j^T + \mathbf{P}_j^+ - \mathbf{f}_{\mathbf{x}j} \mathbf{P}_{j-1}^+ \mathbf{f}_{\mathbf{x}j}^T \right] \tag{33}$$

where $\mathbf{P}_j^+ - \mathbf{f}_{\mathbf{x}j} \mathbf{P}_{j-1}^+ \mathbf{f}_{\mathbf{x}j}^T$ covers the change in the covariances between time steps, and $\Delta \mathbf{x}_k$ is the state correction sequence. Since the state vector in the errorEKF, defined in Equation (10), is the error in the variables, and the a priori state vector is null,

$$\Delta \mathbf{x}_k = \hat{\mathbf{x}}_k^+ - \mathbf{0} \tag{34}$$

The same procedure can be followed to estimate the measurement noise covariance matrix. Thus, assuming that $\mathbf{\Sigma}^P$ is known and independent of α, and that $\mathbf{\Sigma}^S$ is diagonal,

$$\hat{\mathbf{\Sigma}}^S{}_k = \frac{1}{N} \sum_{j=j_0}^{k} \left[\mathbf{y}_j \mathbf{y}_j^T - \mathbf{h}_{\mathbf{x}j} \mathbf{P}_j^- \mathbf{h}_{\mathbf{x}j}^T \right] \tag{35}$$

Note that both estimations (Equations (33) and (35)) are considering that the transition matrix, $\mathbf{f}_{\mathbf{x}}$, and observation matrix, $\mathbf{h}_{\mathbf{x}}$, are time-varying.

Furthermore, the shape of both matrices for the errorEKF with force estimation is known. The information of the matrices shape is useful to help the algorithm to reach the convergence. Thus, the estimated PNCM and MNCM are modified in order to adapt them to their theoretical shape. The MNCM is a diagonal matrix composed by the covariance of each sensor. The PNCM, according to [9], is

$$\hat{\mathbf{\Sigma}}^P = \begin{bmatrix} \mathbf{0} & \mathbf{0} & \mathbf{0} \\ \mathbf{0} & \mathbf{0} & \mathbf{0} \\ \mathbf{0} & \mathbf{0} & diag(\hat{\sigma}^2) \end{bmatrix} \tag{36}$$

where $diag(\hat{\sigma}^2)$ is a diagonal matrix containing the variance of all the components of the discrete plant noise at acceleration level.

A last consideration is that the covariance matrices must be semi-definite positive. With this algorithm, this cannot be guaranteed. Having a negative definite matrix would lead the filter to failure. To ensure the semi-definite positiveness of the matrices, the negative values are set to their absolute value [39].

2.3. Methods

All the experiments of this work are performed in a simulation environment in MATLAB®, using the Open Source toolbox MBDE (See https://github.com/MBDS/mbde-matlab (accessed on 12 May 2021)). The simulations are executed in a single core of a PC with an Intel Core i5 at 3.50 GHz and 8 Gb of RAM. Since there is not a physical prototype, three multibody models of each mechanism are employed in order to replicate a real scenario through simulation. All the simulations are launched with a time step of 5 milliseconds, that is a frequency of 200 Hz. They are detailed hereafter:

- *Real mechanism*: This model represents the real version of the mechanism, providing the *ground truth*. It is equivalent to the physical mechanism. The sensor measurements are obtained from this model. The sensor data is gathered from the kinematics of the *real mechanism*. Since the sensor data is obtained from a simulation, it is required to add white noise to the simulated measurements in order to represent the noise properties of real sensors.
- *Model*: This model represents the modeling of the *real mechanism*. The *model* can be affected by several uncertainties. Even though multibody modeling can represent with high accuracy a real mechanism, it is always subjected to modeling errors. While the geometry of a system can be accurately determined, the force models present a high level of uncertainty. Hence, this model is modified, creating a discrepancy between model and real mechanism. It should be mentioned that the errors in the mass or the mass distribution will lead to a similar accuracy level regarding kinematics magnitudes, since both force and mass errors result in acceleration errors. In addition, while the mass usually remains constant, force models are prone to change during a maneuver. Thus, it is of interest to test the estimator under errors in force models. The modeling error introduced is of $1 \, \text{m/s}^2$ in gravity acceleration. In addition, the initial value of the crank angle has an offset of $\pi/16 \, \text{rad}$.
- *Observer*: The estimations are computed based on this model. Regarding the modeling, it is the same as the *model*. The difference is that it is combined with the filter. Thus, its motion and dynamics are corrected with the information of the sensors installed on the *real mechanism*.

For the estimation, different sensors are considered. Although there are multiple possible configurations, for a reliable comparative with the results of [9], the same four combinations are considered. The four-bar and five-bar linkage share the same configurations for the sensors, duplicating the number of sensors in the case of the five-bar linkage, since it has two degrees of freedom. The sensor models are provided in [9]. The characteristics of each sensor are presented in Table 3. They are obtained from the technical data sheet of similar sensors. The four considered configurations are listed next:

- Encoder on the crank for measuring the crank angle, which is the degree of freedom (Figure 3a).
- Gyroscope on the crank for measuring the angular rates (Figure 3b).
- Gyroscope in the coupler of the mechanism (Figure 3c).
- Pair of accelerometers at the end of the crank, in such a manner that there is one sensitive axis parallel and another perpendicular to the crank (Figure 3d).

Table 3. Characteristics of the sensors considered in this work.

	Encoder	**Gyroscope**	**Accelerometers**
Standard deviation	$1.745 \times 10^{-2} \, \text{rad}$	$9.839 \times 10^{-4} \, \text{rad/s}$	$5.638 \times 10^{-2} \, \text{m/s}^2$
Sampling frequency (Hz)	200	200	200

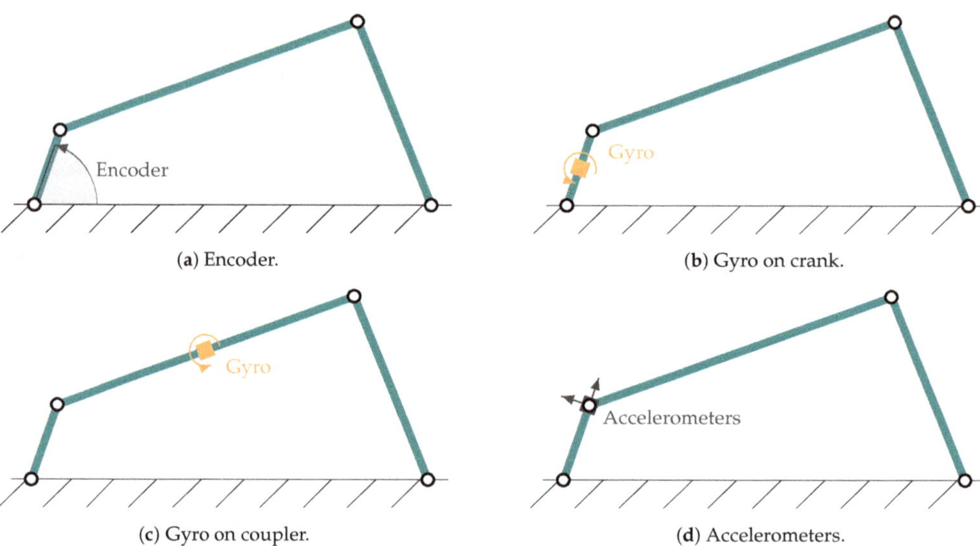

(**a**) Encoder.

(**b**) Gyro on crank.

(**c**) Gyro on coupler.

(**d**) Accelerometers.

Figure 3. Sensor configurations considered in the four-bar linkage.

The performance of the proposed estimator is evaluated in terms of accuracy and robustness. For the accuracy, a set of tests is executed under different initial noise covariance matrices. It is expected that the filter converges to a similar accuracy level independently from the initial noise covariance matrices, due to the adaptive equations. With respect to the robustness, it is of interest that the estimator can cope with unexpected events during the simulations without causing a divergence in the estimations.

The numerical results of each test are analyzed using two different metrics: the root-mean square error (RMSE) and the confidence interval of the estimation errors. The RMSE is used to measure the accuracy of the estimations in position, velocity, acceleration and force acting in the degree of freedom of each mechanism. The RMSE can be obtained following Equation (37).

$$RMSE = \sqrt{\frac{\sum_{i=0}^{N}(Virtual_i - Real_i)^2}{N}} \tag{37}$$

being N the number of samples gathered during the simulation. *Virtual* refers to the estimated value of the variable and *Real* to the reference value obtained from the *real mechanism*.

The confidence intervals of the errors give a different insight with respect to the RMSE on the accuracy of the estimator. For being reliable, the error of the estimations must be inside the confidence interval. Thus, the confidence interval of the estimation errors in position, velocity and acceleration are also provided. A confidence interval is calculated as a function of the standard deviation of a variable, which is related to the diagonal elements of the covariance matrix, **P** (Equation (21)). In this work, a confidence interval of 95% is selected. It can be calculated through Equation (38).

$$CI = \bar{E} \pm 1.96\sqrt{\sigma_i} \tag{38}$$

where $\bar{E}$ is the mean estimation error of the ith variable and $\sqrt{\sigma_i}$ the standard deviation of the ith variable.

3. Results

To asses the performance of the proposed adaptive Kalman filter, different tests are launched. During the first test, the mechanisms are in the vertical plane and they are only affected by the gravity force. A batch of tests is executed under these conditions for different initial values of the PNCM. The results of each simulation are compared against the results obtained from the conventional errorEKF with force estimation. In a second test, the mechanisms are modified with the addition of a torsional spring in the crank. During the simulation, the spring is removed from the *real mechanism*, simulating a failure on a real machine. The AerrorEKF-FE is expected to overcome this new modeling error (together with the error in the gravity acceleration and initial position) in the *observer* and adapt the covariance matrices to the new situation.

The simulations are executed with a time step of 5 milliseconds, that is a frequency of 200 Hz. The measurements are gathered at 200 Hz. Hence, there is one measurement available at each time step. The errors introduced in the model for each test are of $1\,\mathrm{m/s^2}$ in gravity acceleration and of an offset in the initial value of the crank angle $\pi/16\,\mathrm{rad}$. In addition, for testing the robustness, a torsional spring is introduced in the crank. During the simulation, this spring is removed only from the *real mechanism*. The *observer* should detect this event through the sensor measurements and perform the required corrections.

3.1. Accuracy Test

For testing the convergence of the adaptive filter, different initial values of the PNCM are selected. While the MNCM can be approximated by testing the sensors, an initial value for the process noise is more difficult to obtain. Due to this inconvenience, it is of interest to test the robustness of the proposed filter under different initial values of the PNCM. The results are compared with the ones obtained from the errorEKF-FE. This would show the error reduction when using the AerrorEKF-FE if the initial PNCM is not close to the true value.

Following Equation (36), for the four-bar linkage, only one value is required. For the five-bar linkage, since it has two degrees of freedom, two elements of the PNCM matrix should be instantiated. For simplicity, both elements are set to the same value. The different initial values are summarized in Table 4.

Table 4. Initial values of the PNCM.

Test	1	2	3	4	5	6
σ_0^2	0	0.0001	0.0023	0.1	1	10

Note that the initial values of both noise covariance matrices are used for the errorEKF-FE and AerrorEKF-FE. For the errorEKF-FE, these values are constant during the simulation, since it assumes that the PNCM is constant. Meanwhile, for the AerrorEKF-FE, these are the initial values of the PNCM, which are modified by the adaptive method. Hence, a total of six simulations for each sensor configuration are launched for each filter in order to compare their performance.

From the initial tests, an undesired behavior is detected. If both the PNCM and MNCM are estimated at the same time, the filter shows a trend to diverge. In Figure 4, the evolution of the estimated PNCM in the four-bar linkage is represented and compared with its evolution when the MNCM is not included in the estimation. Except for the case of the accelerometers, there is a trend to a continuous growth of the PNCM. This implies that the filter is reducing its trust on the model in exchange of the measurements.

The different behavior within configuration relies on the nature of the sensors. In the scenarios without accelerometers, there is a trend of divergence in the acceleration error. In addition, when using only an encoder, both the velocity and acceleration error diverge. Once that the estimator detects that the error is increasing through the innovation sequence, it starts to reduce the confidence in the model and trusting more in the sensors. However, correcting the acceleration error from measurements in position or velocity is not accurate. Hence, the error keeps growing causing the filter to diverge. At some point, this issue will lead to a failure of the filter. This, together with the non-linearities of the mechanisms, can result in unpredictable behaviors as the case of the gyroscope in the crank. Hence, for the rest of the experiments, only the PNCM is estimated.

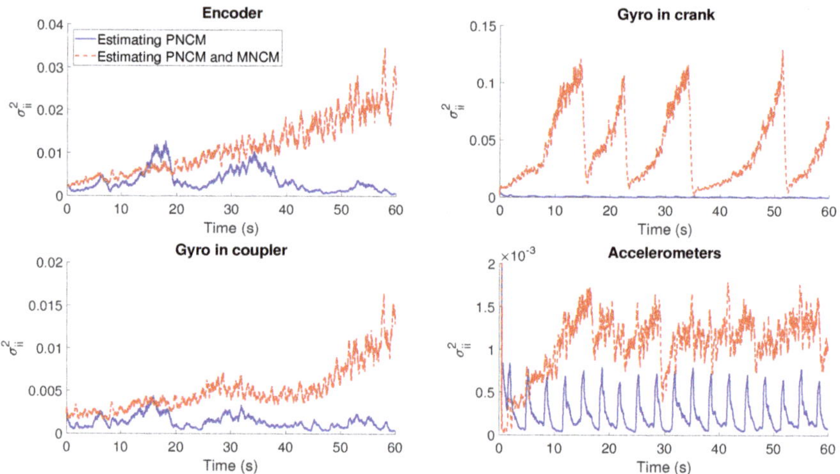

Figure 4. Estimated PNCM element in the four-bar linkage test for the case of estimating PNCM and MNCM simultaneously and estimating only the PNCM.

Regarding the size of the sliding window for the innovation sequence, it should be noted that a large window leads to low mean errors in smooth maneuvers. However, a large window also implies slow corrections to the covariance matrix. Hence, in some tests, the delay in the correction can result in an error which is not corrected. This is the case shown in Figure 5. It corresponds to a simulation of the four-bar linkage with a gyroscope on the coupler and an initial process covariance noise of $\sigma_{ii,0}^2 = 10$, and a window length of 100 time steps. The same maneuver is performed with a window length of 50 samples. The innovation sequences are shown in Figure 5a. The value of the crank angle for each slide window size is shown in Figure 5b.

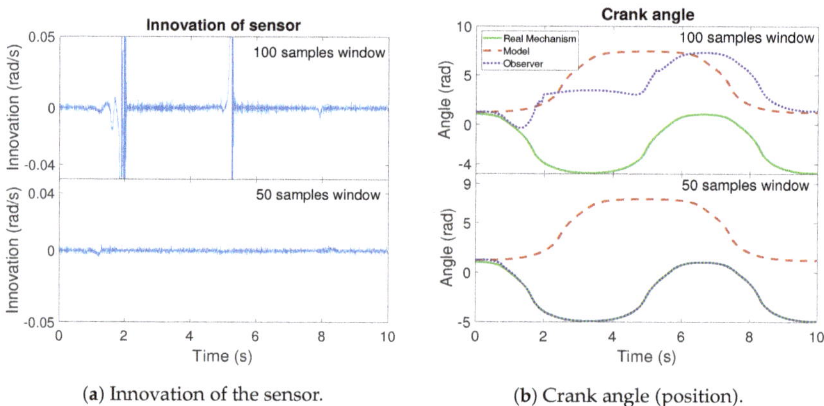

(**a**) Innovation of the sensor.

(**b**) Crank angle (position).

Figure 5. Comparison of innovation and crank angle for the simulation of the four-bar linkage with a gyroscope on the coupler for different window lengths.

As can be seen in Figure 5a, with a window length of 100 samples, there are a notable difference between the estimated and measured angular rate of the coupler at particular moments of the maneuvers. In these points of the maneuver, the mechanism reaches a position where the crank and the rocker are parallel. In this situation, the possible velocities of the coupler are limited. Thus, for an error in position, the error in velocity can be incorrigible. This is a consequence of the non-linearity of the system.

Comparing the results between different window sizes, it is possible to appreciate the effects of the samples considered for the estimation. With a large window, the filter cannot correct with immediacy the error. Although it is able to correct the error in terms of crank angular rate, as is shown in the innovation sequence, a permanent offset appears in the position of the crank angle. Meanwhile, with a short window length, the influence of previous values is reduced and the filter can perform the correction properly. Hence, the tests of this work are executed with a window of 50 samples for the innovation sequence.

There is no a general criteria for selecting the length of this window. It should be adjusted depending on the nature of the maneuver aiming to include only the relevant information of past events. The main idea is to avoid the effect of past events which do not have relation with the actual state of the mechanism. If a maneuver is close to steady state, a large window can be selected. On the other hand, a short window length is adequate in cases where the system changes.

In order to analyze the performance of the proposed adaptive filter, the RMSE for the position, velocity, acceleration and force acting in the degree of freedom of each mechanism is evaluated. From Figures 6–9, the RMSE between the AerrorEKF-FE and errorEKF-FE are compared for the four-bar mechanism. Results show that the adaptive version of the errorEKF-FE leads in most of the test to a lower error in all the measured magnitudes. In fact, in some configurations, the RMSE is almost constant. This behavior is what could be initially expected: despite the different initial PNCM, the AerrorEKF-FE is capable of finding an optimal covariance that minimizes the errors in all tests.

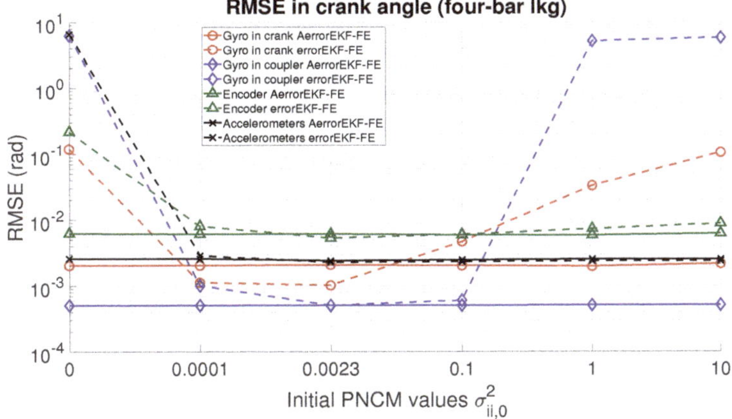

Figure 6. RMSE in the crank angle (position) provided by the observers for the four-bar linkage.

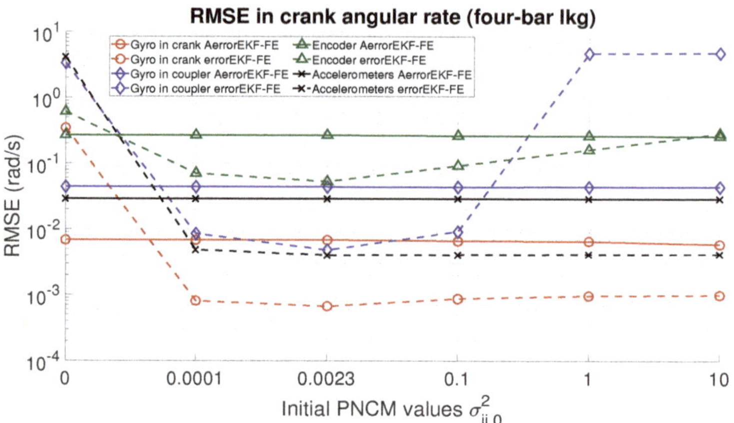

Figure 7. RMSE in the crank angular rate (velocity) provided by the observers for the four-bar linkage.

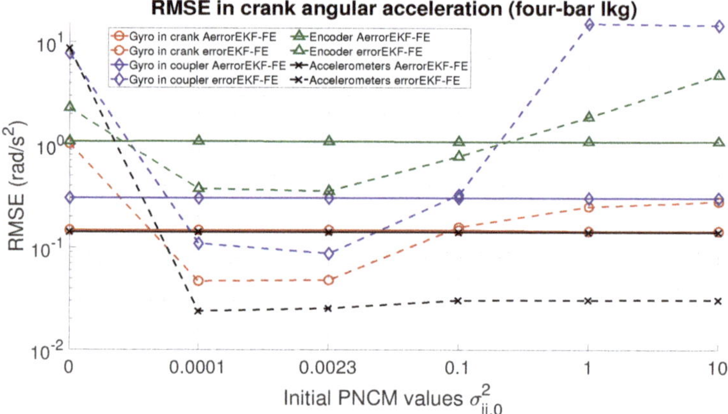

Figure 8. RMSE in the crank angular acceleration provided by the observers for the four-bar linkage.

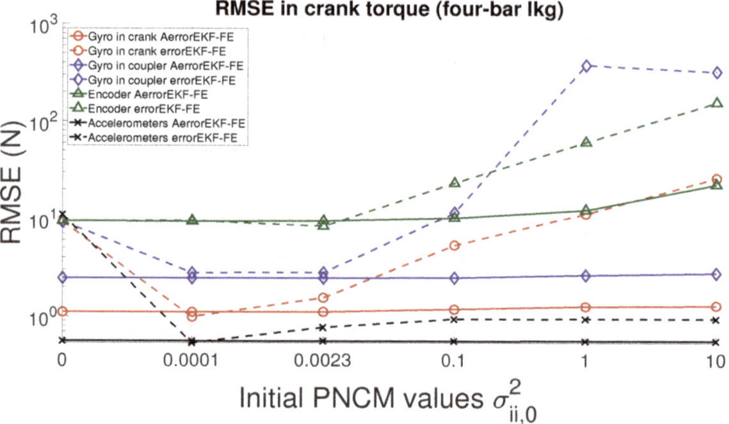

Figure 9. RMSE in the forces acting on the crank provided by the observers for the four-bar linkage.

The confidence intervals of the error in position, velocity and acceleration are shown in Figure 10 for one of the simulations. It can be seen how the errors converge to the true value and that the confidence intervals are consistent with the evolution of the error. Considering that the confidence interval is of 95%, it can be asserted that only the 5% of the estimations will have an error higher than 1.96 times the standard deviation of the variable (i.e., the square root of the diagonal elements of the covariance matrix, $\mathbf{P}$, of Equation (21)). These plots also provide useful information about the observability of the system. Since the confidence interval is based on the PNCM elements, if the system were not observable, it could be expected an unlimited growth of the covariance of the states.

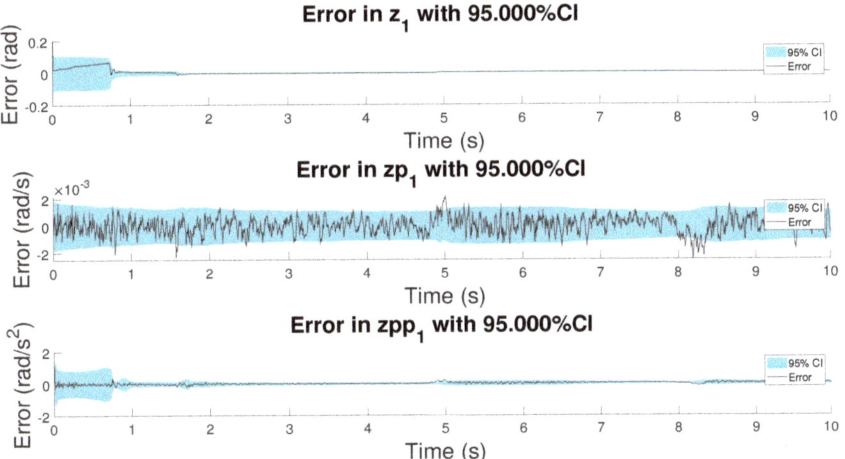

Figure 10. Error and confidence interval of the position, velocity and acceleration of the crank angle in the configuration which considers a gyroscope on the coupler.

With respect to the five-bar linkage, a similar trend can be appreciated from Figures 11–14. Except for the crank angle and angular rate when using gyroscopes in couplers, the rest of the experiments show that the AerrorEFK-FE converges to accurate solutions in every tested case. Thus, it is able to estimate the value of the PNCM which minimizes the error.

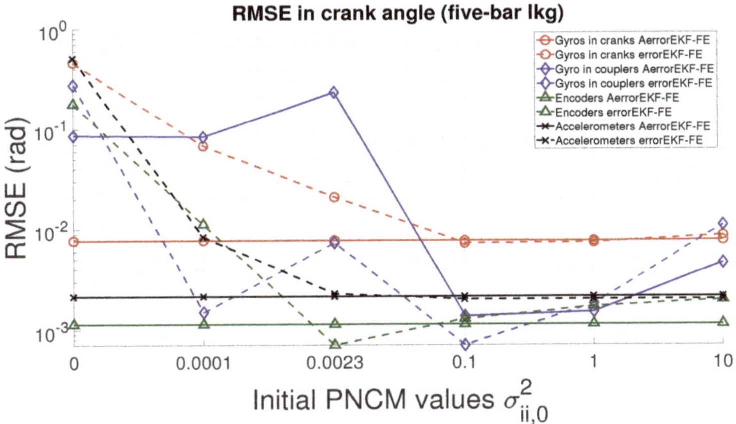

Figure 11. RMSE of the norm of the two crank angles (position) provided by the observers for the five-bar linkage.

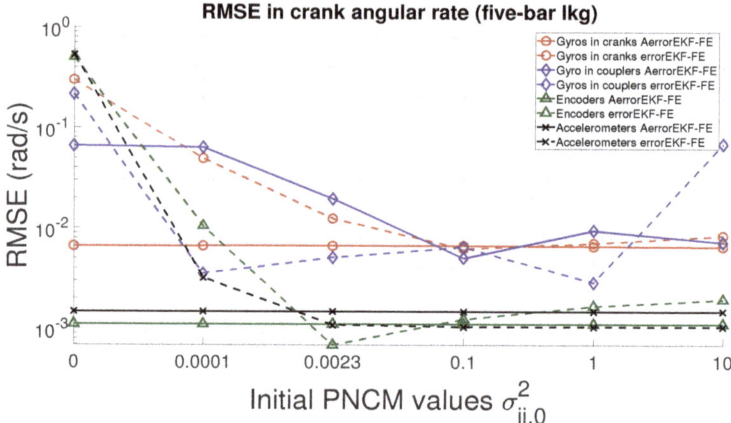

Figure 12. RMSE of the norm of the two crank angular rates (velocity) provided by the observers for the five-bar linkage.

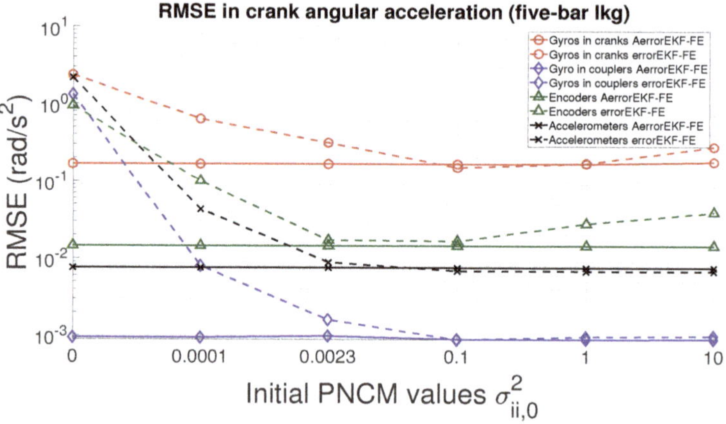

Figure 13. RMSE of the norm of the two crank angular accelerations provided by the observers for the five-bar linkage.

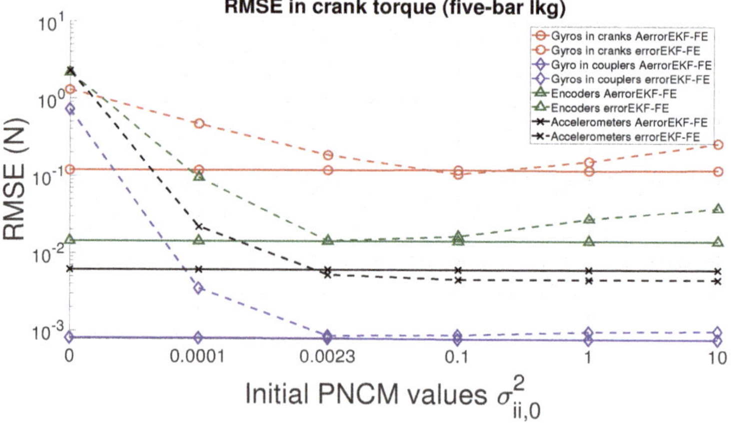

Figure 14. RMSE of the norm of the forces acting on the two cranks provided by the observers for the five-bar linkage.

3.2. Robustness Test

For evaluating the robustness of the filter, the four-bar linkage is equipped with a torsional spring actuating over the crank angle. In order to replicate a failure, at 5 s, the spring breaks. In this new scenario, the *real mechanism* continues the maneuver without the spring, while the *observer* is still considering its existence for the dynamics. This test is useful to analyze the response of the filter to unexpected changes. The *observer* is also under the errors of previous tests: $1 \, \text{m/s}^2$ in gravity acceleration. In addition, the initial value of the crank angle has an offset of $\pi/16 \, \text{rad}$.

The results in terms of force estimation are shown in Figure 15. The estimated force can be understood as the torque, which would have to be applied to the crank to compensate the errors. It is equivalent to the difference between the force applied in the *real mechanism* and the *model*. It can be seen how, at the time of 5 s, the *observer* has to subtract the force in the crank. Since there is no spring in the *real mechanism*, the *observer* needs to apply a torque equal in magnitude but different in sign to the torque applied by its spring.

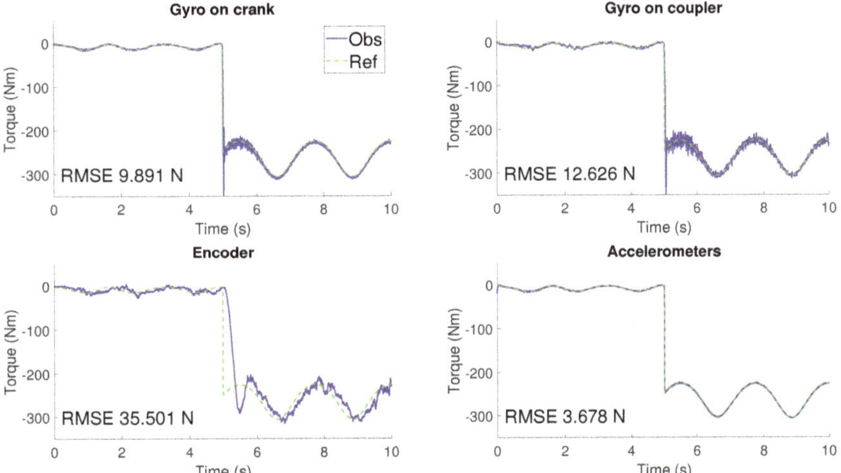

Figure 15. Torque estimation for the four-bar linkage for the spring failure scenario. *Obs* refers to the estimated torque and *Ref* refers to the theoretical value.

It can be seen how, except for the encoder configuration, the *observer* tracks the reference values. In addition, the filter responses with immediacy to the spring failure. This time response can be achieved by adjusting the innovation window length, seeking for a compromise between accuracy and time response. For this case, the window length is of 50 samples. Regarding the encoder configuration, the behavior is similar to what could be appreciated in [9]. There is a delay in the estimation and the noise is noticeable. It can be explained by the fact that obtaining acceleration from position information.

With respect to the other three configurations, the estimations present a reduced noise, being the use of accelerometers the most accurate solution. The use of gyroscopes shows a good behavior before the spring breaks. Once this event takes place, both tests show a remarkable overshooting which is quickly corrected at expenses of higher noise. However, in both cases, it can be appreciated how the noise is being reduced with time. This is in line with what is shown in Figure 16.

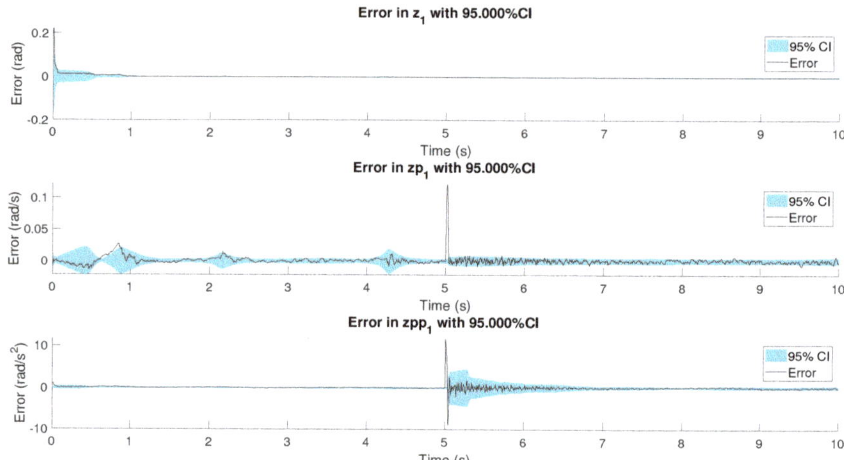

Figure 16. Error and confidence interval of the position, velocity and acceleration of the crank angle in the configuration which considers a gyroscope on the coupler.

It can be seen how the confidence interval becomes wider when the spring breaks. In terms of the PNCM, it means that the AerrorEKF-FE has detected the change in the *real mechanism*. To address this new modeling error, the filter increases the values of the PNCM giving more relevance to the sensor measurements. Once that the *observer* tracks the new scenario of the *real mechanism*, the covariances are reduced together with the noise of the estimations.

3.3. Computational Cost

For most of the industrial applications, it is required to achieve real-time performance. In previous works, the errorEKF-FE has proven to be capable of running in real time with complex multibody models [10]. Hence, it is of interest to analyze the increment in computational cost derived by the presented adaptive procedure.

It is important to remark that this work is developed in MATLAB®, which is not oriented to real-time applications. Hence, measuring the computation time of the algorithm is not a fair test. However, since the errorEKF-FE and AerrorEKF-FE are executed under the same conditions, a reference in the increment in computational cost can be derived. It can be seen in Table 5. From the results, it can be concluded that the AerrorEKF requires about the double of time than the error errorEKF-FE for computing the same simulation.

Table 5. Computational cost analysis of the AerrorEKF-FE. The simulations tested correspond to the use of position sensors.

	Simulated Time (s)	Computing Time (s)	
		errorEKF	AerrorEKF
Four-bar linkage	10	1.5931	3.1955
Five-bar linkage	10	2.2551	4.7234

4. Discussion

From the presented results, it can be seen that the adaptive version of the errorEKF-FE solves one of the main drawbacks of the filter: setting the values of the process noise covariance matrix. The tests performed during this work show that, with independence of the initial assumptions on the PNCM, the AerrorEKF-FE converge to an accurate solution. As shown during the accuracy test, the AerrorEKF-FE is able to achieve a similar level of error in the estimations despite of the initial values of the PNCM. This allows to reduce the development time of multibody-based Kalman filters, where the determination of the statistics of the system is a tedious process. In addition, even though the maximum likelihood method depends on a sliding window of the past innovation, the filter has shown an acceptable response to sudden changes in the system. Through the simulation of a sudden failure in the *real mechanism*, the robustness of the filter is tested. The results show that the estimator is able to track the new situation and correct the new error with accuracy, without compromising the convergence of the simulation.

It can also be concluded that, although the measurement noise covariance matrix can also be estimated through the ML method, in this particular case it led to a filter divergence in the absence of accelerometers. When the estimator detects an error, it starts to rely on the corrections in the measurements instead of the model. However, through sensors in position or velocity, the acceleration cannot be obtained with accuracy and hence, the error cannot be corrected. This, together with the non-linearies of the model, leads to a divergence of the filter. In addition, the estimation of the MNCM did not offer an improvement of the estimations. However, as opposite from the PNCM, it is possible to obtain an acceptable initial value for the MNCM after characterizing the sensors or through the information provided by the manufacturer of each sensor.

The length of the sliding window for the innovation sequence is also an important factor and one of the main limitations of the approach. A large window size can result into wrong estimations in maneuvers with high variability. This is one of the known disadvantages of the ML method, since there does not exist a general rule in order to select the most suitable innovation window length. Algorithms such as Sage-Husa are focused on minimizing the impact of the window length by giving different weights to the elements of past events, giving higher weights to the more recent events.

Regarding the computational cost, the AerrorEKF-FE has lower efficiency than the errorEKF-FE due to the adaptive procedure. Estimating the process noise covariance matrix each time step entails an increment of the computations per time step, which is turned into a noticeable increment of computational cost. This limitation can be critical in real-time applications. It is necessary to test the computational cost of the method in each particular application. The efficiency of the estimator not only depends on the algorithm, but on the platform where it is going to be executed. This also implies to explore the possibility of increasing the efficiency of the solution. Since the estimation of the noise covariance matrices is independent from the state estimation each time step, the code execution can be parallelized, increasing the algorithm efficiency and reducing the computational cost of the approach.

5. Conclusions

This work presents an adaptive Kalman filter for state and input estimation based on multibody models (AerrorEKF-FE). The aim is to accurately estimate the noise covariance matrix of the estimator, which has shown to be a critical factor for accurate and robust observer performance. The method is tested on two different planar mechanisms: a four-bar linkage and a five-bar linkage. In addition, four different sensor configurations are considered, increasing the general application of the proposed solution. Several tests are presented in order to analyze the behavior of the filter in terms of accuracy and robustness.

The AerrorEKF-FE combines an error-state Kalman filter with an adaptive method based on the maximum likelihood criteria. This adaptive technique, commonly used for navigation estimation, has several assumptions which are not fulfilled in multibody dynamics. Even though adaptive algorithms have the capacity of overcoming some of the main difficulties of multibody based Kalman filtering, its application has never been explored. In this work, the selected adaptive method is adjusted to fit with multibody particularities, such as time-variant transition and observation matrices. Furthermore, the estimated matrix should be adapted to the shape of the covariance matrix (if known), increasing the accuracy of the observer.

The first test evaluates the accuracy of the filter under different initial values of the noise covariance matrix. The results are compared against the estimations obtained with a non-adaptive version of the filter (errorEKF-FE). The results show an improvement in accuracy with the use of adaptive techniques. In addition, it can be seen how the adaptive method allows to achieve similar results in terms of accuracy in spite of the initial assumption of the covariance noises.

In a second test, the robustness of the filter is studied through a simulation in which an unexpected event changes the system. The filter shows a quick reaction and corrects with accuracy the new error, without loosing the stability of the estimator.

The tests executed in this work and their results show the potential of adaptive Kalman filtering for multibody based estimation. Determining the noise covariance matrix (specially the noise of the process) in multibody based estimators is an actual limitation for the general use of this techniques in multibody applications. Through adaptive estimation, the uncertainties on the process noise covariance matrix can be solved and the robustness and accuracy of the estimators can be increased and guaranteed during different scenarios. However, this method also implies a reduction of the computational efficiency that should be considered in real-time applications.

In future works, the proposed adaptive filter should be applied in systems of more complexity, such as vehicle dynamics, where the state and input estimation is of high interest. In addition, techniques for reducing the associated increment in computational cost can be studied.

Author Contributions: Conceptualization, A.J.R., E.S., R.P. and M.Á.N.; methodology, A.J.R., E.S. and M.Á.N.; software, A.J.R. and E.S.; validation, A.J.R., E.S. and M.Á.N.; formal analysis, A.J.R., E.S. and M.Á.N.; investigation, A.J.R., E.S. and M.Á.N.; resources, A.J.R., E.S., R.P. and M.Á.N.; data curation, A.J.R.; writing—original draft preparation, A.J.R.; writing—review and editing, A.J.R., E.S., R.P. and M.Á.N.; visualization, A.J.R.; supervision, M.Á.N.; project administration, M.Á.N.; funding acquisition, M.Á.N. All authors have read and agreed to the published version of the manuscript.

Funding: This research was partially financed by the Spanish Ministry of Science, Innovation and Universities and EU-EFRD funds under the project "Técnicas de co-simulación en tiempo real para bancos de ensayo en automoción" (TRA2017-86488-R), and by the Galician Government under grant ED431C2019/29.

Institutional Review Board Statement: Not applicable.

Informed Consent Statement: Not applicable.

Data Availability Statement: Not applicable.

Conflicts of Interest: The authors declare no conflict of interest. The funders had no role in the design of the study; in the collection, analyses, or interpretation of data; in the writing of the manuscript, or in the decision to publish the results.

References

1. Kalman, R. A New Approach to Linear Filtering and Prediction Problems. *J. Basic Eng.* **1960**, *82*, 35–45. [CrossRef]
2. Simon, D. *Optimal State Estimation: Kalman, H Infinity, and Nonlinear Approaches*; John Wiley & Sons: Hoboken, NJ, USA, 2006.
3. García de Jalón, J.; Bayo, E. *Kinematic and Dynamic Simulation of Multibody Systems: The Real Time Challenge*; Springer: Berlin/Heidelberg, Germany, 1994.
4. Pastorino, R. *Model-Based System Testing*; Technical Report; Siemens Digital Industries Software: Plano, TX, USA, 2019.
5. Cuadrado, J.; Dopico, D.; Barreiro, A.; Delgado, E. Real-time state observers based on multibody models and the extended Kalman filter. *J. Mech. Sci. Technol.* **2009**, *23*, 894–900. [CrossRef]
6. Cuadrado, J.; Dopico, D.; Perez, J.A.; Pastorino, R. Automotive observers based on multibody models and the extended Kalman filter. *Multibody Syst. Dyn.* **2012**, *27*, 3–19. [CrossRef]
7. Pastorino, R.; Richiedei, D.; Cuadrado, J.; Trevisani, A. State estimation using multibody models and non-linear Kalman filter. *Int. J. Non-Linear Mech.* **2013**, *53*, 83–90. [CrossRef]

8. Sanjurjo, E.; Naya, M.Á.; Blanco-Claraco, J.L.; Torres-Moreno, J.L.; Giménez-Fernández, A. Accuracy and efficiency comparison of various nonlinear Kalman filters applied to multibody models. *Nonlinear Dyn.* **2017**, *88*, 1935–1951. [CrossRef]
9. Sanjurjo, E.; Dopico, D.; Luaces, A.; Naya, M.A. State and force observers based on multibody models and the indirect Kalman filter. *Mech. Syst. Signal Process.* **2018**, *106*, 210–228. [CrossRef]
10. Rodríguez, A.J.; Sanjurjo, E.; Pastorino, R.; Naya, M.A. State, parameter and input observers based on multibody models and Kalman filters for vehicle dynamics. *Mech. Syst. Signal Process.* **2021**, *155*, 107544. [CrossRef]
11. Mehra, R. Approaches to adaptive filtering. *IEEE Trans. Autom. Control* **1972**, *17*, 693–698. [CrossRef]
12. Fraser, C.; Ulrich, S. Adaptive extended Kalman filtering strategies for spacecraft formation relative navigation. *Acta Astronaut.* **2021**, *178*, 700–721. [CrossRef]
13. Filho, J.O.A.L.; Fortaleza, E.L.F.; Silva, J.G.; Campos, M.C.M.M. Adaptive Kalman filtering for closed-loop systems based on the observation vector covariance. *Int. J. Control* **2020**, 1–16. [CrossRef]
14. Mohamed, A.H.; Schwarz, K.P. Adaptive Kalman Filtering for INS/GPS. *J. Geod.* **1999**, *73*, 193–203. [CrossRef]
15. Karasalo, M.; Hu, X. An optimization approach to adaptive Kalman filtering. *Automatica* **2011**, *47*, 1785–1793. [CrossRef]
16. Rodríguez, A.J.; Pastorino, R.; Carro-Lagoa, Á.; Janssens, K.; Naya, M.Á. Hardware acceleration of multibody simulations for real-time embedded applications. *Multibody Syst. Dyn.* **2021**, *55*, 455–473. [CrossRef]
17. Woo, R.; Yang, E.J.; Seo, D.W. A Fuzzy-Innovation-Based Adaptive Kalman Filter for Enhanced Vehicle Positioning in Dense Urban Environments. *Sensors* **2019**, *19*, 1142. [CrossRef] [PubMed]
18. Liu, Y.; Fan, X.; Lv, C.; Wu, J.; Li, L.; Ding, D. An innovative information fusion method with adaptive Kalman filter for integrated INS/GPS navigation of autonomous vehicles. *Mech. Syst. Signal Process.* **2018**, *100*, 605–616. [CrossRef]
19. Wu, Q.; Jia, Q.; Shan, J.; Meng, X. Angular velocity estimation based on adaptive simplified spherical simplex unscented Kalman filter in GFSINS. *Proc. Inst. Mech. Eng. Part G J. Aerosp. Eng.* **2014**, *228*, 1375–1388. [CrossRef]
20. Sage, A.P.; Husa, G.W. Algorithms for sequential adaptive estimation of prior statistics. In Proceedings of the 1969 IEEE Symposium on Adaptive Processes (8th) Decision and Control, University Park, PA, USA, 17–19 November 1969; p. 61. [CrossRef]
21. Wang, J.; Xu, T.; Wang, Z. Adaptive Robust Unscented Kalman Filter for AUV Acoustic Navigation. *Sensors* **2020**, *20*, 60. [CrossRef]
22. Luo, Z.; Fu, Z.; Xu, Q. An Adaptive Multi-Dimensional Vehicle Driving State Observer Based on Modified Sage–Husa UKF Algorithm. *Sensors* **2020**, *20*, 6889. [CrossRef] [PubMed]
23. Sarkka, S.; Nummenmaa, A. Recursive Noise Adaptive Kalman Filtering by Variational Bayesian Approximations. *IEEE Trans. Autom. Control* **2009**, *54*, 596–600. [CrossRef]
24. Davari, N.; Gholami, A. Variational Bayesian adaptive Kalman filter for asynchronous multirate multi-sensor integrated navigation system. *Ocean Eng.* **2019**, *174*, 108–116. [CrossRef]
25. Wang, S.Y.; Yin, C.; Duan, S.K.; Wang, L.D. A Modified Variational Bayesian Noise Adaptive Kalman Filter. *Circuits Syst. Signal Process.* **2017**, *36*, 4260–4277. [CrossRef]
26. Huang, Y.; Zhang, Y.; Wu, Z.; Li, N.; Chambers, J. A Novel Adaptive Kalman Filter with Inaccurate Process and Measurement Noise Covariance Matrices. *IEEE Trans. Autom. Control* **2018**, *63*, 594–601. [CrossRef]
27. Pan, C.; Gao, J.; Li, Z.; Qian, N.; Li, F. Multiple fading factors-based strong tracking variational Bayesian adaptive Kalman filter. *Measurement* **2021**, *176*, 109139. [CrossRef]
28. Dong, X.; Chisci, L.; Cai, Y. An adaptive variational Bayesian filter for nonlinear multi-sensor systems with unknown noise statistics. *Signal Process.* **2021**, *179*, 107837. [CrossRef]
29. Shan, C.; Zhou, W.; Yang, Y.; Jiang, Z. Multi-Fading Factor and Updated Monitoring Strategy Adaptive Kalman Filter-Based Variational Bayesian. *Sensors* **2021**, *21*, 198. [CrossRef] [PubMed]
30. Cuadrado, J.; Gutiérrez, R.; Naya, M.A.; Morer, P. A comparison in terms of accuracy and efficiency between a MBS dynamic formulation with stress analysis and a non-linear FEA code. *Int. J. Numer. Methods Eng.* **2001**, *51*, 1033–1052. [CrossRef]
31. Cuadrado, J.; Dopico, D.; Naya, M.A.; Gonzalez, M. Real-Time Multibody Dynamics and Applications. In *Simulation Techniques for Applied Dynamics*; Springer: Vienna, Austria, 2008; Volume 507, pp. 247–311. [CrossRef]
32. Dopico, D.; Luaces, A.; Saura, M.; Cuadrado, J.; Vilela, D. Simulating the anchor lifting maneuver of ships using contact detection techniques and continuous contact force models. *Multibody Syst. Dyn.* **2019**, *46*, 147–179. [CrossRef]
33. Parra, A.; Rodríguez, A.J.; Zubizarreta, A.; Pérez, J. Validation of a Real-Time Capable Multibody Vehicle Dynamics Formulation for Automotive Testing Frameworks Based on Simulation. *IEEE Access* **2020**, *8*, 213253–213265. [CrossRef]
34. Jaiswal, S.; Rahikainen, J.; Khadim, Q.; Sopanen, J.; Mikkola, A. Comparing double-step and penalty-based semirecursive formulations for hydraulically actuated multibody systems in a monolithic approach. *Multibody Syst. Dyn.* **2021**, *52*, 169–191. [CrossRef]
35. Cuadrado, J.; Dopico, D.; Gonzalez, M.; Naya, M.A. A Combined Penalty and Recursive Real-Time Formulation for Multibody Dynamics. *J. Mech. Des.* **2004**, *126*, 602. [CrossRef]
36. Sanjurjo, E.; Blanco, J.L.; Torres, J.L.; Naya, M.A. Testing the efficiency and accuracy of multibody-based state observers. In Proceedings of the ECCOMAS Thematic Conference on Multibody Dynamics, Barcelona, Spain, 29 June–2 July 2015; pp. 1595–1606.
37. Grewal, M.; Andrews, A. *Kalman Filtering: Theory and Practice Using MATLAB®*; John Wiley & Sons: Hoboken, NJ, USA, 2008.

38. Dopico, D.; Zhu, Y.; Sandu, A.; Sandu, C. Direct and Adjoint Sensitivity Analysis of Ordinary Differential Equation Multibody Formulations. *J. Comput. Nonlinear Dyn.* **2014**, *10*, 011012. [CrossRef]
39. Myers, K.; Tapley, B. Adaptive sequential estimation with unknown noise statistics. *IEEE Trans. Autom. Control* **1976**, *21*, 520–523. [CrossRef]

sensors

MDPI

Article

A Factor-Graph-Based Approach to Vehicle Sideslip Angle Estimation

Antonio Leanza [1], Giulio Reina [1] and José-Luis Blanco-Claraco [2,*

[1] Department of Mechanics, Mathematics, and Management, Polytechnic of Bari, via Orabona 4, 70126 Bari, Italy; antonio.leanza@unisalento.it (A.L.); giulio.reina@poliba.it (G.R.)
[2] Department of Engineering, Campus de Excelencia Internacional Agroalimentario, University of Almería, ceiA3, 04120 Almería, Spain
* Correspondence: jlblanco@ual.es

Abstract: Sideslip angle is an important variable for understanding and monitoring vehicle dynamics, but there is currently no inexpensive method for its direct measurement. Therefore, it is typically estimated from proprioceptive sensors onboard using filtering methods from the family of the Kalman filter. As a novel alternative, this work proposes modeling the problem directly as a graphical model (factor graph), which can then be optimized using a variety of methods, such as whole-dataset batch optimization for offline processing or fixed-lag smoothing for on-line operation. Experimental results on real vehicle datasets validate the proposal, demonstrating a good agreement between estimated and actual sideslip angle, showing similar performance to state-of-the-art methods but with a greater potential for future extensions due to the more flexible mathematical framework. An open-source implementation of the proposed framework has been made available online.

Keywords: vehicle dynamics estimation; sideslip angle estimation; factor graph; graphical models; Kalman filtering

check for updates

Citation: Leanza, A.; Reina, G.; Blanco-Claraco, J.-L. A Factor-Graph-Based Approach to Vehicle Sideslip Angle Estimation. *Sensors* **2021**, *21*, 5409. https://doi.org/10.3390/s21165409

Academic Editors: Javier Cuadrado and Miguel Ángel Naya Villaverde

Received: 21 July 2021
Accepted: 9 August 2021
Published: 10 August 2021

Publisher's Note: MDPI stays neutral with regard to jurisdictional claims in published maps and institutional affiliations.

1. Introduction

A large body of research work has accumulated during the past three decades regarding sideslip estimation, which is a fundamental feature of vehicle dynamics [1]. Despite its central role, the direct measurement of sideslip angle over time is often impractical during vehicle motion, for both technical and economic reasons. For instance, a cheap but few practical solution is to use measures coming from the Global Positioning System (GPS) that can provide the position of the receiver without any numerical integration and then derive the velocity of the vehicle using Doppler measurements [2]. Nevertheless, GPS receivers have issues such as temporary signal unavailability due to signal blocking by trees, buildings, or urban canyons, as well as (typically) a much lower measuring rate than other onboard sensors involved in vehicle dynamics control such as accelerometers or gyroscopes. On the other hand, direct measurement of the vehicle sideslip angle can be achieved by high-precision optical sensors, but they are sophisticated, still in an early research and development (R&D) stage, and expensive for production vehicles [3].

For these reasons, sideslip angle estimation continues to pose a significant challenge in vehicle dynamics research, attracting noticeable interest in both the academic and industrial worlds [4–6]. Several methods have been developed and described throughout the scientific literature which make use of different models and estimators. The most common approach relies on model-based observers [7–11], which make use of a vehicle reference model for state and parameter estimators. Different levels of sophistication can be obtained in order to achieve a phenomenon description that is as accurate as possible. One of the most common combinations that can be found in the literature is the bicycle (single-track) model as an observer for a Kalman filter (KF). This arrangement allows the estimation of states and parameters at the same time. For instance, in [12,13] a linear bicycle

203

model is used in combination with an extended Kalman filter (EKF) for the estimation of the sideslip angle and vehicle parameters at the same time. The former estimates lateral velocity and vehicle mass by also correcting the bias of the gyroscope, while the latter estimates the front and rear cornering stiffness and the sideslip angle directly as a dynamic state. In [14], the same linear model is exploited but using a cubature Kalman filter (CKF) as the estimator for obtaining the sideslip angle and other states from the input noisy measurements (i.e., accelerations, vehicle longitudinal velocity, etc.). In [15], a combined approach between kinematic and dynamic models is carried out, since the kinematic model performs well during transient maneuvers but fails in steady-state conditions. Therefore, the information provided by the kinematic formulation is exploited to update the single-track model parameters (i.e., the tire cornering stiffness), while the dynamic state observer is used in the nearly quasi-state condition. The steady-state or transient conditions are discriminated via a fuzzy-logic procedure. Furthermore, the bicycle model can also be coupled with nonlinear tire models, as in [16], where the authors attempt to estimate the sideslip angle of a heavy-duty vehicle by using a rational tire model and an EKF as the estimator. The study showed good estimation performance, but at the cost of an overparametrized and complex model. On the other hand, four-contact models provide a better description of vehicle dynamics, but obviously at the cost of a greater number of parameters and an increase in the complexity of the systems. For instance, the authors in [17] use an EKF applied to a four-contact vehicle model with a Dugoff tire model in order to estimate the sideslip angle and the tire/road forces. The authors in [18] again use a four-contact model, but with a semiphysical nonlinear tire model called "Unitire", and apply a reduced-order sliding mode observer (SMO), evaluating the performance of the proposed method by means of simulation and experiments. In [19] the Magic Formula (or Pacejka tire model [20]) is exploited, where a preliminary filtering on vertical forces is performed via linear KF in order to estimate the roll angle and then an EKF is applied to the four-contact vehicle model to achieve the sideslip angle. Again, the Magic Formula with a four-contact vehicle model is used in [21], where a dual estimation scheme was adopted: the sideslip angle and tire/road forces by means of a dual unscented Kalman filter (UKF) algorithm and the Pacejka tire parameters by solving a nonlinear least squares (LS) problem. However, the sophistication of these models might be unpractical in real applications. To overcome this issue, some authors (e.g., [22]) rely on direct causality equations without the need of any explicit tire model. In the cited research, different estimators are developed, such as the standard EKF, the CKF, and particle filtering (PF), and the results are evaluated using a vehicle model with 14 degrees of freedom (DOFs) which performs standard maneuvers. Authors in [23] perform vehicle sideslip angle and road bank angle estimation via a simple algebraic relationship in real time, based on two online parameter identification techniques, combining a single-track model with roll and tire slip models and force due to bank angle. Moreover, with the use of a lateral G sensor signal and by including road bank angle effect, the front and rear cornering stiffness and vehicle sideslip angle are identified, in the absence of any a priori knowledge of the road bank angle. In order to completely overcome the need for a vehicle model of any kind and its related complex set of parameters, different approaches based on artificial neural networks (ANNs) [24–27] are now widely investigated, since they are suitable for modeling complex systems using their ability to identify relationships from input–output data.

In this paper, we present a novel estimation technique grounded in factor graphs (FGs) theory. To the best of our knowledge, this is the first time an FG-based observer is proposed in the automotive field. An FG is one of the possible types of probabilistic graphical models that can be used to describe the structure of an estimation problem [28,29]. Factor graphs are bipartite graphs comprising two kinds of nodes: variable and factor nodes. Variable nodes represent the unknown data to be estimated, while factors represent cost functions to be minimized, modeling relationships between the variables. Each factor node is connected to only those variables that appear in its cost function; this allows an FG to explicitly model the key property of how sparse a problem is, depending on the connection pattern between its variables. As shown below, by applying this estimation approach to the classical linear bicycle model, good estimation accuracy can be achieved even in the nonlinear regions of the tire behavior compared to the celebrated Kalman filtering, while preserving the inherent simplicity of the linear model and the use of few parameters (i.e., the front and rear cornering stiffness).

The paper is organized as follows. In Section 2 the equations related to the linear bicycle model are recalled and reorganized in order to directly obtain the sideslip angle. Section 3 introduces the estimation problem with the use of FGs, applied to the case at hand with details on its implementation. Results obtained with the proposed method are evaluated by means of real data gathered in the Stanford database [30] in Section 4. Finally, conclusions are drawn in Section 5.

2. Vehicle Dynamic Model

Lateral dynamics represents a basic challenge in vehicle behavior, because lateral velocity is usually not directly measurable for practical and/or economic reasons. Nonetheless, it represents fundamental information, since it influences the entire vehicle dynamics, especially the direction of motion. In fact, the total velocity of the vehicle center of gravity (CoG) is the vectorial sum of the longitudinal and lateral velocities u and v, respectively. Figure 1 shows the model used in this paper for dealing with the vehicle's lateral behavior. It is known as a "bicycle" or "single-track" model, and it is based on the following simplifications [31]: equal internal and external dynamics so that the tires of the same axle can be fused, leading to the model's bicycle-like appearance; linear range of the tires, thus lateral forces are a linear function of slip angles by means of a specific coefficient named cornering stiffness; rear-wheel drive; negligible motion resistance; small-angle approximation to preserve the linearity of the model; constant longitudinal velocity u, which is guaranteed within a small time-step Δt. This model is characterized by two degrees of freedom (DoFs)—namely, the vehicle lateral velocity v and the yaw rate r. Hence, two equations of motion are sufficient to fully describe its behavior over time:

$$\dot{v} = -\left(\frac{C_f + C_r}{mu}\right)v - \left(\frac{C_f l_f - C_r l_r}{mu} + u\right)r + \frac{C_f \delta}{m}$$

$$\dot{r} = -\left(\frac{C_f l_f - C_r l_r}{J_z u}\right)v - \left(\frac{C_f l_f^2 + C_r l_r^2}{J_z u}\right) + \frac{C_f l_f \delta}{J_z} \tag{1}$$

where m is the vehicle mass, J_z is the moment of inertia with respect to the yaw axis, C_f and C_r are the front and rear cornering stiffness, respectively, l_f and l_r are respectively the distance of the CoG from the front and rear axle, and δ is the front steering angle.

Usually, the sideslip angle β is of interest in the study of lateral dynamics, in place of the lateral velocity v; therefore, one can consider as DoFs β and r in place of the couple v and r. From Figure 1, the relation between v and β is the following:

$$\beta = \arctan\left(\frac{v}{u}\right) \approx \frac{v}{u} \tag{2}$$

where the approximation comes from the assumption of small angles, as stated above. Therefore,

$$v = \beta u \quad \text{and} \quad \dot{v} = \dot{\beta} u + \beta \overset{0}{\dot{u}} \approx \dot{\beta} u, \tag{3}$$

considering the constancy of u within each time-step Δt. The Equation (1) become:

$$\dot{\beta} = -\left(\frac{C_f + C_r}{mu}\right)\beta - \left(\frac{C_f l_f - C_r l_r}{mu^2} + 1\right)r + \frac{C_f \delta}{mu}$$

$$\dot{r} = -\left(\frac{C_f l_f - C_r l_r}{J_z}\right)\beta - \left(\frac{C_f l_f^2 + C_r l_r^2}{J_z u}\right) + \frac{C_f l_f \delta}{J_z} \tag{4}$$

By considering Equation (4), β and r are variables of the vehicle lateral dynamics and the other terms are considered to be known parameters.

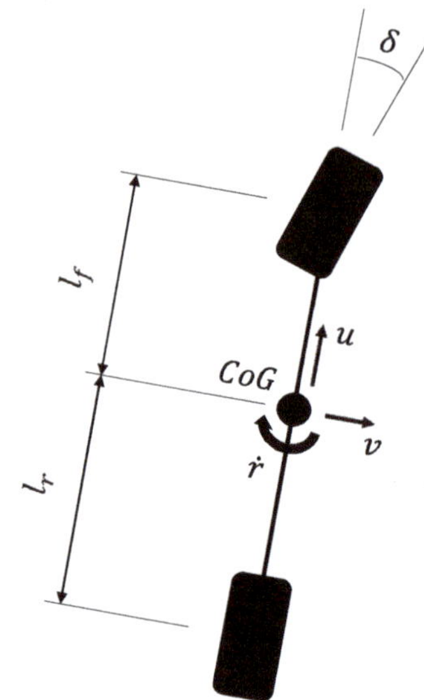

Figure 1. Single-track model.

It is worth noting that the yaw rate r can be directly measured by means of a vertical gyroscope, while the sideslip angle β usually has to be estimated. Although β can also be measured via GPS and Equation (2), this type of measurement is often unstable and unreliable due to the presence of shady areas of the GPS signal, such as tunnels or mountains. This leads to the need to indirectly estimate the sideslip angle. To achieve this estimation, the lateral acceleration a_y is considered as a further measurement. Both a_y and r are generally already available onboard common vehicles via the ESP system. Therefore, the following set of measurements is here considered:

$$\dot{\varphi} = r$$
$$a_y = \dot{v} + ur = u(\dot{\beta} + r) \tag{5}$$

As a matter of practicality, Equations (4) and (5) in their continuous-time form need to be converted into the following discrete-time representations, obtained from the forward Euler integration with time step $\Delta t = t_k - t_{k-1}$:

$$\beta_k = \beta_{k-1} + \Delta t \left[-\frac{C_f + C_r}{m u_{k-1}} \beta_{k-1} - \left(\frac{C_f l_f - C_r l_r}{m u_{k-1}^2} + 1 \right) r_{k-1} + \frac{C_f \delta_{k-1}}{m u_{k-1}} \right] \tag{6}$$

$$r_k = r_{k-1} + \Delta t \left(-\frac{C_f l_f - C_r l_r}{J z} \beta_{k-1} - \frac{C_f l_f^2 + C_r l_r^2}{J z u_{k-1}} r_{k-1} + \frac{C_f l_f \delta_{k-1}}{J z} \right) \tag{7}$$

$$\dot{\varphi}_k = r_k a_{yk} = -\frac{C_f + C_r}{m} \beta_k - \frac{C_f l_f - C_r l_r}{m u_k} r_k + \frac{C_k \delta_k}{m} \tag{8}$$

3. Factor Graph for Vehicle Lateral Dynamics

3.1. The Estimation Problem

The discrete-time Equation (8) refer to the vehicle system and to the measurements gathered for a correct estimate of the DoFs, in particular for the estimation of β, since r is directly measured. The most common method followed throughout the literature relies on a state-space representation of the above equations, writing two compacted matrix equations: one for the propagation of the model over time and one relating to the measures, both as a function of suitable state variables. Therefore, a recursive routine, often based on Kalman filtering, is performed to achieve the correct estimation of the unknown state variables [22].

In this paper, an alternative approach is suggested for the estimation of the unknown variables, without passing through the state-space form—namely, the factor graphs (FGs), a method belonging to the family of probabilistic graphical models. The objective is to minimize given cost functions in each time step. The FG relative to the problem at hand is displayed in Figure 2 for a generic time-step k. Each factor (filled circle in the figure) is connected with the unknown variables involved in the cost function that the factor minimizes. The cost functions come directly from Equation (8):

$$e_\beta^{k-1} = \beta_k - \beta_{k-1} - \Delta t \left[-\frac{C_f + C_r}{m u_{k-1}} \beta_{k-1} - \left(\frac{C_f l_f - C_r l_r}{m u_{k-1}^2} + 1 \right) r_{k-1} + \frac{C_f \delta_{k-1}}{m u_{k-1}} \right] \tag{9}$$

$$e_r^{k-1} = r_k - r_{k-1} - \Delta t \left(-\frac{C_f l_f - C_r l_r}{J z} \beta_{k-1} - \frac{C_f l_f^2 + C_r l_r^2}{J z u_{k-1}} r_{k-1} + \frac{C_f l_f \delta_{k-1}}{J z} \right) \tag{10}$$

$$e_\varphi^k = \dot{\varphi} - r_k \tag{11}$$

$$e_{a_y}^k = a_{yk} + \frac{C_f + C_r}{m} \beta_k + \frac{C_f l_f - C_r l_r}{m u_k} r_k - \frac{C_f \delta_k}{m} \tag{12}$$

The cost functions e_β^{k-1} and e_r^{k-1} are handled, respectively, by the black and gray ternary factors (Figure 2). On the other hand, the unary and binary factors are referred to the measurements: the unary factor minimizes the difference between the actual yaw rate $\dot{\varphi}$ and the associated unknown r, whilst the binary factor minimizes the difference between the actual lateral acceleration a_y and the estimated one.

In principle, the value of the error functions (12) has to be equal to zero, but actually an uncertainty is associated to each factor due to the stochastic nature of the estimation problems. First, measurements are collected through the use of sensors with inherent precision and accuracy. The uncertainties associated to the model include mathematical approximations and uncertainties of the inputs u and δ.

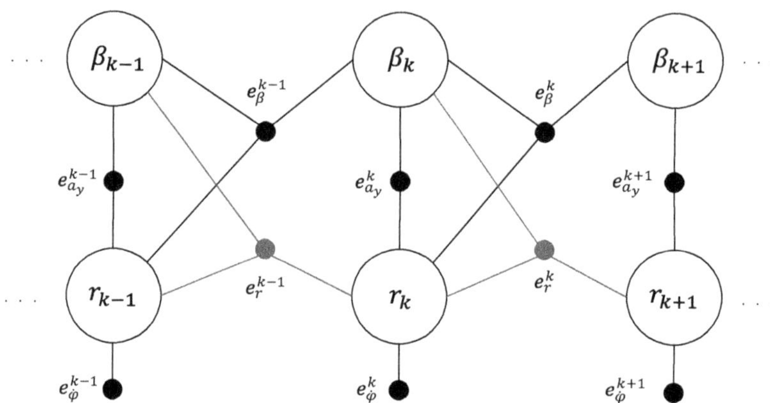

Figure 2. Factor graph (FG) relative to the linear bicycle model.

Without loss of generality, with the assumption of a zero-mean Gaussian probability density function (pdf) associated to the uncertainty in each factor, the minimization problem can be reduced to a linear least squares problem. Since the unknown variables $\mathbf{X}_k$ to minimize at the k-th instant are $[\beta_k \quad r_k]^T$, Equation (12) can be written in compact matrix form as:

$$\begin{bmatrix} e_\beta \\ e_r \\ e_{\dot\varphi} \\ e_{a_y} \end{bmatrix}_k = \mathbf{H}_k \mathbf{X}_k + \mathbf{C}_k \mathbf{h}_k \tag{13}$$

where

$$\mathbf{H}_k = \begin{bmatrix} -\left(1 - \Delta t\dfrac{C_f + C_r}{mu_{k-1}}\right) & \Delta t\left(\dfrac{C_f l_f - C_r l_r}{mu_{k-1}^2} + 1\right) & 1 & 0 \\[2ex] \Delta t\dfrac{C_f l_f - C_r l_r}{J_z} & -\left(1 - \Delta t\dfrac{C_f l_f^2 + C_r l_r^2}{J_z u_{k-1}}\right) & 0 & 1 \\[2ex] 0 & 0 & 0 & -1 \\[2ex] 0 & 0 & \dfrac{C_f + C_r}{m} & \dfrac{C_f l_f - C_r l_r}{mu_k} \end{bmatrix}$$

$$\mathbf{C}_k = \mathrm{diag}\left(\dfrac{C_f}{mu_{k-1}} \quad \dfrac{C_f l_f}{J_z} \quad 0 \quad -\dfrac{C_f}{m} \right)$$

$$\mathbf{h}_k = [\delta_{k-1} \quad \delta_{k-1} \quad 0 \quad \delta_k]^T$$

By defining the *state update vector* as $\boldsymbol{\Delta}_k := \mathbf{X}_k - \mathbf{X}_{k-1}$, the next equation holds:

$$\mathbf{H}_k \mathbf{X}_k = \mathbf{H}_k \boldsymbol{\Delta}_k + \mathbf{H}_k \mathbf{X}_{k-1} \tag{14}$$

The target is the estimation of the state update vector Δ_k via least squares. It is important to remark that the problem at hand is linear; therefore, the following weighted least squares (WLS) problem is set:

$$\hat{\Delta} = \underset{\Delta}{\mathrm{argmin}} \sum_k \left\| \mathbf{H}_k \Delta_k - (\mathbf{z}_k - \mathbf{H}_k \mathbf{X}_{k-1} - \mathbf{C}_k \mathbf{h}_k) \right\|_{\mathbf{Q}_k}^2 \tag{15}$$

where $\mathbf{z}_k$ is the k-th vector of measures, $\mathbf{Q}_k$ is the error covariance matrix and the parenthesis $(\mathbf{z}_k - \mathbf{H}_k \mathbf{X}_{k-1} - \mathbf{C}_k \mathbf{h}_k)$ is the *prediction error*. The argument of the sum is the Mahalanobis norm:

$$\left[\mathbf{Q}_k^{-\frac{1}{2}} [\mathbf{H}_k \Delta_k - (\mathbf{z}_k - \mathbf{H}_k \mathbf{X}_{k-1} - \mathbf{C}_k \mathbf{h}_k)] \right]^T \left[\mathbf{Q}_k^{-\frac{1}{2}} [\mathbf{H}_k \Delta_k - (\mathbf{z}_k - \mathbf{H}_k \mathbf{X}_{k-1} - \mathbf{C}_k \mathbf{h}_k)] \right] \tag{16}$$

and from the following replacements:

$$\mathbf{Q}_k^{-\frac{1}{2}} \mathbf{H}_k = \mathbf{A}_k \tag{17}$$

$$\mathbf{Q}_k^{-\frac{1}{2}} (\mathbf{z}_k - \mathbf{H}_k \mathbf{X}_{k-1} - \mathbf{C}_k \mathbf{h}_k) = \mathbf{b}_k \tag{18}$$

one obtains the following simple least squares problem:

$$\hat{\Delta} = \underset{\Delta}{\mathrm{argmin}} \left\| \mathbf{A}\Delta - \mathbf{b} \right\|_2^2 = \left(\mathbf{A}^T \mathbf{A} \right)^{-1} \mathbf{A}^T \mathbf{b} \tag{19}$$

where $\mathbf{A}$ is a large matrix collecting all matrices $\mathbf{A}_k$. From the calculation of the state update vector: $\hat{\mathbf{X}}_k = \hat{\mathbf{X}}_{k-1} + \hat{\Delta}_k$, where $\mathbf{A}$, $\mathbf{b}$, and Δ grow over time.

3.2. Implementation

Figure 3 displays the first three steps for the estimation problem at hand. The first step, surrounded by the closed dashed black line, includes two prior factors (the unitary gray ones) in place of the dynamic factors. The prior factors minimize the difference between the first estimates and the initial guess of the unknown variables, enforcing known initial conditions; therefore, a reliable initial guess is assumed, with values very close to the correct ones. Their error functions are:

$$e_{p\beta}^0 = \beta_1 - \beta_0 \tag{20}$$

$$e_{pr}^0 = r_1 - r_0 \tag{21}$$

The other steps follow the logic of the second step, contained in the closed dashed-dotted gray line. This line encloses four unknown variables and four factors: two belonging to the dynamic model and two for the measures. By looking at the first dynamic step enclosed in the gray line, the dynamic model is managed by the ternary factors e_β^1 and e_r^1 for the estimation of the sideslip angle and the yaw rate, respectively, while factors $e_{a_y}^2$ and e_ϕ^2 introduce new measurements and minimize the difference between what is effectively measured and the what is expected from the model.

In order to understand how to move from the graphical representation of factors to the minimization problem, the $\mathbf{A}$ matrix and vectors $\mathbf{b}$ and Δ are shown for the first three steps of Figure 3.

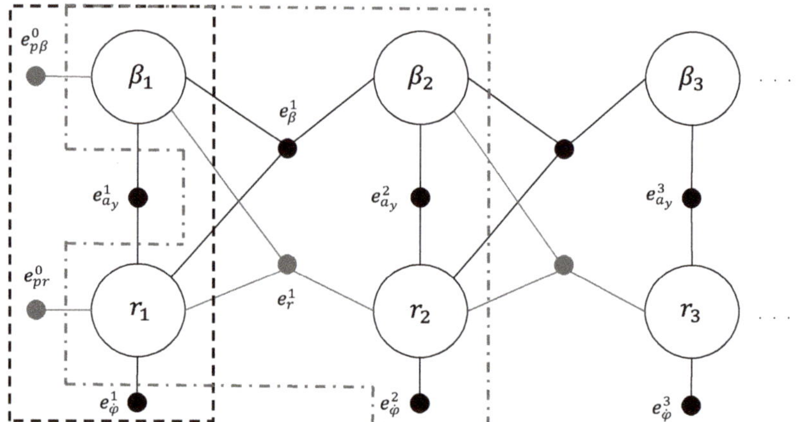

Figure 3. FG relative to the first three steps of the estimation problem: the first step with priors is shown in the black dashed window, and the generic step is shown in the gray dashed-dotted window, with the involved factors.

In Table 1 matrix **A** and vector **b** are reported, and the Δ vector for the first three steps is that in Equation (22).

$$
\begin{bmatrix} \Delta_1 \\ \Delta_2 \\ \Delta_3 \\ \Delta_4 \\ \Delta_5 \\ \Delta_6 \end{bmatrix} = \begin{bmatrix} \beta_1 - \beta_0 \\ r_1 - r_0 \\ \beta_2 - \beta_1 \\ r_2 - r_1 \\ \beta_3 - \beta_2 \\ r_3 - r_2 \end{bmatrix} \tag{22}
$$

Table 1. Matrix **A** and vector **b** for the first three steps of estimation.

Factor	Δ_1	Δ_2	Δ_3	Δ_4	Δ_5	Δ_6	b
p1	$A_{1,1}$						b_1
p2		$A_{2,2}$					b_2
m1		$A_{3,2}$					b_3
m2	$A_{4,1}$	$A_{4,2}$					b_4
d1	$A_{5,1}$	$A_{5,2}$	$A_{5,3}$				b_5
d2	$A_{6,1}$	$A_{6,2}$		$A_{6,4}$			b_6
m3				$A_{7,4}$			b_7
m4			$A_{8,3}$	$A_{8,4}$			b_8
d3			$A_{9,3}$	$A_{9,4}$	$A_{9,5}$		b_9
d4			$A_{10,3}$	$A_{10,4}$		$A_{10,6}$	b_{10}
m5						$A_{11,6}$	b_{11}
m6					$A_{12,5}$	$A_{12,6}$	b_{12}

Matrix **A** in Table 1 is composed of blocks, each one corresponding to a single time-step of the FG, as shown in Figure 3. The first block corresponds to the black dashed window containing the two priors, two measures, and the first two unknown variables β_1 and r_1. Hence, the whitened matrix $\mathbf{A}_1$ is a 4×2 matrix, where each row is associated to a specific factor of Figure 3. The whitened matrix $\mathbf{A}_2$ is composed of a square matrix containing the first and second blocks together, and so forth. Vectors **b** and Δ grow accordingly. It is worth noting that the resulting matrix is a sparse matrix, with elements concentrated near the diagonal. This characteristic can be exploited in the minimization problem of Equation (19). For further details on the elements of matrix **A** and vector **b** in Table 1, the interested reader can refer to the Appendix A.

Matrix $\mathbf{Q}_k$ collects the uncertainty associated to each factor. Therefore, $\mathbf{Q}_k^{-\frac{1}{2}}$ contains the weights of the factors, representing the tuning parameters of this estimation problem. The greater the confidence given to a factor, the higher the values associated with its weight and the lower the corresponding standard deviation. In this case, by looking at Equation (12) it is clear that the factor associated to the yaw rate measurement has a large weight, being precisely measured; therefore, the standard deviation associated to the error of Equation (11) is very small. In this work, a value of $\sigma_{\dot{\phi}} = 10^{-8}$ rad/s is assumed. Instead, the error associated to the Equation (12) relative to the lateral acceleration measurement is taken to be larger because a linear model is compared with the actual lateral acceleration; therefore, a smaller weight is associated to that factor by assuming a $\sigma_{a_y} = 10^{-2}$ m/s^2. Regarding the error relative to the model equations, a high weight is associated to the corresponding factors for the well-known reliability of the model, with standard deviations equal to $\sigma_{\beta} = 10^{-5}$ rad and $\sigma_r = 10^{-4}$ rad/s for the sideslip angle model and yaw rate model, respectively.

During vehicle motion, a large amount of data can be collected. Two estimator implementations are explored: performing a batch estimation on the whole dataset, and a fixed-lag-smoother. The latter considers a sliding window, which contains a fixed number of samples M. Hence, the minimization is achieved on these samples and the most reliable estimate is retained as a set of priors for the next estimation. The length of the sliding window is a tuning parameter. In this work, a window length of $M = 5$ samples is heuristically found to be a a good trade-off between computational speed and goodness of estimation.

By looking at Figure 3, this window will be composed of one closed dashed black line with the priors consisting of the previous estimate (or of the initial conditions for the first window) and by five closed dashed-dotted gray lines, where the indices of every factor change accordingly with the time step. Of course, the matrix $\mathbf{A}$ will be composed of six blocks following the rationale of Table 1, and vectors $\mathbf{b}$ and $\boldsymbol{\Delta}$ grow accordingly.

4. Results

This section collects results obtained with the proposed approach and evaluated by means of real data acquired by an instrumented Ferrari 250 LM Berlinetta GT and made publicly available by Center for Automotive Research at Stanford [30]. A global navigation satellite system (GNSS)-aided inertial navigation system provides overall vehicle body motion, resulting in centimeter-level position accuracy and direct slip-angle measurement. The 2014 Targa Sixty-Six event served as the data collection venue that took place at the Palm Beach International Raceway, a 3.3 km-long track featuring 10 turns and a 1 km straight. Experiments have been carried out using an open-source implementation of the proposed framework (Available online at https://github.com/MBDS/sideslip-angle-vehicle-estimation (accessed on 9 August 2021)).

For a thorough understanding of the advantages of the proposed method, the results obtained with the celebrated linear KF are first shown, which are well known in the automotive field. For the sake of brevity, the KF equations and the state-space form of the problem at hand is here omitted, since it is widespread in the literature (e.g., [13]). Figure 4 shows the estimation of the vehicle sideslip angle β and yaw rate r obtained via linear Kalman filtering during some laps of the Palm Beach International circuit. As expected, the unknown state associated to the yaw rate is estimated very well, since it is directly measured. In fact, only the sensor noise is filtered out, leading to an RMSE value of 0.27 deg/s. Note as well the relatively good estimation of β (RMSE of 0.87 deg) despite the approximation of assuming a linearized model. This is explained by the relatively small range of β values observed in practice and, in particular, for this race dataset.

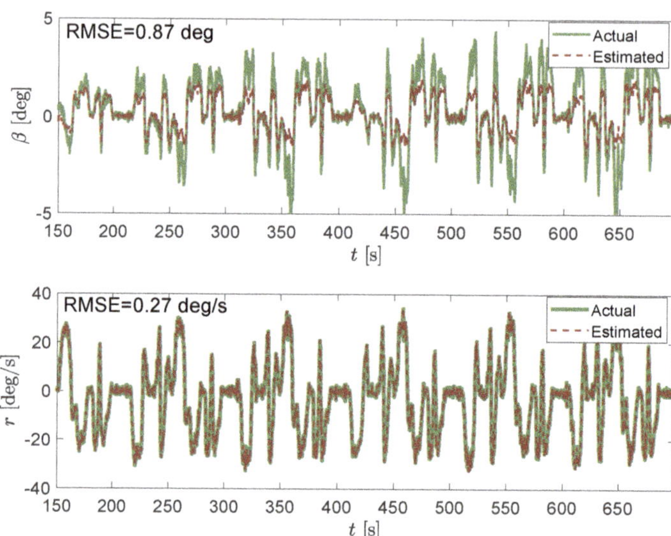

Figure 4. Vehicle sideslip angle β estimate at the top and yaw rate r at the bottom obtained by using a KF-based observer applied to the linear bicycle model.

In Figure 5 the estimation of the sideslip angle β and yaw rate r is shown, but this time obtained from the FG-based observer. As seen from the figure, the estimator can track β more closely and even for higher values than KF, although it is based on a linear model. A corresponding RMSE of 0.57 deg is achieved that is a 34% improvement over the KF implementation. Note that the numerical estimators used for the FG are capable of optimally handling strongly nonlinear models due to their iterative nature, but in this particular case this advantage is not exploited due to the linearity of the model. In fact, the adopted nonlinear solver (Gauss–Newton) only ran a single iteration before detecting that it reached the optimum.

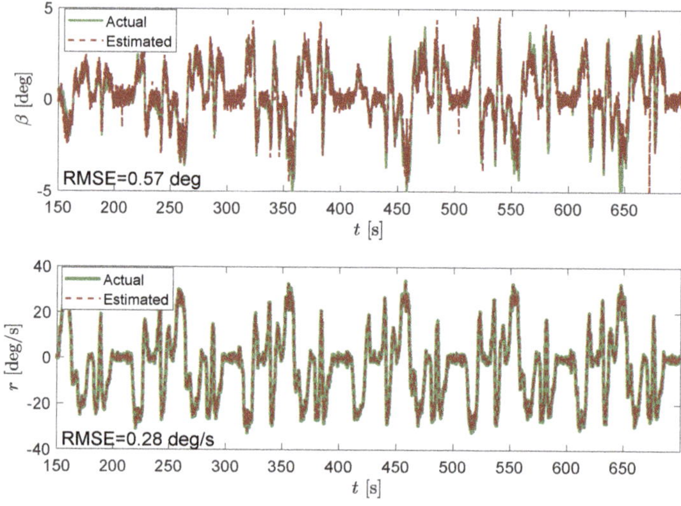

Figure 5. Vehicle sideslip angle β estimate at the top and yaw rate r at the bottom obtained by using FG applied to the linear bicycle model.

On the other hand, one fundamental difference between the FG solver and the KF is the use of a sliding-window estimator for the former, which is then capable of smoothing out sensor noise much more effectively than KF is able to by sequentially processing time steps one by one. However, for very small β angles (see the whole comparison in Figure 6 and detailed view in Figure 7), KF seems to provide a less noisy output. This effect may be explained by the use of different tuning parameters in both approaches.

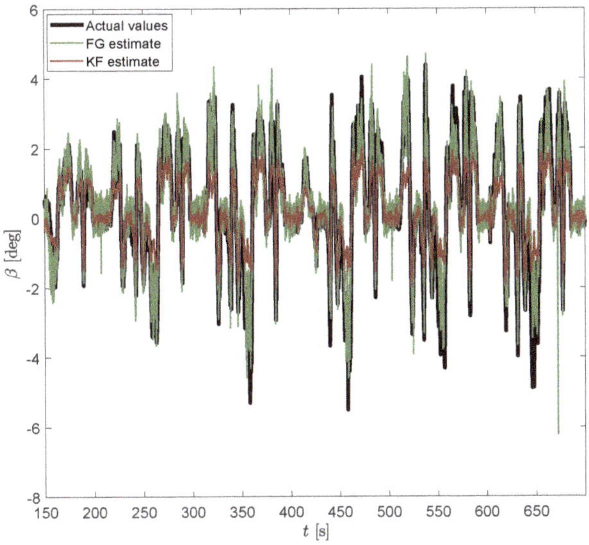

Figure 6. Estimation of vehicle sideslip angle β by considering both KF and FG estimators with a window of 5 samples.

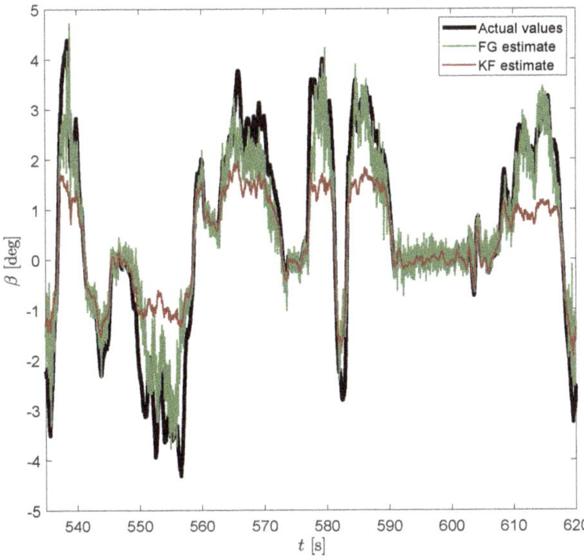

Figure 7. Estimation of vehicle sideslip angle β by considering both KF and FG estimators for small values of the sideslip angle.

For completeness, another experiment performed on a different set of real data gathered in a different race session is displayed in Figure 8. Again, the KF shows the best performance in the linear region characterized by small β values, but fails for higher values of sideslip angle. In contrast, the FG is noisier for small β values but well estimates the actual sideslip angle for higher values, which are the most interesting for vehicle handling and passenger safety.

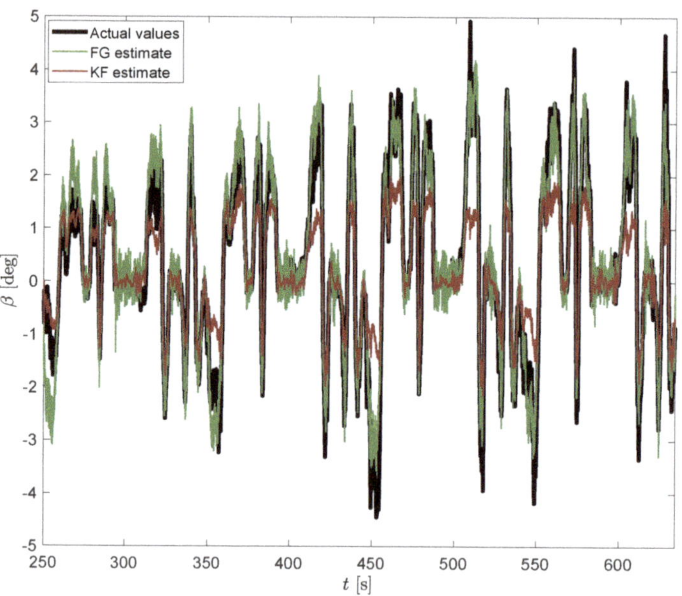

Figure 8. Estimation of vehicle sideslip angle β by considering both KF and FG estimators with a window of 5 samples, for another set of real data.

In order to show the potential of the proposed method to cope with large datasets by exploiting the variable sparsity, the batch estimation is reported in Figure 9 on the total number of samples acquired. As one can see, both the advantages of KF and FG with sliding window method are obtained, consisting of a good estimation for high values of β beyond the linear region and a very smooth estimate, especially in the linear one, similar to the KF, where a better visualization is achieved at the bottom of the figure in a shorter period of time, associated to a single lap. Moreover, Figure 10 displays the absolute error of the sideslip angle estimation performed with KF (red dashed line) and FG (green solid line) with respect to the ground truth, as further proof of the validity of the proposed approach. As can be seen, the error associated with the KF is almost always higher than the FG error.

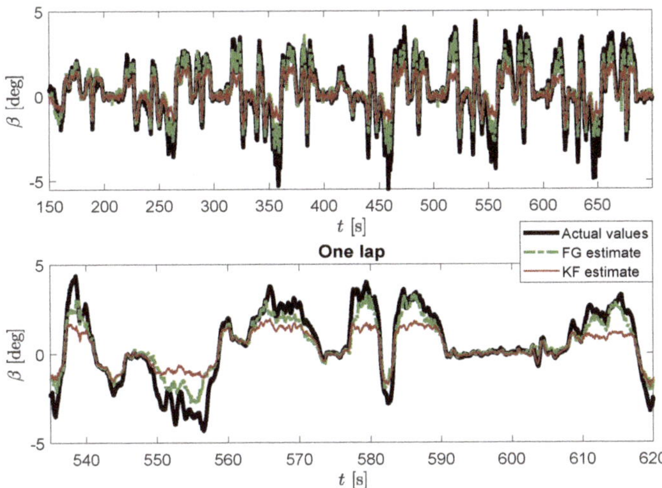

Figure 9. Estimation of vehicle sideslip angle β by considering both KF and FG batch estimators.

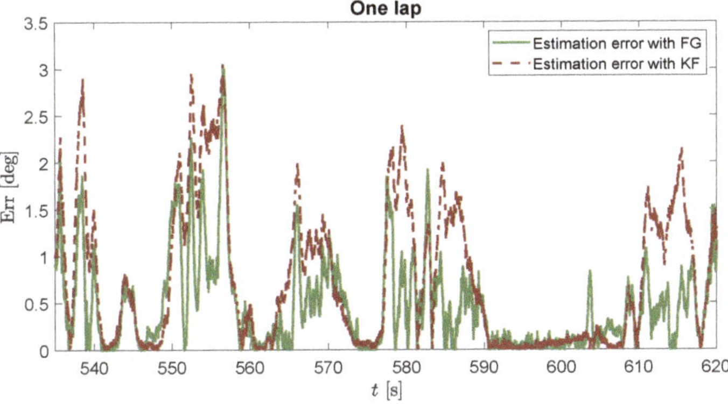

Figure 10. Absolute error in the estimation of sideslip angle β by considering KF against the FG batch estimator.

Finally, for the sake of completeness, the path followed by the vehicle during a single lap is shown in Figure 11, effectively obtaining a path shape that matches the Palm Beach International Raceway. The black path is the actual one obtainable from GPS data or by exploiting the actual yaw rate, sideslip angle, and longitudinal and lateral velocities provided by the dataset. As one can see from the figure, the green path representing the shape of the circuit estimated via FG is closer to the black path than is the red one corresponding to the path estimated by using KF for most of the route, especially for sudden curves, in agreement with the previous results.

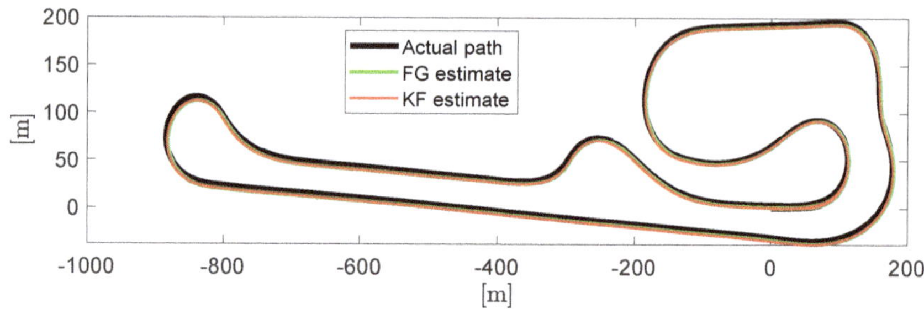

Figure 11. Path followed by the vehicle during a single lap: black is ground truth, red represents KF, and green indicates FG.

5. Conclusions

In this paper an estimator grounded in the factor graph theory is applied for the first time to the estimation of the sideslip angle of a vehicle during motion. A linear single-track model is considered in this research as the vehicle model for the estimator. The performance of the proposed estimation approach is evaluated by using real data and contrasted with the performance obtained from the standard Kalman filter. It is demonstrated that the proposed method accurately estimates the sideslip angle even when its value exceeds the linearity range, which represents the more critical situation in terms of vehicle handling and passenger safety, and thus the knowledge of the correct β value becomes more important. On the other hand, for small values of the sideslip angle, the Kalman filtering preserves its superiority, as expected. Nonetheless, a batch estimation leads to a good estimate of the sideslip angle with a level of noise similar to that achieved using the KF. However, it is worth pointing out that higher values of the sideslip angle can be correctly estimated by keeping a linear bicycle model with very few parameters with respect to other more complex vehicle models, representing an important aspect in terms of practical implementation in real commercial vehicles.

Next steps will take into account FGs applied to more complex (nonlinear) vehicle models, in order to investigate the effectiveness and the ability to achieve even more accurate estimates, as well as other aspects related to vehicle dynamics that cannot be represented with the bicycle model.

Author Contributions: Conceptualization, J.-L.B.-C. and G.R.; software, A.L. and J.-L.B.-C.; writing—original draft preparation, A.L.; writing—review and editing, A.L., G.R. and J.-L.B.-C. All authors have read and agreed to the published version of the manuscript.

Funding: The financial support of the projects Agricultural Interoperability and Analysis System (ATLAS), H2020 (Grant No. 857125), and Multimodal Sensing for Individual Plant Phenotyping in Agriculture Robotics (ANTONIO), ICT-AGRI-FOODCOFUND (Grant No. 41946) is gratefully acknowledged.

Institutional Review Board Statement: Not applicable.

Informed Consent Statement: Not applicable.

Conflicts of Interest: The authors declare no conflict of interest.

Appendix A

In this Appendix, details about matrix **A** and vector **b** in Table 1 are provided for the convenience of the interested reader. Starting with matrix **A**, two different kinds of block matrix can be found: one associated to the first step with priors and one associated to the other steps with the dynamic model. By recalling that

$$\mathbf{Q}^{-\frac{1}{2}} = \mathrm{diag}\left(\frac{1}{\sigma_\beta}, \quad \frac{1}{\sigma_r}, \quad \frac{1}{\sigma_\phi}, \quad \frac{1}{\sigma_{a_y}}\right) \tag{A1}$$

the former kind of block matrix is a 4×2 matrix where:

$$\begin{bmatrix} A_{1,1} & 0 \\ 0 & A_{2,2} \\ 0 & A_{3,2} \\ A_{4,1} & A_{4,2} \end{bmatrix} = \mathbf{Q}^{-\frac{1}{2}} \begin{bmatrix} 1 & 0 \\ 0 & 1 \\ 0 & -1 \\ \dfrac{C_f + C_r}{m} & \dfrac{C_f l_f - C_r l_r}{m u_k} \end{bmatrix} \tag{A2}$$

The longitudinal velocity u_k is that measured at the specific time step on which the priors are applied—for instance, for the first time step with priors associated to initial guess, $u_k = u(k = 1)$. The other terms are constant in this work. The latter kind of block matrix is a 4×4 matrix obtained as in the following:

$$\begin{bmatrix} A_{1,1} & A_{1,2} & A_{1,3} & 0 \\ A_{2,1} & A_{2,2} & 0 & A_{2,4} \\ 0 & 0 & 0 & A_{3,4} \\ 0 & 0 & A_{4,3} & A_{4,4} \end{bmatrix} = \mathbf{Q}^{-\frac{1}{2}} \mathbf{H}_k \tag{A3}$$

with $\mathbf{H}_k$ depending on the k-th time step. A smoother with a window comprising 5 samples has an $\mathbf{A}$ matrix composed of one matrix (A2) and five matrices (A3) arranged on the diagonal.

Vector $\mathbf{b}$ grows accordingly with matrix $\mathbf{A}$. By looking at Equation (18), the parentheses enclose the prediction error between the actual value and the estimated one. Therefore:

- For priors, the actual value is the guess and the function is the difference between these two values; hence:

$$b_1 = \left(\beta_0 - \beta_0 + \beta_0 \right)/\sigma_\beta \tag{A4}$$

and b_2 follows the same rationale.

- For dynamic factors, the actual value is zero, since it is the difference between the forward value and the same obtained by integrating the differential equation; for instance:

$$b_5 = \left[0 - \beta_1 + \left(1 - dt\frac{C_f + C_r}{mu_1} \right)\beta_0 - dt\left(\frac{C_f l_f - C_r l_r}{mu_1^2} + 1 \right)r_0 + dt\frac{C_f \delta_1}{mu_1} \right] \Big/ \sigma_r \tag{A5}$$

and b_6 follows the same rationale.

- Finally, measures follow this rationale (only yaw rate is taken for the sake of brevity):

$$b_3 = \left(\dot{\varphi}_1 + r_0 \right)/\sigma_{\dot{\varphi}} \tag{A6}$$

where, naturally, all indices change according to the time step.

References

1. Chindamo, D.; Lenzo, B.; Gadola, M. On the vehicle sideslip angle estimation: A literature review of methods, models, and innovations. *Appl. Sci.* **2018**, *8*, 355. [CrossRef]
2. Tin Leung, K.; Whidborne, J.F.; Purdy, D.; Dunoyer, A. A review of ground vehicle dynamic state estimations utilising GPS/INS. *Veh. Syst. Dyn.* **2011**, *49*, 29–58. [CrossRef]
3. Caroux, J.; Lamy, C.; Basset, M.; Gissinger, G.L. Sideslip angle measurement, experimental characterization and evaluation of three different principles. *IFAC Proc. Vol.* **2007**, *40*, 505–510. [CrossRef]
4. Manning, W.; Crolla, D. A review of yaw rate and sideslip controllers for passenger vehicles. *Trans. Inst. Meas. Control.* **2007**, *29*, 117–135. [CrossRef]
5. Mastinu, G.; Plöchl, M. *Road and Off-Road Vehicle System Dynamics Handbook*; CRC Press: Boca Raton, FL, USA, 2014.
6. Lenzo, B.; Sorniotti, A.; Gruber, P.; Sannen, K. On the experimental analysis of single input single output control of yaw rate and sideslip angle. *Int. J. Automot. Technol.* **2017**, *18*, 799–811. [CrossRef]
7. Best, M.C.; Gordon, T.; Dixon, P. An extended adaptive Kalman filter for real-time state estimation of vehicle handling dynamics. *Veh. Syst. Dyn.* **2000**, *34*, 57–75.

8. Cheli, F.; Melzi, S.; Sabbioni, E. An adaptive observer for sideslip angle estimation: Comparison with experimental results. In Proceedings of the ASME 2007 International Design Engineering Technical Conferences and Computers and Information in Engineering Conference, Las Vegas, NV, USA, 4–7 September 2007; Volume 48043, pp. 1193–1199.

9. Antonov, S.; Fehn, A.; Kugi, A. Unscented Kalman filter for vehicle state estimation. *Veh. Syst. Dyn.* **2011**, *49*, 1497–1520. [CrossRef]

10. Wei, W.; Bei, S.; Zhu, K.; Zhang, L.; Wang, Y. Vehicle state and parameter estimation based on adaptive cubature Kalman filter. *ICIC Express Lett.* **2016**, *10*, 1871–1877.

11. Luo, W.; Wu, G.; Zheng, S. Design of vehicle sideslip angle observer with parameter adaptation based on HSRI tire model. *Automot. Eng.* **2013**, *33*, 249–255.

12. Reina, G.; Paiano, M.; Blanco-Claraco, J.L. Vehicle parameter estimation using a model-based estimator. *Mech. Syst. Signal Process.* **2017**, *87*, 227–241. [CrossRef]

13. Reina, G.; Messina, A. Vehicle dynamics estimation via augmented Extended Kalman Filtering. *Measurement* **2019**, *133*, 383–395. [CrossRef]

14. Xin, X.; Chen, J.; Zou, J. Vehicle state estimation using cubature kalman filter. In Proceedings of the 2014 IEEE 17th International Conference on Computational Science and Engineering, Chengdu, China, 19–21 December 2014; pp. 44–48.

15. Cheli, F.; Sabbioni, E.; Pesce, M.; Melzi, S. A methodology for vehicle sideslip angle identification: Comparison with experimental data. *Veh. Syst. Dyn.* **2007**, *45*, 549–563. [CrossRef]

16. Di Biase, F.; Lenzo, B.; Timpone, F. Vehicle sideslip angle estimation for a heavy-duty vehicle via Extended Kalman Filter using a Rational tyre model. *IEEE Access* **2020**, *8*, 142120–142130. [CrossRef]

17. Dakhlallah, J.; Glaser, S.; Mammar, S. Vehicle side slip angle estimation with stiffness adaptation. *Int. J. Veh. Auton. Syst.* **2010**, *8*, 56–79. [CrossRef]

18. Chen, Y.; Ji, Y.; Guo, K. A reduced-order nonlinear sliding mode observer for vehicle slip angle and tyre forces. *Veh. Syst. Dyn.* **2014**, *52*, 1716–1728. [CrossRef]

19. Syed, U.H.; Vigliani, A. *Vehicle Side Slip and Roll Angle Estimation*; Technical Report; SAE Technical Paper; Politecnico di Torino: Torino, Italy, 2016.

20. Pacejka, H. *Tire and Vehicle Dynamics*; Elsevier: Amsterdam, The Netherlands, 2005.

21. Davoodabadi, I.; Ramezani, A.A.; Mahmoodi-k, M.; Ahmadizadeh, P. Identification of tire forces using Dual Unscented Kalman Filter algorithm. *Nonlinear Dyn.* **2014**, *78*, 1907–1919. [CrossRef]

22. Reina, G.; Leanza, A.; Mantriota, G. Model-based observers for vehicle dynamics and tyre force prediction. *Veh. Syst. Dyn.* **2021**, 1–26. doi:10.1080/00423114.2021.1928245. [CrossRef]

23. You, S.H.; Hahn, J.O.; Lee, H. New adaptive approaches to real-time estimation of vehicle sideslip angle. *Control. Eng. Pract.* **2009**, *17*, 1367–1379. [CrossRef]

24. Chindamo, D.; Gadola, M. Estimation of vehicle side-slip angle using an artificial neural network. *MATEC Web Conf.* **2018**, *166*, 02001. [CrossRef]

25. Wei, W.; Shaoyi, B.; Lanchun, Z.; Kai, Z.; Yongzhi, W.; Weixing, H. Vehicle sideslip angle estimation based on general regression neural network. *Math. Probl. Eng.* **2016**, *2016*, 3107910. [CrossRef]

26. Cheli, F.; Ivone, D.; Sabbioni, E. Smart Tyre Induced Benefits in Sideslip Angle and Friction Coefficient Estimation. In *Sensors and Instrumentation*; Springer: Berlin/Heidelberg, Germany, 2015; Volume 5, pp. 73–83.

27. Melzi, S.; Resta, F.; Sabbioni, E. Vehicle Sideslip Angle Estimation Through Neural Networks: Application to Numerical Data. *Eng. Syst. Des. Anal.* **2006**, *42495*, 167–172.

28. Koller, D.; Friedman, N. *Probabilistic Graphical Models: Principles and Techniques*; MIT Press: Cambridge, MA, USA, 2009.

29. Blanco-Claraco, J.L.; Leanza, A.; Reina, G. A general framework for modeling and dynamic simulation of multibody systems using factor graphs. *Nonlinear Dyn.* **2021**. [CrossRef]

30. Kegelman, J.C.; Harbott, L.K.; Gerdes, J.C. 2014 Targa Sixty-Six. Stanford Digital Repository. 2016. Available online: http://purl.stanford.edu/hd122pw0365 (accessed on 2 July 2021).

31. Jazar, R.N. *Vehicle Dynamics*; Springer: New York, NY, USA, 2014.

sensors

Article

Lunar Surface Fault-Tolerant Soft-Landing Performance and Experiment for a Six-Legged Movable Repetitive Lander

Ke Yin, Songlin Zhou, Qiao Sun and Feng Gao *

Key Laboratory of Mechanical System and Vibration, School of Mechanical Engineering, Shanghai Jiao Tong University, Shanghai 200240, China; jixie_yinke@sjtu.edu.cn (K.Y.); zhousonglin@sjtu.edu.cn (S.Z.); qiaosun1234@gmail.com (Q.S.)
* Correspondence: fengg@sjtu.edu.cn

Abstract: The cascading launch and cooperative work of lander and rover are the pivotal methods to achieve lunar zero-distance exploration. The separated design results in a heavy system mass that requires more launching costs and a limited exploration area that is restricted to the vicinity of the immovable lander. To solve this problem, we have designed a six-legged movable repetitive lander, called "HexaMRL", which congenitally integrates the function of both the lander and rover. However, achieving a buffered landing after a failure of the integrated drive units (IDUs) in the harsh lunar environment is a great challenge. In this paper, we systematically analyze the fault-tolerant capacity of all possible landing configurations in which the number of remaining normal legs is more than two and design the landing algorithm to finish a fault-tolerant soft-landing for the stable configuration. A quasi-incentre stability optimization method is further proposed to increase the stability margin during supporting operations after landing. To verify the fault-tolerant landing performance on the moon, a series of experiments, including five-legged, four-legged and three-legged soft-landings with a vertical landing velocity of −1.9 m/s and a payload of 140 kg, are successfully carried out on a 5-DoF lunar gravity ground-testing platform. The HexaMRL with fault-tolerant landing capacity will greatly promote the development of a next-generation lunar prober.

Keywords: movable repetitive lander; fault-tolerant soft-landing; landing configuration; stability optimization

Citation: Yin, K.; Zhou, S.; Sun, Q.; Gao, F. Lunar Surface Fault-Tolerant Soft-Landing Performance and Experiment for a Six-Legged Movable Repetitive Lander. *Sensors* **2021**, *21*, 5680. https://doi.org/10.3390/s21175680

Academic Editor: Javier Cuadrado

Received: 28 July 2021
Accepted: 20 August 2021
Published: 24 August 2021

Publisher's Note: MDPI stays neutral with regard to jurisdictional claims in published maps and institutional affiliations.

1. Introduction

The moon is the hub and bridge between mankind and the universe, while lunar exploration is the premise and basis of deep space exploration. Nowadays, the separated design of an immovable lander and rover is still the core method of zero-distance exploration on the moon. Many countries have made world-renowned achievements such as the Soviet/Russian Luna-9 [1], the first lander to achieve lunar soft-landing, which absorbs impact energy using four airbags; American Surveyor-1 [2], the first legged lander to reach the lunar surface, which uses three three-branch buffered legs filled with aluminum honeycomb material, providing technical support for Apollo program [3]; Chinese Chang'e 4 [4] reaches the far side of the moon first, which utilizes four similar buffer legs. Notably, all these landers are immovable and are designed to help the rover finish landing, so their exploration capacity is restricted to around the fixed landing site. Two kinds of rover are applied to expand the exploration range, one is a manned lunar rover, like LRV [5], which can carry up to two astronauts, another is the unmanned wheeled rover, like Yutu 1 & 2 [4,6], which can maneuver quickly with scientific instruments. However, the separated design of the lander and rover creates a heavy and complex prober system. The rover only executes exploration in a circle district with the lander as the center and the safety distance as the radius.

Thanks to excellent traversing performance on irregular terrain [7,8], legged robots are promising to accomplish lunar exploration compared with a wheeled rover. On the

Sensors **2021**, *21*, 5680. https://doi.org/10.3390/s21175680 https://www.mdpi.com/journal/sensors

one hand, the leg form is utilized in most existing landers, though they lack mobility on the lunar surface. On the other hand, legged robots are well designed in many fields such as running robots, Bigdog [9], Cheetah-3 [10] and Anymal [11,12]; underwater robots, Crabster [13,14]; heavy-duty robots, Octopus [15]; and exploration robots, Athlete [16], Spaceclimber [17] and Spacebok [18,19]. Furthermore, to combine the excellent speed performance for wheeled robots on even terrain and a great adaptive capacity for legged robots on irregular terrain, the wheel-legged robot [20] adopted a hierarchical framework to control wheel and leg motions; this has drawn a lot of researcher attention. Nevertheless, the current legged or wheel-legged robots are difficult to directly apply to lunar exploration. The hydraulic actuator is widely employed in HyQ2Max [21] or Bigdog [9] to obtain high explosive torque, which is infeasible for extraterrestrial exploration. The leg layout of a running robot cannot withstand the landing impact in all directions. The buffer capacity of the current robot is relatively weak, so the engine nozzle under the lander body will be easily damaged for colliding with the ground.

We have designed a six-legged movable repetitive lander "HexaMRL" in previous work to integrate the function of both lander and rover. IDUs are used to simulate the dynamic characters of an active dissipative system of spring and dampener by impedance control to achieve a buffered landing and protect the leg structure, different from the irreversible deformation of aluminum honeycomb material after landing [22–24]. Hence, the lander can still execute locomotion as a rover. After exploration at a current landing site, it can fly to the next landing site using the engine and repetitively perform buffered landing tasks. This new exploration mode will significantly increase the utilization rate of an individual prober on the moon and greatly extend the exploration district. However, the repetitive work mode has higher requirements for the quality and fault-tolerant landing capacity of the lander.

In the harsh lunar environment (i.e., intense radiation, large temperature difference and ultravacuum), it is hard to repair with the remote operation if some failures occur on the IDUs. Fault-tolerant control (FTC) for the robot has attracted great attention all over the world and is pivotal for the prober to execute exploration tasks. Nowadays, FTC is generally achieved by the following three methods. Firstly, multiple drives are used in the active joints; for example, Zhang et al. [25] employ dual-input/single-output (DISO) to drive the servo press machine. Secondly, the parallel robot could use a redundant drive [26] to eliminate singularities in the workspace. The third one is to increase the DoF of robot end-effector, such as the Canadian space station's remote manipulator system (SSRMS) [27], which adopted seven series joints to improve the workspace. However, the above FTC relies on more drives or more complex mechanisms, which will increase the system mass and complexity that are difficult for the lander to accept. Furthermore, fault-tolerant landing is not generally considered in current landers because most of the existing landers are three-legged or four-legged, which constructively lack fault-tolerant landing capacity when one leg fails. For a three-legged lander like Surveyor-1, the remaining two legs cannot support the lander. As for four-legged landers like Apollo 11 [3] or Chang'e 3, 4, & 5 [4,6,28], the center of mass of the lander will move to the side of the supporting triangle constructed by the remaining three legs, leading to a failed buffer landing on the uneven lunar surface. Therefore, hard strict standards are required for the manufacture and control of such landers and would be abandoned if any failures occur.

The six-legged design in HexaMRL makes the fault-tolerant soft-landing feasible without any supplement of drivers or mechanisms. In this paper, we have systematically studied the fault-tolerant soft-landing performance on the moon for HexaMRL. Firstly, we analyze the classification and stability of the landing configuration and establish the relationship between fault number and landing configuration by the synthesis equation. Secondly, regarding stable configuration, the corresponding fault-tolerant landing algorithms are designed to achieve a buffered landing, and a quasi-incentre stability optimization method is further proposed to increase the stability margin during supported operations. Thirdly, to verify the fault-tolerant landing on the moon, a series of experiments including five-legged, four-legged and three-legged soft-landing with a vertical landing velocity of -1.9 m/s and a payload of 140 kg are successfully carried out on a 5-DoF lunar gravity ground testing platform.

The rest of the paper is organized as follows. Section 2 introduces the lander system. The landing configuration is analyzed in Section 3. Sections 4 and 5 design the fault-tolerant algorithm and optimize the stability margin in supporting the operations, respectively. Section 6 clarifies the fault-tolerant landing experiments, and Section 7 discusses the experiment results. The last section is the conclusion and its expansion.

2. Lander System

The HexaMRL is composed of six identical legs that adopt a hybrid mechanism actuated by three IDUs. During buffer landing, each IDU imitates the dynamic characters of both spring and damper. If some errors occur in the IDUs of the leg under the harsh lunar environment, the leg residual mobility capacities are a great variant in different fault combinations.

2.1. HexaMRL

As shown in Figure 1 the HexaMRL is designed to execute repetitive soft-landing and roving for lunar exploration. Its size is about 1.35 m long, 0.94 m wide and 0.75 m high; it weighs 60 kg, including a 20 kg aluminum shell. The robot consists of a body and six identical legs. Each leg is composed of side IDU, thigh IDU, shank IDU, thigh, rocker, connecting link, sole and shank. All legs are connected to the body that carries the controller, inertial measurement unit (IMU), power system, sensing system, payloads, etc.

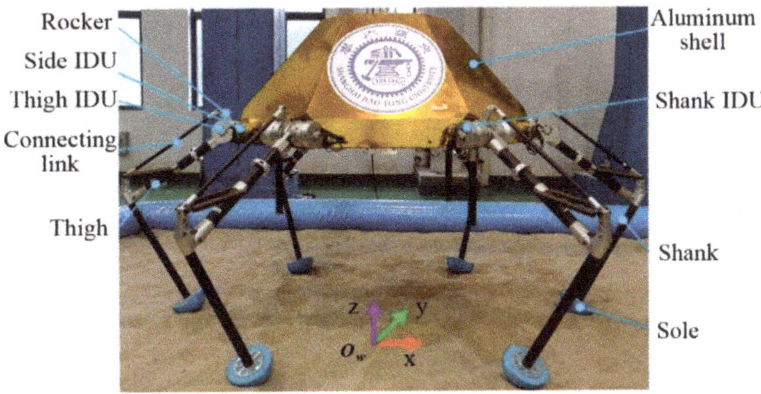

Figure 1. The six-legged movable and repetitive lander (HexaMRL).

The leg distribution design needs to meet landing and roving functions simultaneously. On the one hand, to withstand the impact force uniformly during buffer landing, all legs are arranged as an equilateral polygon like a regular triangle in Surveyor-1 [2] or Square in Apollo program [3] and Chang'e series [4,6,28]. As for the six-legged lander, the angular interval between adjacent legs should be 60°. On the other hand, to achieve quick locomotion, all legs will be laid out in a slender shape as in animals like the cheetah, lion or goat, like the hexapod robot RHex [29] or TUM-walking machine [30]. Eventually, as illustrated in Figure 2, the angle between leg 2 and leg 3 is designed to be 54.2°. Leg 1 (or 2) and leg 5 (or 4) are symmetric about axis-x, while leg 1 (or 5) and leg 2 (or 4) are symmetric about axis-y.

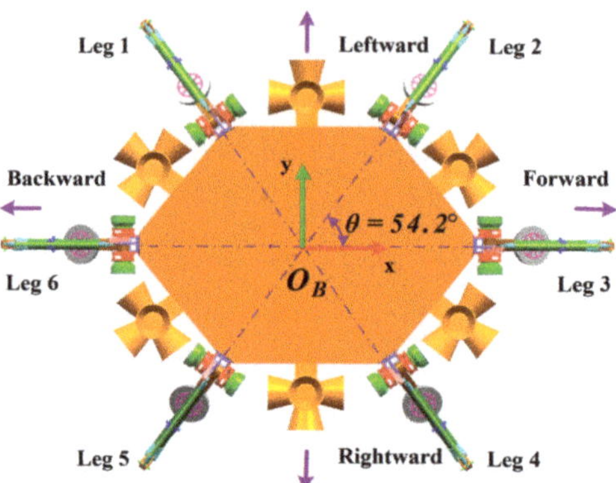

Figure 2. The leg distribution of HexaMRL.

2.2. Leg Mechanism

The leg mechanism is presented in Figure 3. It has three active DoFs that are actuated by side IDU, thigh IDU and shank IDU to control the mobility of the rocker, thigh and shank, respectively. Let $q = \begin{bmatrix} \alpha & \beta & \gamma \end{bmatrix}^T$ denote the generalized coordinate vector where α, β and γ are the joint angles of the side, thigh and shank, respectively. The tiptoe position P_{tip} in the leg coordinate frame can be obtained as follows:

$$P_{tip} = \begin{bmatrix} l_t cos\beta + l_s cos\gamma \\ cos\alpha(l_t sin\beta + l_s sin\gamma) \\ sin\alpha(l_t sin\beta + l_s sin\gamma) \end{bmatrix} \tag{1}$$

where l_t and l_s are the length of the thigh and shank, separately.

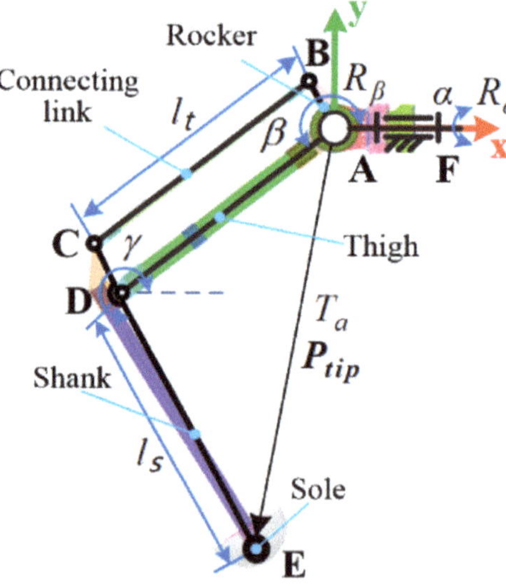

Figure 3. The leg mechanism.

The IDU consists of an encoder, servo motor, torque sensor, harmonic reducer, shell, coupler, bearing, etc. During buffer landing, each IDU imitates the dynamic characters of an active torsion spring and an active torsion damper by the impedance control method, whose control rule can be written as follows:

$$\tau_{act} - \tau_{des} = K_a(\varphi_{des} - \varphi_{act}) + B_a(\dot{\varphi}_{des} - \dot{\varphi}_{act}) \tag{2}$$

where τ_{act} and τ_{des} are the actual torque and desired torque of IDU, K_a and B_a are the coefficients of stiffness and damping of active compliance, φ_{des} and φ_{act} are the desired angle and actual angle of the active joint, $\dot{\varphi}_{des}$ and $\dot{\varphi}_{act}$ are corresponding angular velocity.

2.3. Leg Residual Capacity

In order to reduce the rocket size, all legs need folding, as in Figure 4a, for compact volume when launched from Earth. Before the buffer landing on the moon, all legs need to deploy from the folded state, as in Figure 4b, to obtain better supporting stability and a larger buffer stroke. Since the mobility demand before and after the deploying task is not the same, the fault-tolerant landing capacity is significantly different as well.

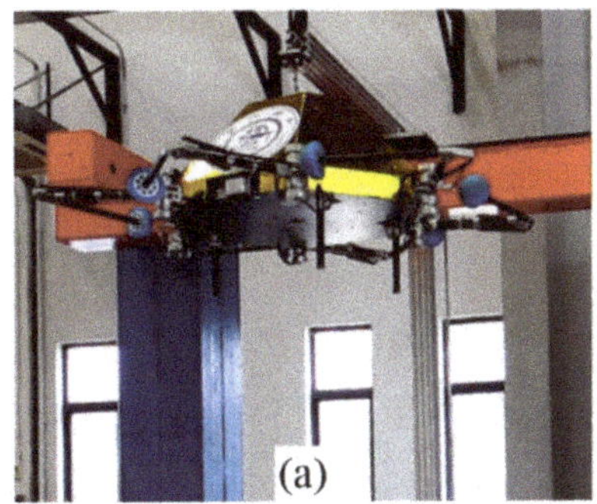

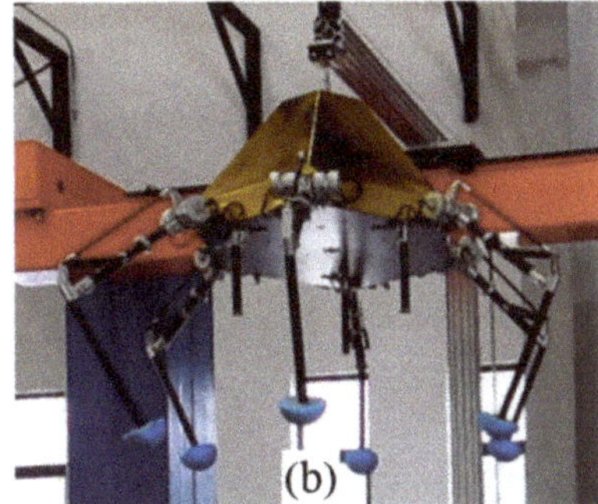

Figure 4. Leg function. (**a**) folding; (**b**) deploying.

Here, because the capacities of fault-tolerant are different, we use LB and LA to distinguish the time of fault occurrence on side IDU before and after deploying operation. On the contrary, as for the thigh or shank IDU, distinguishing separate failure times is unnecessary due to them having the same capacities. The normal IDU of the thigh or shank is denoted by N, while the failed IDU is represented by L. When some IDUs are failed, the residual workspaces are illustrated in Figure 5. There are four cases when one IDU fails. Firstly, if the side IDU fails after deploying, and the other two IDUs are normal, this case can be denoted by LANN. Its residual workspace is the purple vertical plane, and the mobility character is expressed as $G_F^{II}(0, R_\beta, 0; T_a, 0, 0)$ by G_f theory [31]. Where R_β and T_a are the swing movement and the stretching/shrinking movement in the leg sagittal plane. Secondly, if the side IDU fails before deploying and the other two IDUs are normal, denoted by LBNN, the residual workspace is the blue horizontal plane, and the mobility character is expressed as $G_F^{II}(0, R_\beta, 0; T_a, 0, 0)$. Thirdly, if the thigh IDU fails, denoted by NLN, the residual workspace is the red surface, and the mobility character is expressed as $G_F^{II}(R_\alpha, R_\beta, 0; 0, 0, 0)$ where R_α is the abduction/adduction movement. Lastly, if the shank IDU fails, denoted by NNL, the residual workspace is the green surface, and the mobility character is expressed as $G_F^{II}(R_\alpha, R_\beta, 0; 0, 0, 0)$. There are five cases if two IDUs fail, separately denoted by LBNL, LBLN, NLL, LANL and LALN. Their residual workspaces are the dotted curves of green, blue, red, black and purple; their corresponding mobility characters are expressed as $G_F^{II}(0, R_\beta, 0; 0, 0, 0)$ or $G_F^{II}(R_\alpha, 0, 0; 0, 0, 0)$. There are two cases if all three IDUs fail, denoted by LBLL and LALL; their residual workspaces are a fixed point of black and red while the mobility characters are $G_F^{I}(0, 0, 0; 0, 0, 0)$.

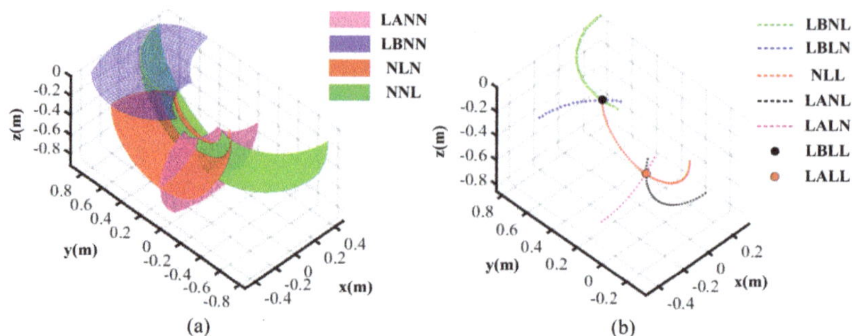

Figure 5. The leg residual workspace. (**a**) One fault; (**b**) two or three faults.

Noticeably, the lander must possess the up/down movement character during the buffer landing period. Therefore, only the fault case LANN satisfies the mobility demand and has the fault-tolerant landing capacity, further symbolized as $E1_1$. The other three cases with one fault don't have fault-tolerant landing capacity and are symbolized as $E1_2$, uniformly. All cases with two or three faults cannot finish fault-tolerant landing, so the fault time before or after deploying operation will not be distinguished, and the symbol LA or LB will be replaced by L. Finally, the two faulted IDUs cases are denoted as E2, while the three faulted IDUs case is written as E3. The detailed fault-tolerant capacity of the leg is shown in Table 1.

Table 1. Fault-tolerant capacity of single leg.

Fault Number	Side	Thigh	Shank	Number	Symbol	G_f Set	Fault Tolerance
	LA	N	N		$E1_1$	$G_F^{II}(0,\ R_\beta,0;T_a,0,0)$	1
1	LB	N	N	4		$G_F^{II}(0,\ R_\beta,0;T_a,0,0)$	0
	N	L	N		$E1_2$	$G_F^{II}(R_\alpha,\ R_\beta,0;0,0,0)$	0
	N	N	L			$G_F^{II}(R_\alpha,\ R_\beta,0;0,0,0)$	0
	L	L	N			$G_F^{II}(0,\ R_\beta,0;0,0,0)$	0
2	L	N	L	3	E2	$G_F^{II}(0,\ R_\beta,0;0,0,0)$	0
	N	L	L			$G_F^{II}(R_\alpha,\ 0,0;0,0,0)$	0
3	L	L	L	1	E3	$G_F^{I}(0,\ 0,0;0,0,0)$	0

3. Landing Configuration Analysis

When more than one IDU fails, different legs may be involved, and their spatial distribution will affect the landing performance. The landing performance will be determined by the configuration that consists of the supporting polygon of the foothold of the remaining normal legs and the position of the lander's center of mass. The stabilities in landing configurations are different, and we systematically assess this by the dimensionless index SAI. Then, we establish a synthesis equation to deduce all possible configurations under a certain number of failed IDUs.

3.1. Classification

Specifically, we need to consider the equivalent leg when the landing configurations are analyzed. As shown in Figure 2, all legs have different effects on landing performance because their interval angles are not equal to 60°. However, legs 1, 2, 4 and 5 are a group of equivalent legs while legs 3 and 6 are the other group, and the influence of equivalent legs on landing performance is the same.

Noticeably, there are three basic properties for landing configuration. (1) The landing performance is different if the landing configuration is different. (2) The configuration is the same if one equivalent leg fails, e.g., leg 1 or 2 failed. (3) If configuration 1 coincides with configuration 2 after rotation and symmetry operations, the two configurations are considered the same, e.g., legs 1 and 3 or legs 3 and 6.

The classification of landing configuration is shown in Figure 6. The red point denotes the center of mass, while the purple points express the footholds. The solid purple lines illustrate the leg position, while the blue dotted lines show the body contour. The supporting polygon constructed by the footholds of remaining normal legs is denoted by the green plane. The relationship between the normal supporting leg and landing configuration is illustrated in Table 2. As for six-legged landing, there is only one configuration ($C_6^6 = 1$) called VI-1. As for five-legged landing, there are six configurations ($C_6^5 = 6$) that are further divided into two groups: V-1 includes supporting leg 2-3-4-5-6, 1-3-4-5-6, 1-2-3-5-6 and 1-2-3-4-6; V-2 consists of supporting leg 1-2-4-5-6 and 1-2-3-4-5. As for four-legged landing, there are fifteen configurations ($C_6^4 = 15$) that are detailly classified into six groups: IV-1 includes supporting leg 3-4-5-6 and 1-2-3-6; IV-2 includes supporting leg 2-4-5-6, 1-3-4-5, 1-2-4-6 and 1-2-3-5; IV-3 includes supporting leg 2-3-5-6 and 1-3-4-6; IV-4 includes supporting leg 2-3-4-6 and 1-3-5-6; IV-5 includes supporting leg 2-3-4-5, 1-4-5-6 and 1-2-5-6; IV-6 includes supporting leg 1-2-4-5. As for three-legged landing, there are twenty configurations ($C_6^3 = 20$) categorized in six groups: III-1 includes supporting leg 1-2-3, 1-2-6, 3-4-5 and 4-5-6; III-2 includes supporting leg 1-2-4, 1-2-5, 1-4-5 and 2-4-5; III-3 includes supporting leg 1-3-4, 1-4-6, 2-3-5 and 2-5-6; III-4 includes supporting leg 1-3-5 and 2-4-6; III-5 includes supporting leg 1-3-6, 2-3-6, 3-4-6 and 3-5-6; III-6 includes supporting leg 2-3-4 and 1-5-6. Here, the analysis of configuration consisting of one or two legs is omitted because the lander cannot finish a fault-tolerant landing.

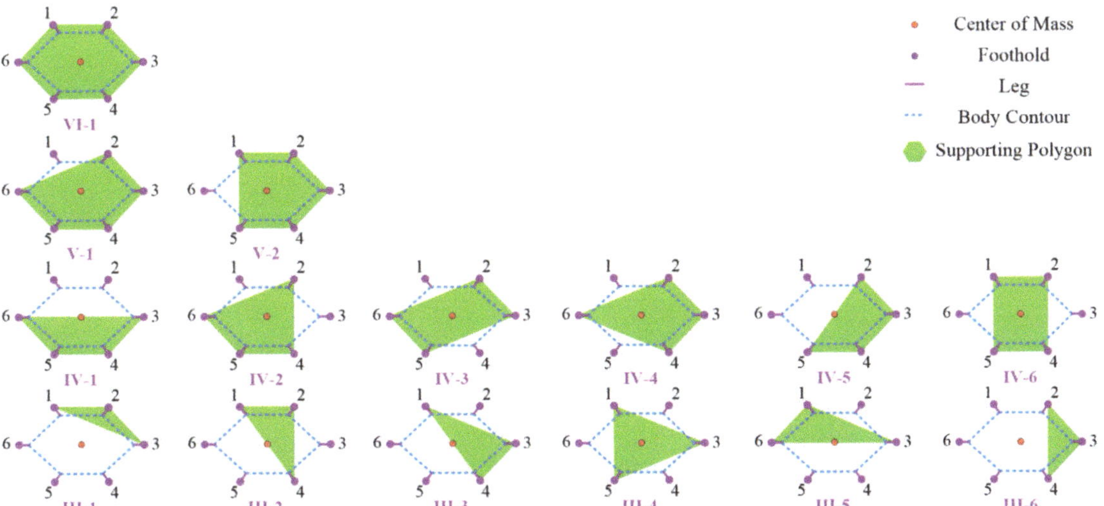

Figure 6. Classification of landing configuration.

Table 2. Analysis table of landing configuration under fault combinations.

Type I	Number	Supporting Leg	Type II
VI	$C_6^6 = 1$	1-2-3-4-5-6	VI-1
V	$C_6^5 = 6$	2-3-4-5-6	V-1
		1-3-4-5-6	
		1-2-3-5-6	
		1-2-3-4-6	
		1-2-4-5-6	V-2
		1-2-3-4-5	
IV	$C_6^4 = 15$	3-4-5-6	IV-1
		1-2-3-6	
		2-4-5-6	IV-2
		1-3-4-5	
		1-2-4-6	
		1-2-3-5	
		2-3-5-6	IV-3
		1-3-4-6	
		2-3-4-6	IV-4
		1-3-5-6	
		2-3-4-5	IV-5
		1-4-5-6	
		1-2-5-6	
		1-2-3-4	IV-6
		1-2-4-5	

Type I	Number	Supporting Leg	Type II
III	$C_6^3 = 20$	1-2-3	III-1
		1-2-6	
		3-4-5	
		4-5-6	
		1-2-4	III-2
		1-2-5	
		1-4-5	
		2-4-5	
		1-3-4	III-3
		1-4-6	
		2-3-5	
		2-5-6	
		1-3-5	III-4
		2-4-6	
		1-3-6	III-5
		2-3-6	
		3-4-6	
		3-5-6	
		2-3-4	III-6
		1-5-6	

3.2. Stability

The influence of different configurations on landing stability is determined by the shortest distance from the ground projection point of the center of mass to each side of the supporting polygon (d) and the area of the supporting polygon (S). The dimensionless index SAI is proposed to evaluate the stability of each configuration and is written as follows:

$$SAI = \frac{d}{d_{VI}} * \frac{S}{S_{VI}} \tag{3}$$

where d_{VI} and S_{VI} are the values of d and S in normal six-legged landing configuration VI, respectively. If the center of mass is within the supporting polygon, then $d > 0$. On the contrary, $d < 0$.

Eventually, the SAI value in each configuration can be calculated because the legs are deployed to the predetermined position from the folded status for the buffer landing. According to the SAI, the configuration stabilities will be divided into three statuses: stable, critical stable and unstable, and can be written as follows:

$$SS = \begin{cases} Stable\ (S), & if\ SAI > 0; \\ Critical\ Stable\ (CS), & if\ SAI = 0; \\ Unstable\ (US), & if\ SAI < 0. \end{cases} \tag{4}$$

As seen in Table 3, the stable configurations have eight cases: VI-1, V-1, V-2, IV-2, IV-3, IV-4, IV-6 and III-4, in which the robot can execute fault-tolerant landing. The critical stable configurations have five cases: IV-1, IV-5, III-2, III-3 and III-5. In these cases, the robot can only perform soft-landing if the landing site is absolutely even. The unstable configurations include III-1 and III-6. Owing to the irregular lunar surface, the lander cannot finish soft-landing under both critical stable and unstable configurations.

Table 3. Stability evaluation results of landing configurations.

Type I	Type II	Stability Radius d(m)	Supporting Area S(m^2)	*SAI*	*SS*
VI	VI − 1	0.5998	1.5233	1	S
V	V − 1	0.3577	1.2640	0.4948	S
	V − 2	0.4324	1.2803	0.6058	S
IV	IV − 1	0	0.7617	0	CS
	IV − 2	0.3577	1.021	0.3997	S
	IV − 3	0.3577	1.0047	0.3933	S
	IV − 4	0.4324	1.0047	0.4754	S
	IV − 5	0	0.7617	0	CS
	IV − 6	0.4324	1.0373	0.4908	S
III	III − 1	−0.3577	0.2593	−0.1015	US
	III − 2	0	0.5187	0	CS
	III − 3	0	0.5023	0	CS
	III − 4	0.3577	0.7617	0.2982	S
	III − 5	0	0.5023	0	CS
	III − 6	−0.4324	0.243	−0.1150	US

3.3. Relationship between Fault Number and Configuration

When the number of faulted IDUs is N_m, all the possible configurations will be concluded by the following equation:

$$\begin{cases} \sum_{i=1}^{3} N_i \cdot i = N_m \\ \sum_{i=1}^{3} N_i = N_L \end{cases} \tag{5}$$

$$s.t. \begin{cases} 0 \leq N_L \leq min\{6, N_m\} \\ 0 \leq N_m \leq 18 \\ 0 \leq N_i \leq 6, i = 1, 2, 3 \end{cases} \tag{6}$$

where N_i is the number of legs with i faulted IDUs, N_L is the total number of failed legs. The solve set $\{ N_1, N_2, N_3 \}$ illustrates the faults distribution. Considering the fault-tolerant capacity of a single leg in Table 1, we can get the possible configuration details. When N_m is 1, the solve set is $\{ 1, 0, 0 \}$. The landing configuration is VI if the failed leg is E1$_1$. On the contrary, the landing configuration is V if the failed leg is E1$_2$. Similarly, when N_m is 2, there are two solve sets. Different fault leg groups will result in VI, V, or IV configuration. Table 4 lists all possible landing configurations when the number of faulted IDUs is less than seven. Lastly, we can get landing configuration type II by substituting the number of failed legs into Table 2. If the faulted number is more than six, the combination results can be obtained by the above Equations (5) and (6), similarly.

Table 4. Fault combination results.

N_m	Solve Sets	Fault Legs Group	Type I	N_m	Solve Sets	Fault Legs Group	Type I
1	{1,0,0}	$E1_1$	VI		{1,0,1}	$E1_1E3$	V
		$E1_2$	V			$E1_2E3$	IV
	{0,1,0}	$E2$	V		{0,2,0}	$E2E2$	IV
2	{2,0,0}	$E1_1E1_1$	VI		{2,1,0}	$E1_1E1_1E2$	V
		$E1_1E1_2$	V	4		$E1_1E1_2E2$	IV
		$E1_2E1_2$	IV			$E1_2E1_2E2$	III
	{0,0,1}	$E3$	VI		{4,0,0}	$E1_1E1_1E1_1E1_1$	VI
	{1,1,0}	$E1_1E2$	V			$E1_1E1_1E1_1E1_2$	V
		$E1_2E2$	IV			$E1_1E1_1E1_2E1_2$	IV
3		$E1_1E1_1E1_1$	VI			$E1_1E1_2E1_2E1_2$	III
	{3,0,0}	$E1_1E1_1E1_2$	V		{0,0,2}	$E3E3$	IV
		$E1_1E1_2E1_2$	IV		{1,1,1}	$E1_1E2E3$	IV
		$E1_2E1_2E1_2$	III			$E1_2E2E3$	III
	{0,1,1}	$E2E3$	IV		{0,3,0}	$E2E2E2$	III
	{2,0,1}	$E1_1E1_1E3$	V		{3,0,1}	$E1_1E1_1E1_1E3$	V
		$E1_1E1_2E3$	IV			$E1_1E1_1E1_2E3$	IV
		$E1_2E1_2E3$	III			$E1_1E1_2E1_2E3$	III
	{1,2,0}	$E1_1E2E2$	IV	6	{2,2,0}	$E1_1E1_1E2E2$	IV
		$E1_2E2E2$	III			$E1_1E1_2E2E2$	III
5	{3,1,0}	$E1_1E1_1E1_1E2$	V		{4,1,0}	$E1_1E1_1E1_1E1_1E2$	V
		$E1_1E1_1E1_2E2$	IV			$E1_1E1_1E1_1E1_2E2$	IV
		$E1_1E1_2E1_2E2$	III			$E1_1E1_1E1_2E1_2E2$	III
	{5,0,0}	$E1_1E1_1E1_1E1_1E1_1$	VI		{6,0,0}	$E1_1E1_1E1_1E1_1E1_1E1_1$	VI
		$E1_1E1_1E1_1E1_1E1_2$	V			$E1_1E1_1E1_1E1_1E1_1E1_2$	V
		$E1_1E1_1E1_1E1_2E1_2$	IV			$E1_1E1_1E1_1E1_1E1_2E1_2$	IV
		$E1_1E1_1E1_2E1_2E1_2$	III			$E1_1E1_1E1_1E1_2E1_2E1_2$	III

4. Fault-Tolerant Landing Algorithms

During the buffer landing period, the force and torque equations of the lander can be written as:

$$
\begin{cases}
F_{ez} = m_b g + \sum_{i=1}^{N} F_{giz} \\
\tau_{exy} = \sum_{i=1}^{N} \left(r_c + r_{bfi} \right) \times F_{giz}
\end{cases}
\tag{7}
$$

where m_b is the mass of the lander, g is the gravity acceleration on the moon, N is the number of normal supporting legs, F_{giz} is the vertical supporting force of the i-th leg, r_c is the vector from the center of mass to the origin O_b of the body coordinate frame, r_{bfi} is the vector from O_b to the i-th foothold, $F_{giz} = \begin{bmatrix} 0 & 0 & F_{giz} \end{bmatrix}^T$ is the vector form of F_{giz}, F_{ez} and $\tau_{exy} = \begin{bmatrix} \tau_{ex} & \tau_{ey} \end{bmatrix}^T$ are dividedly the resultant force and torque.

By adjusting the leg force F_{giz}, we can control the stable body states. After landing, we desire a constant body height and a horizontal body plane. The F_{ez} will be set to zero and the τ_{exy} will be employed to control the angle deviation of pitch and roll. The PID controller will generate the adjusted torque in real-time. Considering the desired stable values of angle and angular velocity are zero, the τ_{exy} will be calculated from the following equation:

$$
\begin{cases}
\tau_{ex} = -k_{px}\theta_{ra} - k_{dx}\dot{\theta}_{ra} - k_{ix} \int \theta_{ra} dt \\
\tau_{ey} = -k_{py}\theta_{pa} - k_{dy}\dot{\theta}_{pa} - k_{iy} \int \theta_{pa} dt
\end{cases}
\tag{8}
$$

where k_{px}, k_{dx}, k_{ix}, k_{py}, k_{dy} and k_{iy} are the proportional, integral and derivative gains in the x-axis and y-axis, respectively. θ_{ra} and θ_{pa} are the actual angles of roll and pitch. $\dot{\theta}_{ra}$ and $\dot{\theta}_{pa}$ are the corresponding velocities.

4.1. VI Configuration

In six-legged normal soft-landing, the force/torque balance equations can be written as follows in matrix form by substituting Equation (8) and $F_{ez} = 0$ into Equation (7).

$$\begin{bmatrix} 1 & 1 & 1 & 1 & 1 & 1 \\ r_{cy} + r_{bf1y} & r_{cy} + r_{bf2y} & r_{cy} + r_{bf3y} & r_{cy} + r_{bf4y} & r_{cy} + r_{bf5y} & r_{cy} + r_{bf6y} \\ r_{cx} + r_{bf1x} & r_{cx} + r_{bf2x} & r_{cx} + r_{bf3x} & r_{cx} + r_{bf4x} & r_{cx} + r_{bf5x} & r_{cx} + r_{bf6x} \end{bmatrix} \cdot \begin{bmatrix} F_{g1z} \\ F_{g2z} \\ F_{g3z} \\ F_{g4z} \\ F_{g5z} \\ F_{g6z} \end{bmatrix} = \begin{bmatrix} -m_b g \\ \tau_{ex} \\ \tau_{ey} \end{bmatrix} \tag{9}$$

However, Equation (9) is an indeterminate equation group and has numerous solutions. It is time-consuming to solve the generalized inverse matrix in a real-time system. Here, a virtual three-legged supporting method (VTLSM) is proposed to allocate the adjusted force F_{giz} quickly. As illustrated in Figure 7, we divide the six supporting legs into two groups: leg 1-3-5 supporting (group A) and leg 2-4-6 supporting (group B). During buffer landing, the legs of each group provide half the adjusted force and torque. Then Equation (9) can be rewritten as follows:

$$\begin{bmatrix} 1 & 1 & 1 \\ r_{cy} + r_{bfky} & r_{cy} + r_{bfmy} & r_{cy} + r_{bfny} \\ r_{cx} + r_{bfkx} & r_{cx} + r_{bfmx} & r_{cx} + r_{bfnx} \end{bmatrix} \cdot \begin{bmatrix} F_{gkz}^{kmn} \\ F_{gmz}^{kmn} \\ F_{gnz}^{kmn} \end{bmatrix} = \frac{1}{2} \begin{bmatrix} -m_b g \\ \tau_{ex} \\ \tau_{ey} \end{bmatrix} \tag{10}$$

where k, m and n are the number of supporting legs in each group. Particularly, $k = 1$, $m = 3$ and $n = 5$ in group A while $k = 2$, $m = 4$ and $n = 6$ in group B. Lastly, the indeterminate Equation (9) is transformed into a two determinate Equation (10) that will generate the adjusted foot force F_{giz} easily in real-time by solving the inverse matrix of a 3×3 matrix.

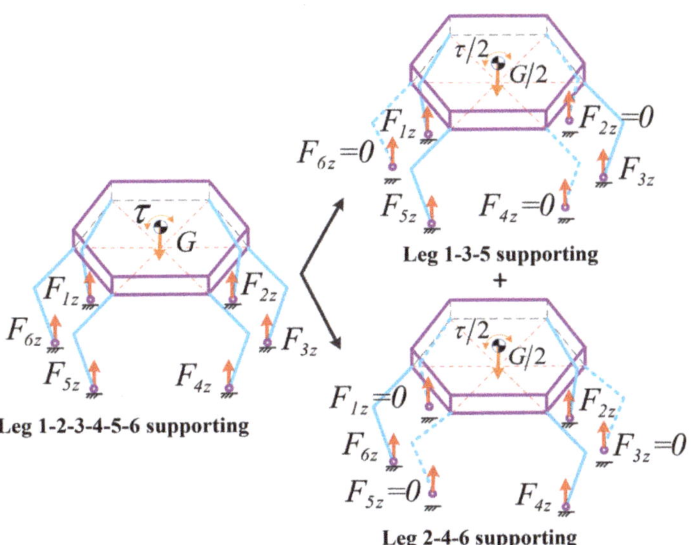

Figure 7. Virtual three-legged supporting method (VTLSM).

The desired torque of the joint impedance controller in Equation (2) comes from the following equation:

$$\tau_{des} = J(q)_f^{-1} F_g \tag{11}$$

where $F_g = \begin{bmatrix} 0 & 0 & F_{gz} \end{bmatrix}^T$ is the vector form of adjusted leg force and $J(q)_f$ is the force Jacobian matrix.

4.2. V Configuration

As for five-legged landing, the 3×6 coefficient matrix in Equation (9) will be transformed into a 3×5 matrix that is still indeterminate. Here we chose configuration V-2 to illustrate the allocated process of leg forces. The force/torque balance equations are written as:

$$\begin{bmatrix} 1 & 1 & 1 & 1 & 1 \\ r_{cy} + r_{bf1y} & r_{cy} + r_{bf2y} & r_{cy} + r_{bf3y} & r_{cy} + r_{bf4y} & r_{cy} + r_{bf5y} \\ r_{cx} + r_{bf1x} & r_{cx} + r_{bf2x} & r_{cx} + r_{bf3x} & r_{cx} + r_{bf4x} & r_{cx} + r_{bf5x} \end{bmatrix} \cdot \begin{bmatrix} F_{g1z} \\ F_{g2z} \\ F_{g3z} \\ F_{g4z} \\ F_{g5z} \end{bmatrix} = \begin{bmatrix} -m_b g \\ \tau_{ex} \\ \tau_{ey} \end{bmatrix} \quad (12)$$

The VTLSM will be again used to quickly solve the foot force. There are ten virtual triangles ($C_5^3 = 10$) constructed by any three supporting legs. They are divided into two cases according to the ground projected position of the center of mass of the lander. If this point is inside the virtual supporting triangles, these triangles are valid, and the number of them is N_v. On the contrary, if the others are invalid and their number is N_i. The supporting legs of the valid triangle are leg 1-2-4, leg 1-2-5, leg 1-3-4, leg 1-4-5, leg 2-3-5 and leg 2-4-5, while the ones of the invalid triangle are leg 1-2-3, leg 2-3-4 and leg 3-4-5. Then the N_v is seven and N_i is three. Meanwhile, the Equation (10) is modified as follows:

$$\begin{bmatrix} 1 & 1 & 1 \\ r_{cy} + r_{bfky} & r_{cy} + r_{bfmy} & r_{cy} + r_{bfny} \\ r_{cx} + r_{bfkx} & r_{cx} + r_{bfmx} & r_{cx} + r_{bfnx} \end{bmatrix} \begin{bmatrix} F_{gkz}^{kmn} \\ F_{gmz}^{kmn} \\ F_{gnz}^{kmn} \end{bmatrix} = \frac{1}{N_v} \begin{bmatrix} -m_b g \\ \tau_{ex} \\ \tau_{ey} \end{bmatrix} \quad (13)$$

Noticeably, the i-th supporting leg provides adjusted force in multiple valid triangles, so we will obtain the eventual foot force by the following sum operation:

$$\begin{cases} F_{g1z} = F_{g1z}^{124} + F_{g1z}^{125} + F_{g1z}^{134} + F_{g1z}^{135} + F_{g1z}^{145} \\ F_{g2z} = F_{g2z}^{124} + F_{g2z}^{125} + F_{g2z}^{235} + F_{g2z}^{245} \\ F_{g3z} = F_{g3z}^{134} + F_{g3z}^{135} + F_{g3z}^{235} \\ F_{g4z} = F_{g4z}^{124} + F_{g4z}^{134} + F_{g4z}^{145} + F_{g4z}^{245} \\ F_{g5z} = F_{g5z}^{125} + F_{g5z}^{135} + F_{g5z}^{145} + F_{g5z}^{235} + F_{g5z}^{245} \end{cases} \quad (14)$$

In configuration V-1, we will just change the leg number to 2, 3, 4, 5 and 6. Then the adjusted force can be found by the similar Equations (13) and (14). The detailed classification of the virtual supporting triangle is listed in Table 5.

Table 5. Classification of virtual supporting triangle of stable configuration in five-legged soft-landing.

Type II	Valid Virtual Supporting Triangle	N_v	Invalid Virtual Supporting Triangle	N_i
V-1	2-3-5, 2-3-6, 2-4-5, 2-4-6, 2-5-6, 3-4-6, 3-5-6	7	2-3-4, 3-4-5, 4-5-6	3
V-2	1-2-4, 1-2-5, 1-3-4, 1-3-5, 1-4-5, 2-3-5, 2-4-5	7	1-2-3, 2-3-4, 3-4-5	3

4.3. IV and III Configurations

As for four-legged landing, the 3×6 coefficient matrix in Equation (9) will be transformed into a 3×4 matrix. For example, in IV-2 configuration, the force/torque balance equations are written as:

$$
\begin{bmatrix}
1 & 1 & 1 & 1 \\
r_{cy} + r_{bf2y} & r_{cy} + r_{bf4y} & r_{cy} + r_{bf5y} & r_{cy} + r_{bf6y} \\
r_{cx} + r_{bf2x} & r_{cx} + r_{bf4x} & r_{cx} + r_{bf5x} & r_{cx} + r_{bf6x}
\end{bmatrix}
\cdot
\begin{bmatrix}
F_{g2z} \\
F_{g4z} \\
F_{g5z} \\
F_{g6z}
\end{bmatrix}
=
\begin{bmatrix}
-m_b g \\
\tau_{ex} \\
\tau_{ey}
\end{bmatrix}
\tag{15}
$$

Similarly, we utilize the VTLSM to calculate the adjusted leg force in real-time. The classification of the virtual supporting triangle is listed in Table 6.

Table 6. Classification of virtual supporting triangle of stable configuration in four-legged soft-landing.

Type II	Valid Virtual Supporting Triangle	N_v	Invalid Virtual Supporting Triangle	N_i
IV-2	2-4-5, 2-4-6, 2-5-6	3	4-5-6	1
IV-3	2-3-5, 2-3-6, 2-5-6, 3-5-6	4		0
IV-4	2-3-6, 2-4-6, 3-4-6	3	2-3-4	1
IV-6	1-2-4, 1-2-5, 1-4-5, 2-4-5	4		0

As for three-legged landing, the 3×6 coefficient matrix in Equation (9) will be transformed into a 3×3 matrix. For configuration III-4, the equation is written as follows:

$$
\begin{bmatrix}
1 & 1 & 1 \\
r_{cy} + r_{bf1y} & r_{cy} + r_{bf3y} & r_{cy} + r_{bf5y} \\
r_{cx} + r_{bf1x} & r_{cx} + r_{bf3x} & r_{cx} + r_{bf5x}
\end{bmatrix}
\cdot
\begin{bmatrix}
F_{g1z} \\
F_{g3z} \\
F_{g5z}
\end{bmatrix}
=
\begin{bmatrix}
-m_b g \\
\tau_{ex} \\
\tau_{ey}
\end{bmatrix}
\tag{16}
$$

Then we can get the adjusted force directly by solving the inverse matrix without the usage of VTLSM.

5. Quasi-Incentre Stability Optimization

After buffer landing, the normal legs need to support the lander to finish the sampling task using the drill or spoon mounted on the body. If the fault does not occur, the ground projected point just locates on the center of the hexagon constructed by all leg footholds. In this case, the robot has the largest stability margin without adjustment demand for the center of mass. However, if several IDUs have failed, the old center of the hexagon does not coincide with the new center of the supporting polygon constructed by the remaining normal leg footholds. We should find a new center that maximizes the supporting stability margin.

5.1. Quasi-Incentre Definition

Here, we define the target center as a quasi-incentre that satisfies the following two/s: firstly, the sum value of distance (d_i) between this point and any sides of the supporting polygon is the maximum; secondly, all the distances d_i should be as consistent as possible. Specifically, for regular polygons, the quasi-incentre is the center of the inscribed circle. As for the regular triangle, the quasi-incentre is just the incentre.

According to the definition of quasi-incentre, we can search the target point q^* by the following function:

$$
\min f(q) = w_1 \cdot \sum_{i=1}^{n} \frac{1}{d_i} + w_2 \cdot \sum_{i=1}^{n} \frac{|d_i - d_{max}|}{d_{max}}
\tag{17}
$$

where n is the number of normal supporting legs, d_{max} is the maximum distance between all d_i, w_1 and w_2 is the weight coefficients, $\sum_{i=1}^{n} \frac{1}{d_i}$ illustrates the first demand while $\sum_{i=1}^{n} \frac{|d_i - d_{max}|}{d_{max}}$ denotes the second demand.

5.2. Quasi-Incentre Searching

Particle Swarm Optimization (PSO) algorithm [32,33] is widely used in function optimization because it is simple to implement and has fewer adjusting parameters. In this algorithm, each particle possesses attributes of position and velocity. They are updated according to the following equation:

$$v_i(k+1) = v_i(k) + c_1 \cdot rand(\ ) \cdot (Popt_i(k) - x_i(k)) + c_2 \cdot rand(\ ) \cdot (Wopt\ (k) - x_i(k)) \quad (18)$$

$$x_i(k+1) = x_i(k) + v_i(k+1) \quad (19)$$

where $x_i(k)$ and $x_i(k+1)$ are the position of the i-th particle at the k-th and (k + 1)-th iteration, $v_i(k)$ and $v_i(k+1)$ are the corresponding velocity, $rand(\ )$ is the random number between zero and one, $Popt_i(k)$ is the best position of i-th particle, $Wopt\ (k)$ is the best position of the swarm, c_1 and c_2 are the learning factors. The actual particle velocity satisfies the following limits:

$$v_i(k+1) = \begin{cases} v_{max}, & if\ v_i(k+1) \geq v_{max}; \\ -v_{max}, & if\ v_i(k+1) \leq v_{max}; \\ v_i(k+1), & else. \end{cases} \quad (20)$$

where v_{max} is the maximum of the searching velocity.

5.2.1. Searching in V Configuration

During five-legged landing, V-1 and V-2 are stable configurations. Here, we chose the V-2 configuration as the searching example. As shown in Figure 8, the numbers 1, 2, 3, 4, 5 and 6 denote the foothold positions of normal landing. All the particle positions at k-th iteration are expressed by the blue circles. The red star is the center of the hexagon that is constructed by all leg footholds and represented by the red dotted line. The new supporting polygon constructed by normal legs is expressed by the solid green line. Each subgraph is arranged in an increment of 5 iterations. At the first iteration, all the particles are scattered in the supporting plane. After 21 iterations, they are quickly concentrated into a small circular area. Then the particles slowly converge to the optimal position.

A similar process of particle convergence can also be found in the curve of target function value vs. iterations in V-2 configuration (Figure 9). At the first iteration, the function value is a larger number of 2.856. After a quick descent, it becomes a relatively stable value at the thirteenth iteration. Lastly, the value converges to a stable value of 1.692.

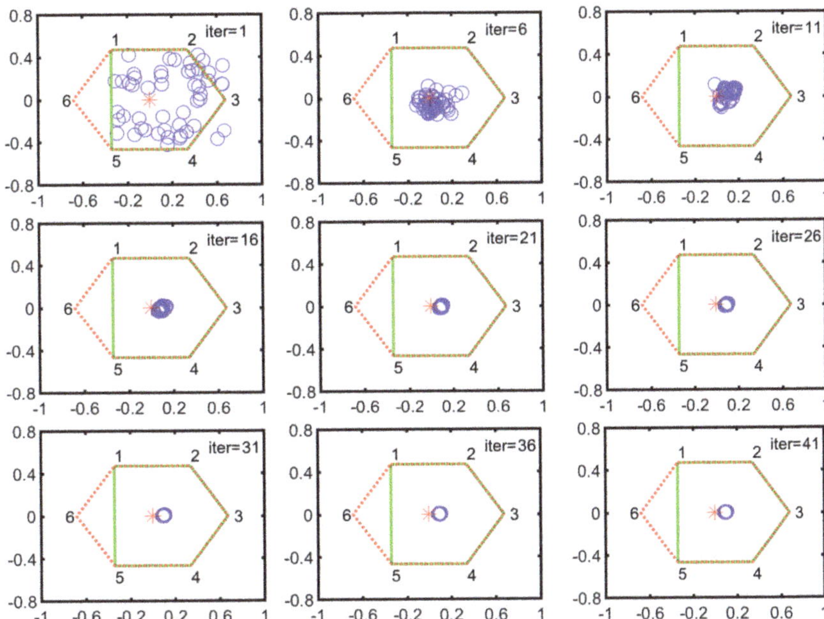

Figure 8. Particle evolution process in V-2 configuration (the horizontal axis denotes the forward/backward direction illustrated in Figure 2 while the vertical axis represents the leftward/rightward direction, respectively).

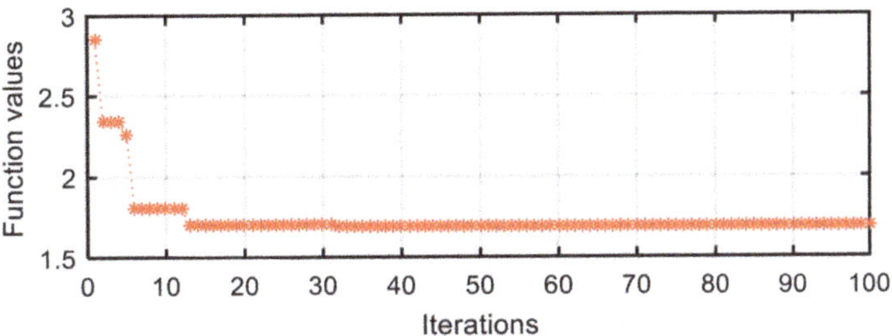

Figure 9. Target function value vs. iterations in V-2 configuration.

5.2.2. Searching in IV Configuration

During four-legged landing, the stable configurations are IV-2, IV-3, IV-4 and IV-6. Here, we take the IV-2 configuration as an example to illustrate the converge process. Figure 10 shows the particle evolution process in IV-2 configuration. All particles are distributed in the supporting plane at the first iteration. After a quick evolution process, they concentrate on a small circle district, then converge to the optimal position with slow velocity.

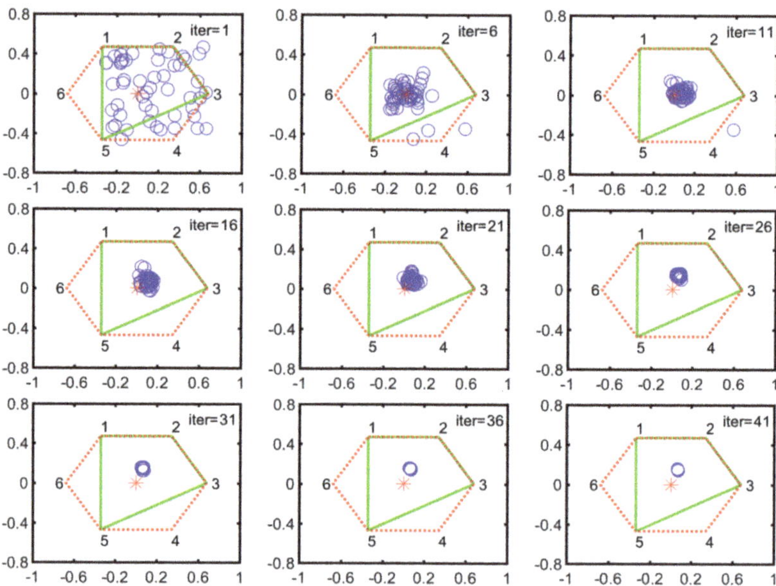

Figure 10. Particle evolution process in IV-2 configuration (the definition of vertical or horizontal axis is same as the one in Figure 8).

At the same time, the converged process is detailly illustrated in the curve of target function value vs. iterations in IV-2 configuration (Figure 11). The initial target function value is 2.124 at the first iteration. After 23 iterations, it reduces to 1.839 rapidly. Eventually, the value converges to the stable value of 1.834 at a slow velocity.

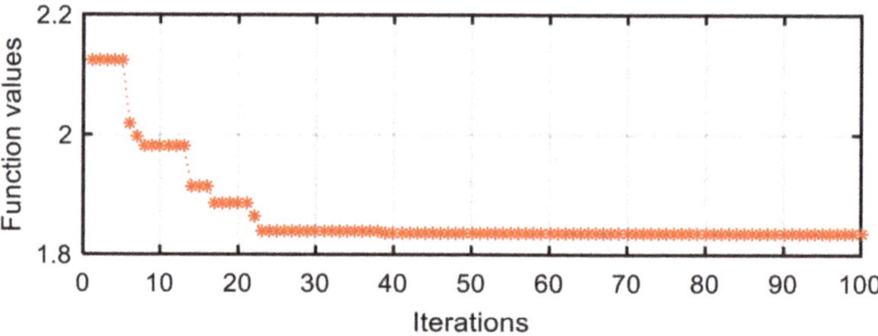

Figure 11. Target function value vs. iterations in IV-2 configuration.

5.2.3. Searching in III Configuration

In a three-legged landing, the stable landing configuration is III-4. As is well-known, the triangle must exist in the center. In the III-4 configuration, its incentre coincides with the old center of the normal supporting hexagon and is set as the origin of the world coordinate frame. It is not necessary to search the quasi-incentre by the PSO algorithm. However, we can verify the target position obtained in the PSO algorithm by comparing it with the incentre. Here, we chose the III-3 configuration to check the searching result. The particle evolution process is shown in Figure 12. The green star is the theoretical incentre of the supporting triangle. After 31 iterations, the scattered particles concentrate to the incentre quickly and are limited in a small circle. Then they converge to the optimal position with a slow velocity.

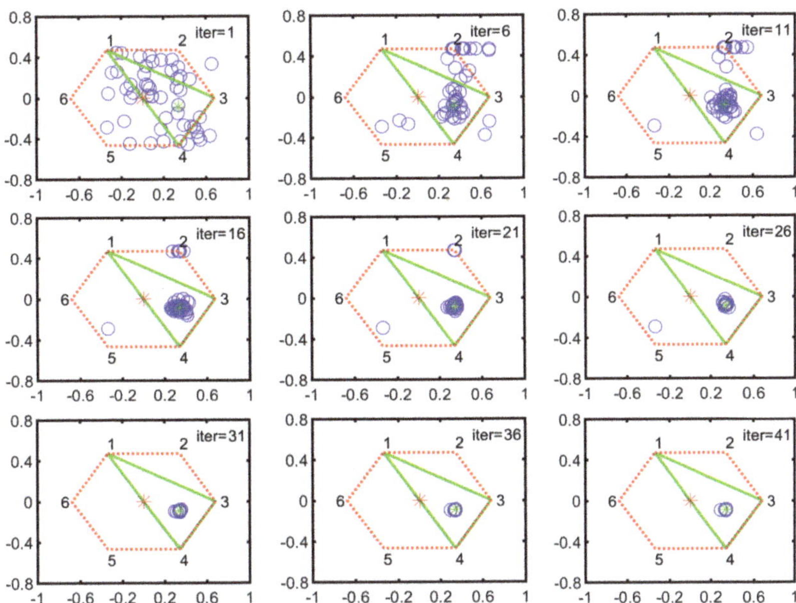

Figure 12. Particle evolution process in III-4 configuration (the definition of vertical or horizontal axis is same as the one in Figure 8 or Figure 10).

As shown in Figure 13, the curve of the target function value vs. iterations in the III-4 configuration can be divided into three terraces. The first terrace is the period from the first iteration to the fourth iteration; the second one is from the fifth iteration to the nineteenth iteration, while the third one is from the twentieth iteration to the final iteration. The corresponding function value reduces from initial 0.1302 to 0.0055, then to zero. The target function value of zero means that the distances between optimal position and any sides of the supporting triangle are equal, and the quasi-incentre is the incentre. Noticeably, we only use the second part in Equation (17) to achieve the searching in a supporting triangle while the weight coefficient of the first part is zero ($w_1 = 0$).

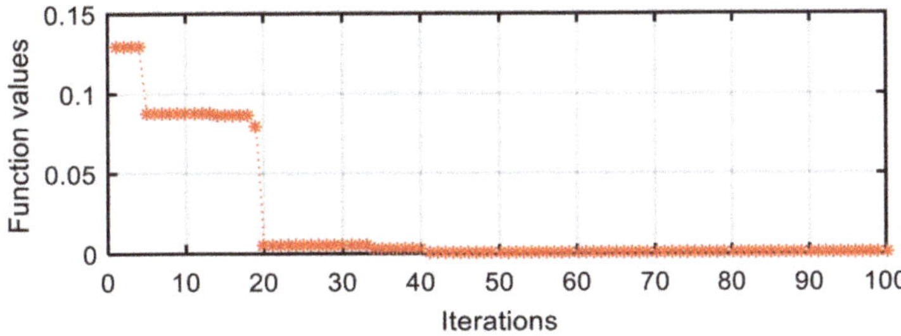

Figure 13. Target function value vs. iterations in III-4 configuration.

6. Experiments

In order to verify the fault-tolerant landing, we design a 5-Dof lunar gravity ground testing platform to provide the experiment scene. There were a series of experiments for stable configurations, including the five-legged, four-legged and three-legged soft-landing experiment conducted to verify the fault-tolerant landing capacity on the platform.

6.1. Experiment Platform

Figure 14a shows the detailed components, while Figure 14b expresses the construction of the counterweight system. The simulation capacity of the platform includes vertical movement z, horizontal x, and 3-Dof spatial rotation of R_x, R_y and R_z. The design principle of counterweight is to make that the resultant force for the lander system in the vertical direction is equal to the one on the lunar surface by trickily dividing the load weight into two parts of counterweight one and counterweight two. In the experiments, we would simulate the soft-landing with a load of 140 kg on the Moon, which means the total mass (m_t) including lander and load is 180 kg, the counterweight one mass is 75 kg while the counterweight two mass is 65 kg.

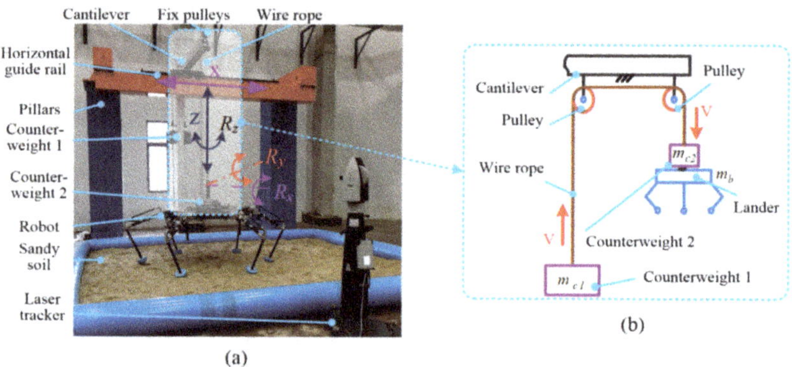

Figure 14. 5-Dof lunar gravity ground testing platform. (**a**) components and Degrees of freedom; (**b**) components of counterweight system.

6.2. Five-Legged Landing

As shown in Figure 15, the lander adopts V-2 configuration to finish soft-landing while the other configuration can bring out similar results. No. 01 is the initial position. No. 02 and No. 03 are declining in the air. No. 04 is the moment of full-touching with the ground. No. 05 is compressing to the lowest position (No. 06). After a damped vibration, the body reaches the stable position No. 10.

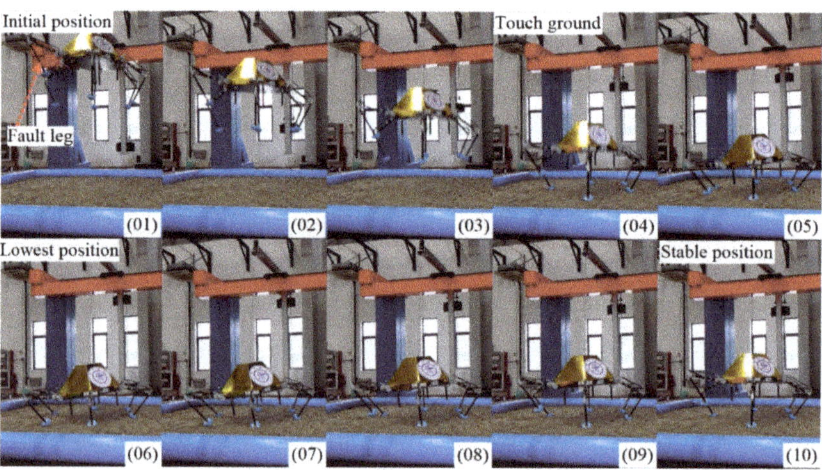

Figure 15. Keyframe snapshots in five-legged fault-tolerant landing.

Owing to the non-centrosymmetric pentagon constructed by the normal supporting leg, the difference of joint torques is great. As illustrated in Figure 16, the maximum of peak joint torques in each leg occurs in leg 1 while the minimum occurs in leg 2. The torque changes dramatically at the moment of touching the ground. The thigh peak torque is 203.7 Nm, while the shank peak torque is -84.14 Nm in leg 1. As for leg 2, the thigh peak torque is 115.1 Nm, while the shank peak torque is -72.47 Nm. The side joint torques are always small. After about a 2 s fluctuation, all joint torques reach a low, stable value. The curves of body angles are shown in Figure 17a. The fluctuation ranges are $-1.8\sim0°$ and $-7.98\sim1.13°$ for roll and pitch angles, respectively. The curves of position and velocity of the body are seen in Figure 17b. The lander touches the ground with a maximum velocity of -1.9 m/s in the z-direction. Next, the body continues to fall with a deceleration. After 0.36 s, the body velocity reduces to zero, and the body reaches the lowest position. Owing to the overturning force/torque caused by eccentricity, extra maximum velocities of 0.355 m/s and 0.04 m/s in the x and y direction are generated. Eventually, the body reaches a stable state after a damping vibration of 1.8 s.

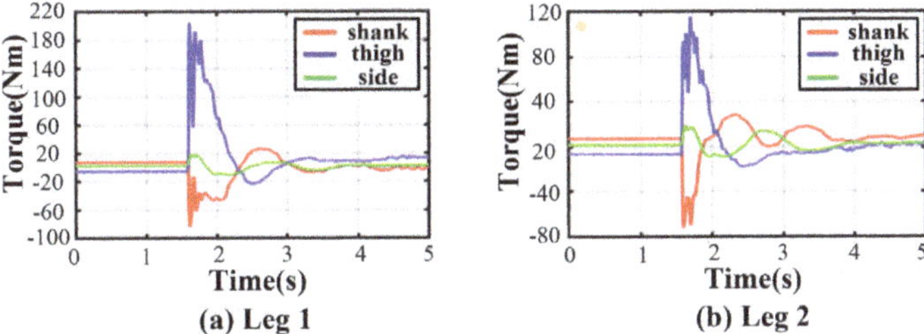

(a) Leg 1 (b) Leg 2

Figure 16. Joint torques in five-legged fault-tolerant landing.

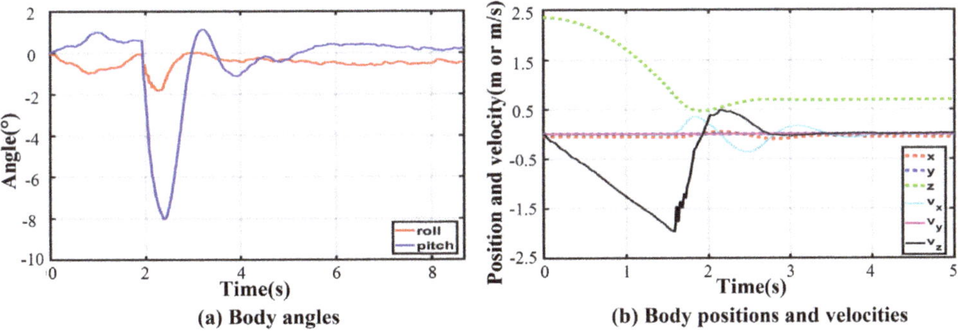

(a) Body angles (b) Body positions and velocities

Figure 17. Body states in five-legged fault-tolerant landing.

6.3. Four-Legged Landing

Here, we chose the IV-6 configuration as the example to present the four-legged soft-landing. As shown in Figure 18, the buffer process in four-legged landing is similar to the one in five-legged landing. No. 01, No. 04, No. 06 and No. 10 are the initial position, touching ground moment, lowest position, and stable position, respectively.

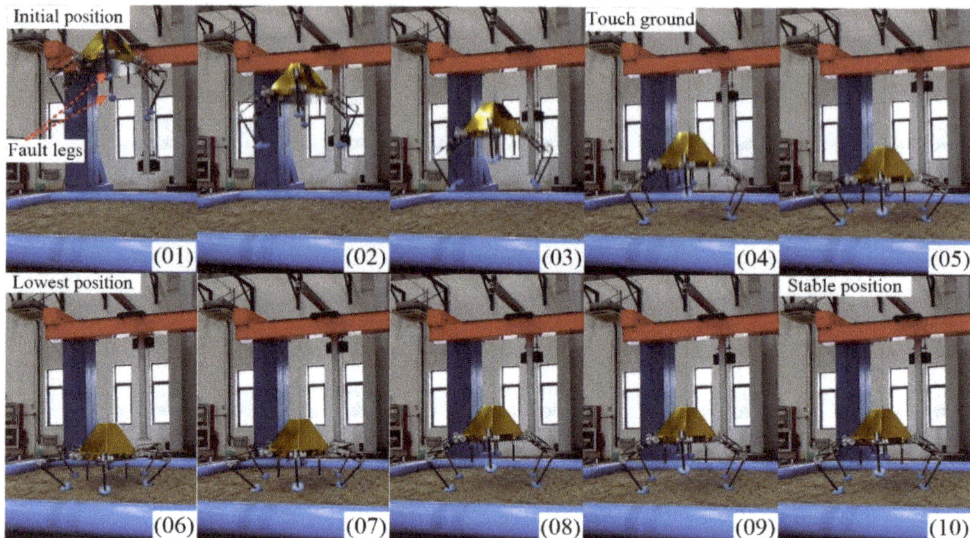

Figure 18. Keyframe snapshots in four-legged fault-tolerant landing.

Thanks to the centrosymmetric supporting rectangle, all curves of joint torque in each leg are basically the same. The maximum peak torque occurs in leg 4. As illustrated in Figure 19, the thigh peak torque is 175.2 Nm, while the shank peak torque is −98.39 Nm. At the moment of touching the ground, the torques of the thigh or shank and the angles of roll or pitch change greatly. As expressed in Figure 20, the maximum roll angle is −1.46° while the one of pitch angle is −0.93°. The touching ground velocity is the same as the one in a five-legged landing, but the fluctuation time is shorter and lasts about 1.6 s. The extra horizontal velocities in the x and y direction are almost zero because of the excellent symmetry of the supporting polygon.

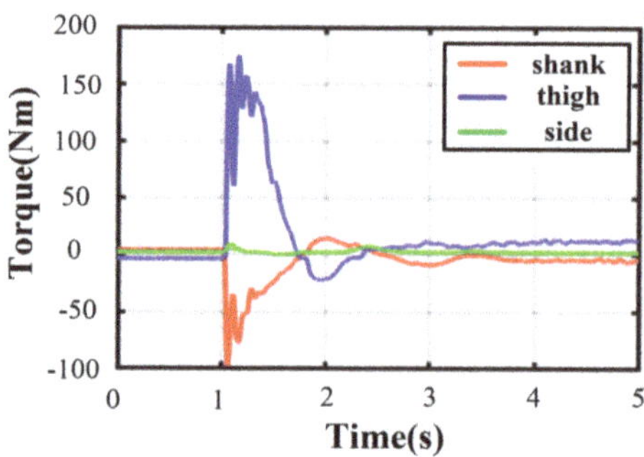

Figure 19. Joint torques in four-legged fault-tolerant landing.

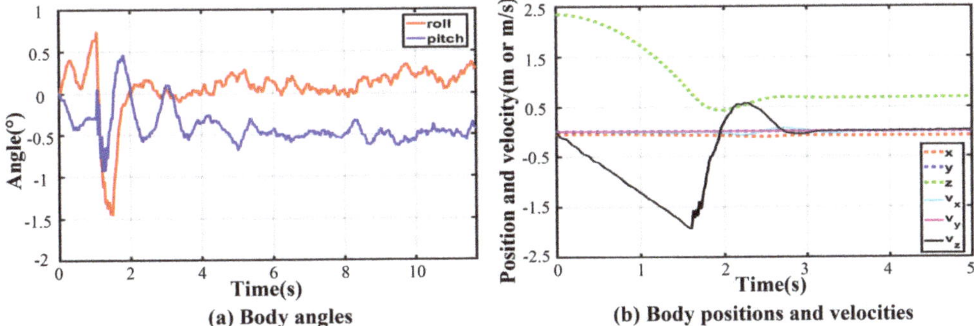

(a) Body angles   (b) Body positions and velocities

Figure 20. Body states in four-legged fault-tolerant landing.

6.4. Three-Legged Landing

As for three-legged landing, there is only one stable configuration denoted by III-4. Figure 21 shows the fluctuation process with damping vibration. The curves of joint torque in each leg are almost consistent, and the maximum peak torque occurs in leg 1. As illustrated in Figure 22, the thigh peak torque is 184.7 Nm, while the shank peak torque is −78.57 Nm. The fluctuation ranges of the angles of roll and pitch are −1.26~0.53° and −1.16~1.08° in Figure 23a, respectively. The maximum velocity in the z-direction is −1.9 m/s at the moment of touching the ground. The body velocity reduces to zero and reaches the lowest position after 0.416 s. Lastly, the body keeps a stable height by a damping vibration of about 2 s.

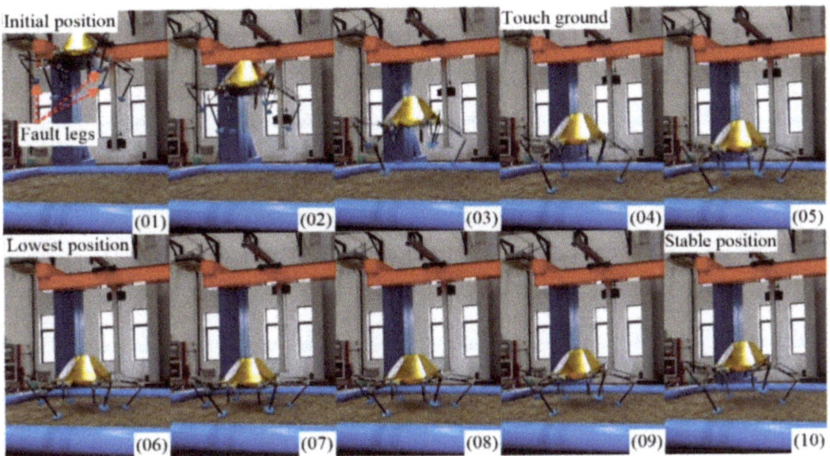

Figure 21. Keyframe snapshots in three-legged fault-tolerant landing.

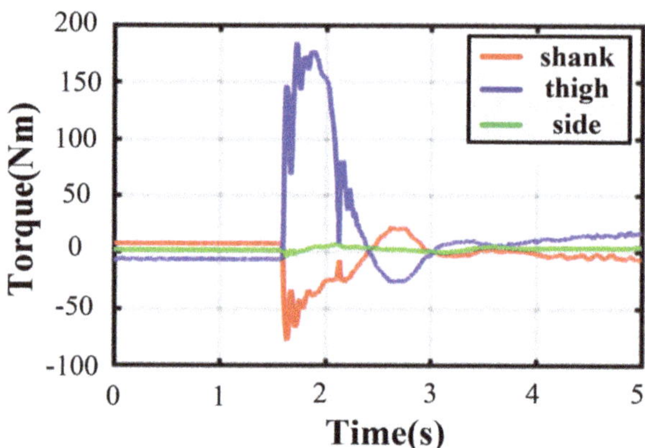

Figure 22. Joint torques in three-legged fault-tolerant landing.

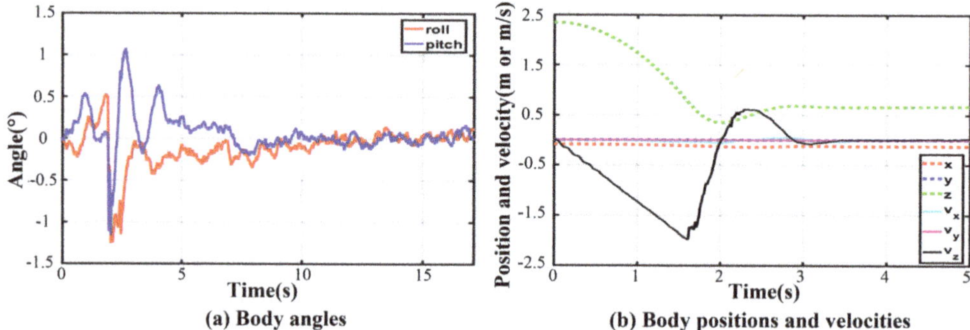

(a) Body angles　　　　　(b) Body positions and velocities

Figure 23. Body states in three-legged fault-tolerant landing.

7. Discussion

As shown in Table 7, the soft-landing performances in different landing configurations with the same touch-ground conditions are obviously numerous. While the number of supporting legs has a great influence on landing performance, the spatial distribution of normal legs also plays an important role. In the four-legged and three-legged landing, all normal legs are evenly distributed, which generates almost no derivative velocity and results in a small angle derivation of roll and pitch ($\approx \pm 1.5°$). Thanks to more normal legs, the peak torque is smaller, and the damping vibration duration is shorter in four-legged landing than the ones in three-legged landing. As for five-legged landing, this case has the most supporting legs than three/four-legged landing, but its landing performance is not very great due to the terrible non-centrosymmetric distribution of normal legs. The indexes of peak torque and angle derivation in five-legged landing are worse than the ones in the centrosymmetric configuration, like four/three-legged landing. Furthermore, a derivative velocity of 0.355 m/s and 0.04 m/s in the x and y direction is separately generated due to the large angle derivation. Thanks to more supporting legs, the damping vibration duration in a five-legged landing is longer than the one in a four-legged landing, but it is still shorter than the one in a three-legged landing.

Table 7. The comparison of the key index in different fault-tolerant landings.

Index	Five-Legged Landing	Four-Legged Landing	Three-Legged Landing
Touch-ground velocity (m/s)	1.9	1.9	1.9
System mass (kg)	180	180	180
Thigh peak torque (Nm)	203.7	175.2	184.7
Shank peak torque (Nm)	−84.14	−98.39	−78.57
Roll angle derivation (°)	−1.8~0°	−1.46~0.74°	−1.26~0.53°
Pitch angle derivation (°)	−7.98~1.13°	−0.93~0.47°	−1.16~1.08°
Damping vibration duration (s)	1.8	1.6	2
Derivative x Velocity (m/s)	0.355	≈ 0	≈ 0
Derivative y Velocity (m/s)	0.04	≈ 0	≈ 0

8. Conclusions

To execute the tasks of landing and roving simultaneously, a six-legged movable repetitive lander is designed and manufactured. Instead of absorbing the landing impact energy by irreversible deformation of aluminum honeycomb material in the current legged lander, a new electric IDU with high power, low weight and small volume is utilized to dissipate the energy actively by simulating the dynamic characters of spring and damper based on impedance control. The leg structure is still intact rather than permanent deformation, so the HexaMRL can perform repetitive exploration like the lander and rover. Fault-tolerant landing capacity is important for adapting this repetitive work mode. The main contributions are as follows:

(1) To achieve the soft-landing as far as possible with a failed IDU, we systematically analyze the classification and stability of landing configurations. Then the relationship between fault number and landing configuration is concluded by equation and listed by table.

(2) As for stable configuration, we have designed corresponding fault-tolerant landing algorithms to achieve buffer landing and further proposed a quasi-incentre stability optimization method to increase the stability margin during supported operations.

(3) A series of experiments including five-legged, four-legged and three-legged soft-landing with a vertical landing velocity of −1.9 m/s and a payload of 140 kg were conducted to verify the fault-tolerant landing capacity on the constructed 5-DoF lunar gravity ground testing platform by means of counterweight.

In future work, we will study the fault-tolerant walking capacity for HexaMRL to further improve the control theory under faults.

Author Contributions: Conceptualization, K.Y. and F.G.; methodology, K.Y.; software, K.Y. and Q.S.; validation, K.Y., S.Z. and Q.S.; formal analysis, K.Y.; investigation, K.Y.; resources, F.G.; data curation, K.Y.; writing—original draft preparation, K.Y.; visualization, K.Y. and S.Z.; supervision, F.G.; project administration, F.G.; funding acquisition, F.G. All authors have read and agreed to the published version of the manuscript.

Funding: This work was funded by the National Natural Science Foundation of China (No. U1613208), the National Key Research and Development Plan of China (No. 2017YFE0112200), the European Union's Horizon 2020 research and innovation programme under the Marie Skłodowska-Curie grant agreement (No. 734575), the Research Program of National Major Research Instruments of China (No. 51927809).

Institutional Review Board Statement: Not applicable.

Informed Consent Statement: Not applicable.

Conflicts of Interest: The authors declared no potential conflict of interest with respect to the research, authorship, and/or publication of this article.

References

1. Efanov, V.; Dolgopolov, V. The moon: From research to exploration (to the 50th anniversary of Luna-9 and Luna-10 Spacecraft). *Sol. Syst. Res.* **2017**, *51*, 573–578. [CrossRef]
2. Behm, H. Results of The Ranger, Lunar 9, and Surveyor 1 missions. *J. Astronaut. Sci.* **1967**, *14*, 101.
3. Doiron, H.; Zupp, G., Jr. Lunar module landing dynamics. In Proceedings of the 41st Structures, Structural Dynamics, and Materials Conference and Exhibit, Atlanta, GA, USA, 3–4 April 2000; p. 1678.
4. Ma, Y.; Liu, S.; Sima, B.; Wen, B.; Peng, S.; Jia, Y. A precise visual localisation method for the Chinese Chang'e-4 Yutu-2 rover. *Photogramm. Rec.* **2020**, *35*, 10–39. [CrossRef]
5. Young, A. *Lunar and Planetary Rovers: The Wheels of Apollo and the Quest for Mars*; Springer Science & Business Media: New York, NY, USA, 2007.
6. Basilevsky, A.T.; Abdrakhimov, A.M.; Head, J.W.; Pieters, C.M.; Wu, Y.; Xiao, L. Geologic characteristics of the Luna 17/Lunokhod 1 and Chang'E-3/Yutu landing sites, northwest mare imbrium of the moon. *Planet. Space Sci.* **2015**, *117*, 385–400. [CrossRef]
7. Lin, R.; Guo, W.; Li, M.; Hu, Y.; Han, Y. Novel design of a legged mobile lander for extraterrestrial planet exploration. *Int. J. Adv. Robot. Syst.* **2017**, *14*, 17298814–17746120. [CrossRef]
8. Luo, J.; Zhao, Y.; Ruan, L.; Mao, S.; Fu, C. Estimation of CoM and CoP trajectories during human walking based on a wearable visual odometry device. *IEEE Trans. Autom. Sci. Eng.* **2020**, 3036530. [CrossRef]
9. Raibert, M.; Blankespoor, K.; Nelson, G.; Playter, R. Bigdog, the rough-terrain quadruped robot. In Proceedings of the 17th World Congress, the International Federation of Automatic Control, Seoul, Korea, 6–11 July 2008; pp. 10822–10823.
10. Bledt, G.; Powell, M.J.; Katz, B.; di Carlo, J.; Wensing, P.M.; Kim, S. MIT Cheetah 3: Design and control of a robust, dynamic quadruped robot. In Proceedings of the 2018 IEEE/Rsj International Conference on Intelligent Robots and Systems, Madrid, Spain, 1–5 October 2018; pp. 2245–2252.
11. Hutter, M.; Gehring, C.; Jud, D.; Lauber, A.; Bellicoso, C.D.; Tsounis, V.; Hwangbo, J.; Bodie, K.; Fankhauser, P.; Bloesch, M.; et al. Anymal-a highly mobile and dynamic quadrupedal robot. In Proceedings of the 2016 IEEE/RSJ International Conference on Intelligent Robots and Systems (IROS), Daejeon, Korea, 9–14 October 2016; pp. 38–44.
12. Hutter, M.; Gehring, C.; Lauber, A.; Gunther, F.; Bellicoso, C.D.; Tsounis, V.; Fankhauser, P.; Diethelm, R.; Bachmann, S.; Bloesch, M.; et al. Anymal-toward legged robots for harsh environments. *Adv. Robot.* **2017**, *31*, 918–931. [CrossRef]
13. Shim, H.; Yoo, S.-y.; Kang, H.; Jun, B.-H. Development of arm and leg for seabed walking robot CRABSTER200. *Ocean. Eng.* **2016**, *116*, 55–67. [CrossRef]
14. Kim, J.-Y. Dynamic balance control algorithm of a six-legged walking robot, little crabster. *J. Intell. Robot. Syst.* **2015**, *78*, 47–64. [CrossRef]
15. Xu, Y.; Gao, F.; Pan, Y.; Chai, X. Method for six-legged robot stepping on obstacles by indirect force estimation. *Chin. J. Mech. Eng.* **2016**, *29*, 669–679. [CrossRef]
16. Wilcox, B.H. ATHLETE: A cargo and habitat transporter for the moon. In Proceedings of the 2009 IEEE Aerospace Conference (AERO), Big Sky, MT, USA, 7–14 March 2009; pp. 1–7.
17. Bartsch, S.; Birnschein, T.; Cordes, F.; Kuehn, D.; Kampmann, P.; Hilljegerdes, J.; Planthaber, S.; Roemmermann, M.; Kirchner, F. Spaceclimber: Development of a six-legged climbing robot for space exploration. In Proceedings of the ISR 2010 (41st International Symposium on Robotics) and ROBOTIK 2010 (6th German Conference on Robotics), Munich, Germany, 7–9 June 2010; pp. 1–8.
18. Arm, P.; Zenkl, R.; Barton, P.; Beglinger, L.; Dietsche, A.; Ferrazzini, L.; Hampp, E.; Hinder, J.; Huber, C.; Schaufelberger, D.; et al. Spacebok: A dynamic legged robot for space exploration. In Proceedings of the 2019 International Conference on Robotics and Automation (ICRA), Montreal, QC, Canada, 20–24 May 2019; pp. 6288–6294.
19. Kolvenbach, H.; Hampp, E.; Barton, P.; Zenkl, R.; Hutter, M. Towards jumping locomotion for quadruped robots on the moon. In Proceedings of the 2019 IEEE/RSJ International Conference on Intelligent Robots and Systems (IROS), Macau, China, 3–8 November 2019; pp. 5459–5466.
20. Wang, S.; Chen, Z.; Li, J.; Wang, J.; Li, J.; Zhao, J. Flexible motion framework of the six wheel-legged robot: Experimental results. *IEEE/ASME Trans. Mechatron.* **2021**, 3100879. [CrossRef]
21. Semini, C.; Barasuol, V.; Goldsmith, J.; Frigerio, M.; Focchi, M.; Gao, Y.; Caldwell, D.G. Design of the hydraulically actuated, torque-controlled quadruped robot HyQ2Max. *IEEE/ASME Trans. Mechatron.* **2016**, *22*, 635–646. [CrossRef]
22. Ugurlu, B.; Havoutis, I.; Semini, C.; Kayamori, K.; Caldwell, D.G.; Narikiyo, T. Pattern generation and compliant feedback control for quadrupedal dynamic trot-walking locomotion: Experiments on RoboCat-1 and HyQ. *Auton. Robot.* **2015**, *38*, 415–437. [CrossRef]
23. Fu, Y.; Luo, J.; Ren, D.; Zhou, H.; Li, X.; Zhang, S. Research on impedance control based on force servo for single leg of hydraulic legged robot. In Proceedings of the 2017 IEEE International Conference on Mechatronics and Automation (ICMA), Takamatsu, Japan, 6–9 August 2017; pp. 1591–1596.
24. Luo, J.; Wang, S.; Zhao, Y.; Fu, Y. Variable stiffness control of series elastic actuated biped locomotion. *Intell. Serv. Robot.* **2018**, *11*, 225–235. [CrossRef]
25. Zhang, J.; Gao, F.; Yu, H.; Zhao, X. Design and development of a novel redundant and fault-tolerant actuator unit for heavy-duty parallel manipulators. Proceedings of the Institution of Mechanical Engineers. *Part C J. Mech. Eng. Sci.* **2011**, *225*, 3031–3044. [CrossRef]

26. Ebrahimi, I.; Carretero, J.A.; Boudreau, R. 3-PRRR redundant planar parallel manipulator: Inverse displacement workspace and singularity analyses. *Mech. Mach. Theory* **2007**, *42*, 1007–1016. [CrossRef]
27. Wertz, J.R.; Bell, R. Autonomous rendezvous and docking technologies: Status and prospects. In *Space Systems Technology and Operations*; International Society for Optics and Photonic: Orlando, FL, USA, 5 August 2003; pp. 20–30.
28. Wang, J.; Zhang, Y.; Di, K.; Chen, M.; Duan, J.; Kong, J.; Xie, J.; Liu, Z.; Wan, W.; Rong, Z.; et al. Localization of the Chang'e-5 Lander Using Radio-Tracking and Image-Based Methods. *Remote Sens.* **2021**, *13*, 590.
29. Saranli, U.; Buehler, M.; Koditschek, D.E. RHex: A simple and highly mobile hexapod robot. *Int. J. Robot. Res.* **2011**, *20*, 616–631. [CrossRef]
30. Pfeiffer, F.; Eltze, J.; Weidemann, H.-J. The TUM-walking machine. *Intell. Autom. Soft Comput.* **1995**, *1*, 307–323. [CrossRef]
31. Gao, F.; Yang, J.; Ge, Q.J. Type synthesis of parallel mechanisms having the second class GF sets and two dimensional rotations. *J. Mech. Robot.* **2011**, *3*, 011003. [CrossRef]
32. Kennedy, J.; Eberhart, R. Particle swarm optimization. In Proceedings of the ICNN'95-international conference on neural networks(ICNN), Perth, WA, Australia, 27 November–1 December 1995; pp. 1942–1948.
33. Shi, Y.; Eberhart, R.C. Empirical study of particle swarm optimization. In Proceedings of the 1999 Congress on Evolutionary Computation-CEC99 (Cat. No. 99TH8406), Washington, DC, USA, 6–9 July 1999; pp. 1945–1950.

MDPI

St. Alban-Anlage 66

4052 Basel

Switzerland

Tel. +41 61 683 77 34

Fax +41 61 302 89 18

www.mdpi.com

Sensors Editorial Office

E-mail: sensors@mdpi.com

www.mdpi.com/journal/sensors